Soil Management for Sustainable Agriculture and Ecosystem Services

Soil Management for Sustainable Agriculture and Ecosystem Services

Editors

Chiara Piccini
Rosa Francaviglia

Basel • Beijing • Wuhan • Barcelona • Belgrade • Novi Sad • Cluj • Manchester

Editors

Chiara Piccini
Council for Agricultural
Research and Economics
(CREA), Research Centre for
Agriculture and Environment
Rome, Italy

Rosa Francaviglia
Council for Agricultural
Research and Economics
(CREA), Research Centre for
Agriculture and Environment
Rome, Italy

Editorial Office
MDPI
St. Alban-Anlage 66
4052 Basel, Switzerland

This is a reprint of articles from the Special Issue published online in the open access journal *Land* (ISSN 2073-445X) (available at: https://www.mdpi.com/journal/land/special_issues/Soil_management2).

For citation purposes, cite each article independently as indicated on the article page online and as indicated below:

Lastname, A.A.; Lastname, B.B. Article Title. *Journal Name* **Year**, *Volume Number*, Page Range.

ISBN 978-3-0365-9574-0 (Hbk)
ISBN 978-3-0365-9575-7 (PDF)
doi.org/10.3390/books978-3-0365-9575-7

Cover image courtesy of Chiara Piccini

© 2023 by the authors. Articles in this book are Open Access and distributed under the Creative Commons Attribution (CC BY) license. The book as a whole is distributed by MDPI under the terms and conditions of the Creative Commons Attribution-NonCommercial-NoDerivs (CC BY-NC-ND) license.

Contents

About the Editors vii

Preface ix

Radwa A. El Behairy, Ahmed A. El Baroudy, Mahmoud M. Ibrahim, Elsayed Said Mohamed, Dmitry E. Kucher and Mohamed S. Shokr
Assessment of Soil Capability and Crop Suitability Using Integrated Multivariate and GIS Approaches toward Agricultural Sustainability
Reprinted from: *Land* 2022, 11, 1027, doi:10.3390/land11071027 1

Mohamed A. E. AbdelRahman, Mohamed M. Metwaly, Ahmed A. Afifi, Paola D'Antonio and Antonio Scopa
Assessment of Soil Fertility Status under Soil Degradation Rate Using Geomatics in West Nile Delta
Reprinted from: *Land* 2022, 11, 1256, doi:10.3390/land11081256 19

Daniela De Benedetto, Emanuele Barca, Mirko Castellini, Stefano Popolizio, Giovanni Lacolla and Anna Maria Stellacci
Prediction of Soil Organic Carbon at Field Scale by Regression Kriging and Multivariate Adaptive Regression Splines Using Geophysical Covariates
Reprinted from: *Land* 2022, 11, 381, doi:10.3390/land11030381 43

Valeria Medoro, Giacomo Ferretti, Giulio Galamini, Annalisa Rotondi, Lucia Morrone, Barbara Faccini and Massimo Coltorti
Reducing Nitrogen Fertilization in Olive Growing by the Use of Natural Chabazite-Zeolitite as Soil Improver
Reprinted from: *Land* 2022, 11, 1471, doi:10.3390/land11091471 61

Kristina Ivashchenko, Emanuela Lepore, Viacheslav Vasenev, Nadezhda Ananyeva, Sofiya Demina, Fluza Khabibullina, et al.
Assessing Soil-like Materials for Ecosystem Services Provided by Constructed Technosols
Reprinted from: *Land* 2021, 10, 1185, doi:10.3390/land10111185 81

Ariel Freidenreich, Sanku Dattamudi, Yuncong Li and Krishnaswamy Jayachandran
Influence of Leguminous Cover Crops on Soil Chemical and Biological Properties in a No-Till Tropical Fruit Orchard
Reprinted from: *Land* 2022, 11, 932, doi:10.3390/land11060932 97

Nina Noreika, Tailin Li, Julie Winterova, Josef Krasa and Tomas Dostal
The Effects of Agricultural Conservation Practices on the Small Water Cycle: From the Farm- to the Management-Scale
Reprinted from: *Land* 2022, 11, 683, doi:10.3390/land11050683 115

Homayra Asima, Victoria Niedzinski, Frances C. O'Donnell and Jack Montgomery
Comparison of Vegetation Types for Prevention of Erosion and Shallow Slope Failure on Steep Slopes in the Southeastern USA
Reprinted from: *Land* 2022, 11, 1739, doi:10.3390/land11101739 131

Siyu Tang, Chong Du and Tangzhe Nie
Inversion Estimation of Soil Organic Matter in Songnen Plain Based on Multispectral Analysis
Reprinted from: *Land* 2022, 11, 608, doi:10.3390/land11050608 151

Donata Drapanauskaitė, Kristina Bunevičienė, Regina Repšienė, Danutė Karčauskienė, Romas Mažeika and Jonas Baltrusaitis
The Effect of Pelletized Lime Kiln Dust Combined with Biomass Combustion Ash on Soil Properties and Plant Yield in a Three-Year Field Study
Reprinted from: *Land* **2022**, *11*, 521, doi:10.3390/land11040521 . . . . **169**

Adrielle Rodrigues Prates, Karen Cossi Kawakami, Aline Renée Coscione, Marcelo Carvalho Minhoto Teixeira Filho, Orivaldo Arf, Cassio Hamilton Abreu-Junior, et al.
Composted Sewage Sludge Sustains High Maize Productivity on an Infertile Oxisol in the Brazilian Cerrado
Reprinted from: *Land* **2022**, *11*, 1246, doi:10.3390/land11081246 . . . . **189**

Beatrice Farda, Rihab Djebaili, Matteo Bernardi, Loretta Pace, Maddalena Del Gallo and Marika Pellegrini
Bacterial Microbiota and Soil Fertility of *Crocus sativus* L. Rhizosphere in the Presence and Absence of *Fusarium* spp.
Reprinted from: *Land* **2022**, *11*, 2048, doi:10.3390/land11112048 . . . . **203**

Yudha Kristanto, Suria Tarigan, Tania June, Enni Dwi Wahjunie and Bambang Sulistyantara
Water Regulation Ecosystem Services of Multifunctional Landscape Dominated by Monoculture Plantations
Reprinted from: *Land* **2022**, *11*, 818, doi:10.3390/land11060818 . . . . **217**

Aikaterini Voudouri, Evgenia Chaideftou and Athanassios Sfougaris
Topsoil Seed Bank as Feeding Ground for Farmland Birds: A Comparative Assessment in Agricultural Habitats
Reprinted from: *Land* **2021**, *10*, 967, doi:10.3390/land10090967 . . . . **237**

About the Editors

Chiara Piccini

Chiara Piccini has been a geologist and senior technologist at the Council for Agricultural Research and Economics, Research Centre for Agriculture and Environment (CREA-AA) in Rome, Italy, since 1998. Her main research topics include the spatialization of experimental data using geostatistics, digital soil mapping techniques and satellite imagery; application of crop simulation models for soil nitrogen and carbon dynamics; soil quality and environmental concerns; monitoring of nutrients leaching from soil; nitrate contamination; statistical data processing; monitoring of soil chemical features; soil chemical, physical and hydrological characterization; soil chemical fertility; soil water dynamics; irrigation water quality and irrigation with non-optimal water quality.

Rosa Francaviglia

Rosa Francaviglia, with expertise in agronomy, was a senior researcher at the Council for Agricultural Research and Economics, Research Centre for Agriculture and Environment (CREA-AA) in Rome, Italy, from 1981 (now retired). Her main research topics include the effect of climate change on agriculture; carbon sink and agricultural soils; soil organic carbon simulation models; soil fertility; conservation agriculture; crop diversification; agro-environmental evaluations; soil quality indicators; good agro-environmental conditions (GAEC) under the EU Common Agricultural Policy.

Preface

Soil ecosystem services include people's direct and indirect benefits from soils. These include provisioning services, such as food, feed, fiber, and fresh water; regulating services, such as flood and disease control and climate regulation; and supporting services, such as soil formation, water and nutrient cycling, the production of atmospheric oxygen, and the provisioning of habitats.

Utilizing soil for agriculture inevitably changes soil properties, such as nutrient status, pH, organic matter content, and physical characteristics. In many cases, changes that are beneficial for food production are detrimental to other ecosystem services. Core farming practices, such as soil tillage, crop residue management, nutrient management, and pest management, impact a range of soil functions and ecosystem services, including water availability for crops, weed control, insect and pathogen control, soil quality and functioning, soil erosion control, soil organic carbon pool, environmental pollution control, greenhouse gas emissions, and crop yield productivity.

Since prevailing farming paradigms perceive high crop yields and low environmental impact as being in conflict, it is crucial to define an environmentally sound range of agronomic activities that would be considered tolerable at a certain extent of intensity. Sustainable agriculture mainly focuses on increasing the productivity of the soil and reducing the harmful effects of agricultural practices on climate, soil, water, environment, and human health. Increases in soil fertility, water protection, and biodiversity protection need to be considered.

Following sustainable agriculture principles, soil must be protected and developed, water and water resources must be protected, natural control of pests and diseases should be adopted, and different agricultural products should be cultivated. Managing soil organic carbon is central because soil organic matter influences numerous soil properties relevant to ecosystem functioning and crop growth.

Chiara Piccini and Rosa Francaviglia
Editors

Article

Assessment of Soil Capability and Crop Suitability Using Integrated Multivariate and GIS Approaches toward Agricultural Sustainability

Radwa A. El Behairy [1], Ahmed A. El Baroudy [1], Mahmoud M. Ibrahim [1], Elsayed Said Mohamed [2,3], Dmitry E. Kucher [3] and Mohamed S. Shokr [1,*]

1. Soil and Water Department, Faculty of Agriculture, Tanta University, Tanta 31527, Egypt; radwa126710stud_pg@agr.tanta.edu.eg (R.A.E.B.); drbaroudy@agr.tanta.edu.eg (A.A.E.B.); mahmoud.abouzaid@agr.tanta.edu.eg (M.M.I.)
2. National Authority for Remote Sensing and Space Sciences, Cairo 11843, Egypt; elsayed.salama55@mail.ru
3. Department of Environmental Management, Institute of Environmental Engineering, People's Friendship University of Russia (RUDN University), 6 Miklukho-Maklaya Street, 117198 Moscow, Russia; kucher-de@rudn.ru
* Correspondence: mohamed_shokr@agr.tanta.edu.eg

Citation: El Behairy, R.A.; El Baroudy, A.A.; Ibrahim, M.M.; Mohamed, E.S.; Kucher, D.E.; Shokr, M.S. Assessment of Soil Capability and Crop Suitability Using Integrated Multivariate and GIS Approaches toward Agricultural Sustainability. *Land* 2022, *11*, 1027. https://doi.org/10.3390/land11071027

Academic Editors: Chiara Piccini and Rosa Francaviglia

Received: 2 June 2022
Accepted: 4 July 2022
Published: 6 July 2022

Publisher's Note: MDPI stays neutral with regard to jurisdictional claims in published maps and institutional affiliations.

Copyright: © 2022 by the authors. Licensee MDPI, Basel, Switzerland. This article is an open access article distributed under the terms and conditions of the Creative Commons Attribution (CC BY) license (https://creativecommons.org/licenses/by/4.0/).

Abstract: Land evaluation has an important role in agriculture. Developing countries such as Egypt face many challenges as far as food security is concerned due to the increasing rates of population growth and the limited agriculture resources. The present study used multivariate analysis (PCA and cluster analysis) to assess soil capability in drylands, Meanwhile the Almagra model of Micro LEIS was used to evaluate land suitability for cultivated crops in the investigated area under the current (CS) and optimal scenario (OS) of soil management with the aim of determining the most appropriate land use based on physiographic units. A total of 15 soil profiles were selected to characterize the physiographic units of the investigated area. The results reveal that the high capability cluster (C1) occupied 31.83% of the total study area, while the moderately high capability (C2), moderate capability (C3), and low capability (C4) clusters accounted for 37.88%, 28.27%, and 2.02%, respectively. The limitation factors in the studied area were the high contents of $CaCO_3$, the shallow soil depth, and the high salinity and high percentage of exchangeable sodium (% ESP) in certain areas. The application of OS enhanced the moderate suitability (S3) and unsuitable clusters (S5) to the suitable (S2) and marginally suitable (S4) categories, respectively, while the high suitability cluster (S1) had increased land area, which significantly affected the suitability of maize crop. The use of multivariate analysis for mapping and modeling soil suitability and capability can potentially help decision-makers to improve agricultural management practices and demonstrates the importance of appropriate management to achieving agricultural sustainability under intensive land use in drylands.

Keywords: soil capability index; PCA; GIS; land capability and suitability; cluster analysis; sustainable agriculture

1. Introduction

Worldwide food insecurity is currently one of the most significant challenges facing humanity. Demand for food is expected to rise by 70.00% by 2050, and agricultural productivity is a crucial component of global food security [1]. Rapid population growth has exacerbated global human food insecurity, thus necessitating long-term evaluation of natural resources. It is thought that the world population will be more than nine billion by 2050 [2,3]. As such, it is anticipated that there might be shortages in both agricultural resources and land [4,5]. One possible solution to compensate for this shortage is to encourage increasing crop yields. However, this entails using pesticides and fertilizers that may affect the environment negatively. Another possible solution is to import more crops to fill the food gap [4,6]. If properly managed, soil is one of the most significant natural resources that can abet in bridging the food demand gap to achieve food security [7]. Agricultural fields in the Nile Valley and Delta, Egypt, account for about 4.00% of the country's total land area [8]. The growth of the agricultural sector in Egypt is considered an important long-term development backbone. The agricultural sector contributes about 14.50% of the gross national product in Egypt and 30.00% of foreign revenue from the export of agricultural products, and has led to a 41.00% decrease in unemployment [9]. Agricultural growth on arable land strives to accomplish long-term agricultural development by the integration of soil, water, and environmental factors [10,11]. The term "land evaluation" refers to the performance rate of the land and its ability for crop production, with the capacity varying according to climate, geographical location, and physiochemical characteristics [12]. Land evaluation can enable decision-makers to select the best-performing crops based on soil properties [11,13]. The soil limiting factors for crop suitability vary in different areas in Egypt, with soil salinity, poor drainage, and compaction as the most common factors in the northern Nile Delta [14–17]. Agriculture is the greatest user of water in Egypt, especially in the northwestern Nile Delta; thus, determining and controlling surface water quality in such areas is vital for protecting water resources and ensuring long-term sustainable agriculture [18]. Soil property characterization, modeling, and mapping at various spatial and temporal scales are required for the study of diverse environments [19] The Geographic Information System (GIS) technique has accelerated spatial variability studies of different environmental phenomena [20]. Thus, integrating GIS and geostatistical analysis to map and detect the spatial variation of soil parameters in previously unstudied areas might be beneficial. For instance, inverse distance weighted is an interpolation procedure that uses known values with corresponding weighted values to estimate unknown values in a study location [21]. Land capability assessment has a vital role in adequate planning, particularly in arid climate zones [22]. Combining the properties of soil in order to evaluate its capability is limited by the intricate nature of the soil system. Consequently, multivariate analysis has been identified as an appropriate tool for soil capability zone evaluation owing to its ability to perform systematic modeling in unclear and indistinct scenarios [23–25]. PCA and cluster analysis are multivariate procedures that are widely used for soil data recognition, classification, and modeling [26]. Models of soil evaluation, theoretical agricultural management scenarios, and spatial analyses are valuable tools used by land managers and decision-makers to achieve sustainability of land use and management for different studied areas [27,28]. The Micro Land Evaluation Information System (Micro-LEIS) has been widely used to assess land suitability around the world [29]. The Micro-LEIS system is based on an integrated soil, climate, and agricultural management databases for assessing land, and contains two models related to land vulnerability and suitability [30]. The Almagra model was designed for land suitability assessment and is one of the major components of Micro-LEIS DSS [31]. The main aim of this work is to use multivariate analysis to assess soil capability in the dryland areas of the northwestern Nile Delta in Egypt. In addition, land suitability for cultivated crops in the study area under CS and OS of soil management was evaluated to determine the most appropriate land use based on physiographic units.

2. Materials and Methods

2.1. The Site Description

The study area was in the northwest Nile Delta in Egypt. It lies between longitudes 30°15′0″–30°40′0″ E and latitudes 31°7′15″–31°30′45″ N, with a total area of 797.00 km^2 (Figure 1). The area is categorized by a Mediterranean climate based on the mean climatic parameters for a period of 50.00 years from 1960 to 2011 [32]. A relatively high average maximum temperature of 30 °C is usually recorded during the dry season in August. The mean minimum temperature in January is 13 °C. Precipitation is naturally light and drizzly from November to February, with a mean rainfall of about 17.23 mm/year. The lowest evaporation rates are noticed in January and December owing to low temperatures, while the highest rates are observed in June and September owing to relatively high temperatures.

The annual mean rate of evaporation ranges from 3.3 to 4.8 mm/day. The lowest percentage of relative humidity of 51% is observed in April, while the highest proportion of 58.4% is observed in December. The area has a torric and thermal soil moisture and temperature regime [33]. Geologically, the western Nile Delta is formed from sedimentary deposits that vary in age from the Late Cretaceous to Quaternary. The eastern and western parts of the study area are covered with Holocene clay and Quaternary sediments, respectively [34]. Surface irrigation is the most commonly used system, in which water is pumped from irrigation canals and drained in furrows and basins [18].

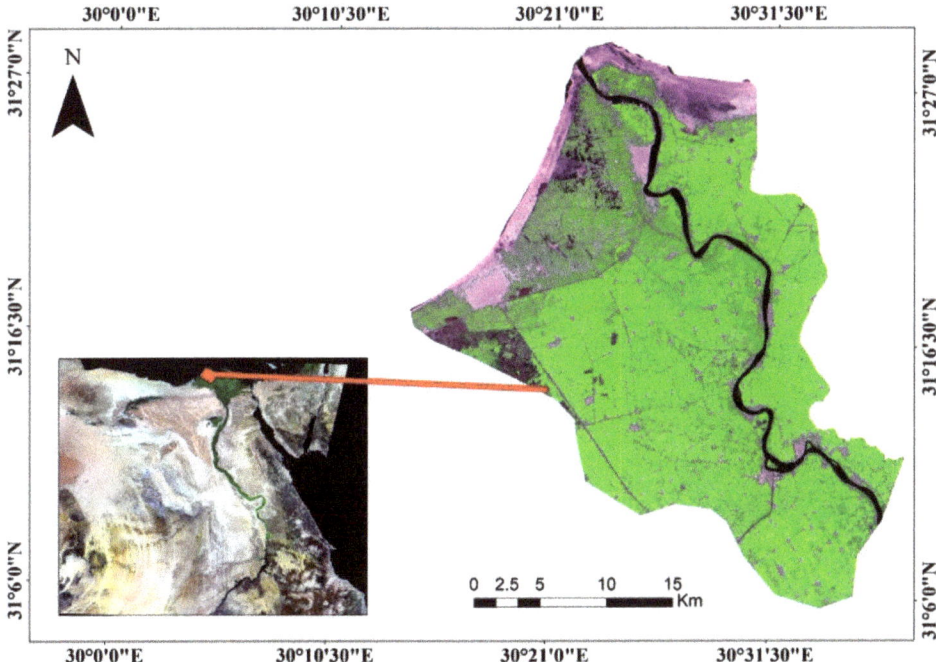

Figure 1. Location of investigated area.

2.2. Extraction of Physiographic Units

In this study, a SENTINEL-2 image acquired in August 2020 under clear-sky conditions was utilized to create landforms and digital soil map features of the study area with the aid of a digital elevation model (DEM). The Sentinel application platform (SNAP) and Environment for Visualizing Images (ENVI 5.4) software were used to process the spectral subset, radiometric calibration, atmospheric, and geometric corrections of the image [35]. Remote sensing (RS) and geographic information system (GIS) are effective for identifying geomorphological units [36]. Thirteen geomorphological units were recognized as representing different geomorphological features within the study area. Subsequently, the image obtained was used as the base map, and each geomorphic unit was homogeneous with the natural land properties [37]. The stepwise methodology for evaluating soil relied on the integrated soil data, remote sensing data, and GIS utilizing multivariate analysis, as illustrated in Figure 2.

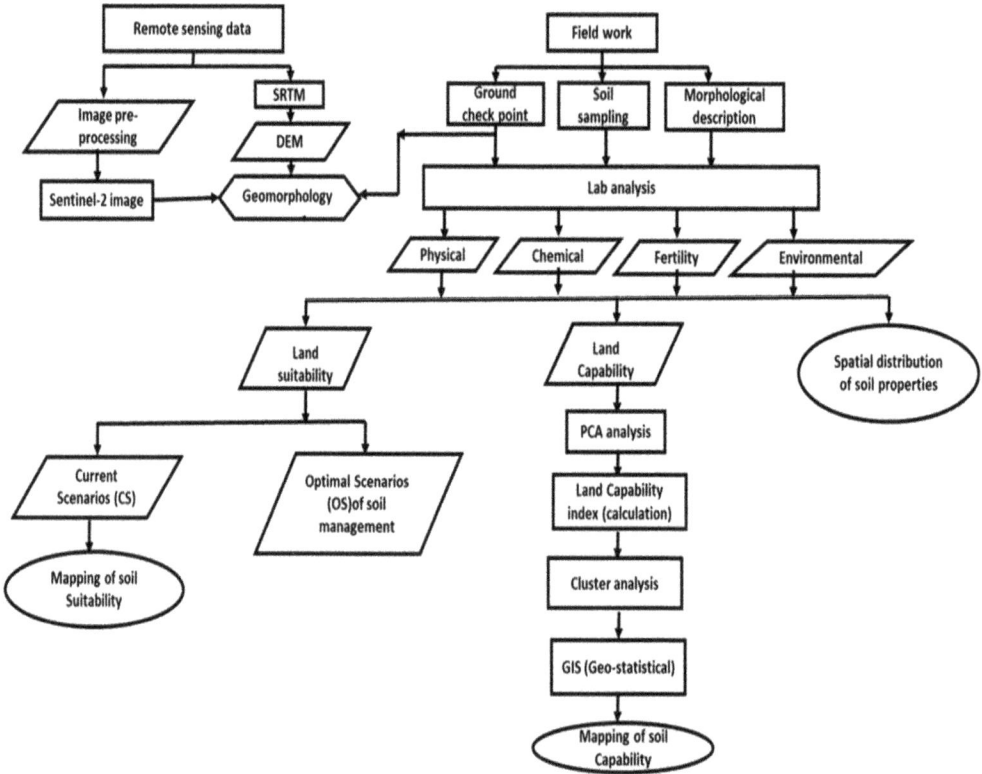

Figure 2. Flow chart illustrating methodology of current work.

2.3. Sample Collection and Lab Analysis

A total of 15 soil profiles were geo-referenced based on geomorphological field mapping of the research area using the Global Positioning System. These profiles were selected from three sampling areas spanning about 80 km² to represent the identified geomorphology and landscape units of the area in Figure 3. Morphological description and classification of soil profiles were carried out according to FAO [12] and USDA [33], respectively. Soil profiles were dug to 150 cm depth or until the water table appeared. Thus, the soil profiles range from 80–150 cm depth. The soil physiochemical parameters (61 soil samples) were analyzed in an ISO/IEC 17025 (2017)-compliant and accredited soil, water, and plant laboratory at the Faculty of Agriculture, Tanta University. Chemical analyses, including salinity (EC), soil reaction (pH), cation exchange capacity (CEC), calcium carbonate percentage ($CaCO_3$), exchangeable sodium percentage (ESP), and trace elements (As, Co, Cu, Ni, and Zn), were conducted to determine the Irrigation Water Quality Index (IWQI). Trace elements and heavy metals in irrigation water are responsible for soil contamination, and are key indicators of irrigation water quality [38]. In addition, analysis of soil physical characteristics, including bulk density, particle size distribution, and fertility as defined by percentage soil organic matter content (SOM%) and available soil nitrogen (N), phosphorus (P), and potassium (K) was conducted [39–43].

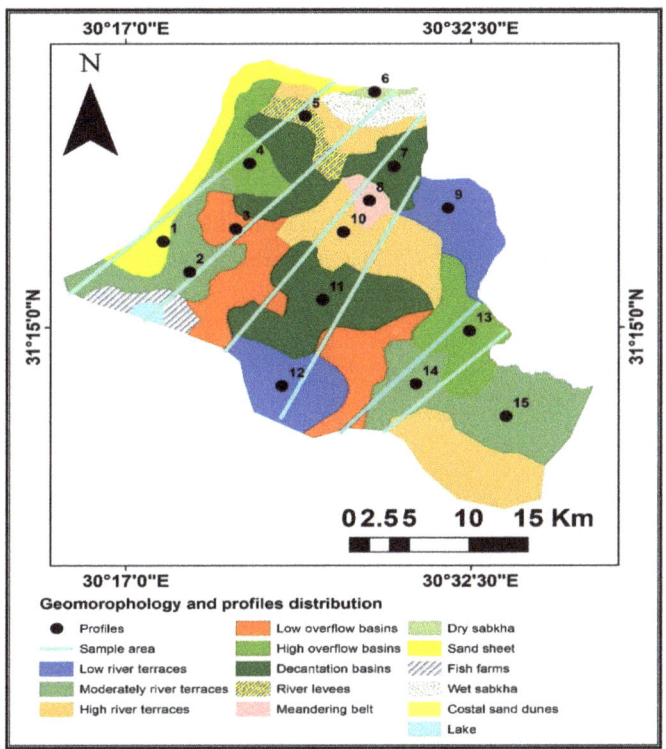

Figure 3. The distribution of soil profiles and sampling areas in this study.

2.4. Determination of IWQI Values

The nature and severity of problems caused by poor irrigation water quality are widely considered to differ based on a variety of factors, such as soil type and crops, the regional environment, and how water is used by farmers. Generally, five measures are used to assess irrigation water quality, including salinity level, infiltration and permeability hazard, and the level of toxic chemicals in water [44,45].

The proposed IWQI, which evaluates the mutual effect of quality parameters, was calculated using Equations (1) and (2):

$$G = \frac{w}{N} \sum_{k=1}^{N} r_k \qquad (1)$$

where k is an incremental index, w is the weight of each hazard, N is the total number of parameters, and r is the rating value of each parameter.

$$IWQ_{index} = \sum_{i=1}^{5} G_i \qquad (2)$$

where i is an incremental index and G is the contribution of each water quality parameter, (salinity, infiltration, specific ion toxicity, trace element toxicity, and miscellaneous effects).

2.5. Statistical Analysis

Descriptive statistics of the studied soil characteristics, including the minimum, maximum, arithmetic mean, standard deviation, and coefficient of variation, were computed using SPSS version 25. PCA was used to reduce the dataset into principal component (PC) variables and to avoid multi-collinearity between the original variables. Prior to PCA, the Pearson correlation coefficient was utilized to verify linear relationships among the soil variables. The Kaiser–Meyer–Olkin (KMO) method was used to assess adequacy of samples for the whole data set, with KMO values larger than 0.5 indicating the suitability of the data for PCA. In addition, data fitness was determined using the Bartlett test, and the results revealed a $p < 0.05$, which further confirmed the data fitness for PCA [46]. SPSS software version 25 was used to perform all statistical analyses. The soil profiles were considered as objects for evaluating soil capability, and were divided into dissimilar clusters utilizing agglomerative hierarchical clustering (AHC) in PCA.

2.6. Soil Capability Assessment Based on PCA

The Weighted Additive method was used according to Equation (3):

$$WAI = \sum_{i=1}^{n} W_i \times S_i \quad (3)$$

where WAI is the Weighted Additive index, S_i is the score, n is the number of indicators, and W_i is the weight of indicators.

All parameters were weighted based on the communality of indicators, which were computed statistically or obtained using factor analysis (IBM, SPSS Statics 25). The weighted value of each parameter was either calculated by dividing each parameter value by the overall sum of their values or reported as a ratio [47]. Each parameter was analyzed using four indicators, namely, chemical (CI), physical (PI), fertility (FI), and environmental (EI) indices, and scores ranging from 0.2 to 1.0 were obtained (Table S1). The final index values were classified into high capability (C1), moderately high capability (C2), moderate capability (C3), and low capability (C4) categories (Table S2). The range of values for each index was divided by the number of categories obtained (4), and the results were subsequently used as the width of each category. The resulting values were successively added to the lowest values of each index to obtain the upper limits of each category. Soil capability assessment depends on defining soil properties and their relationship with agricultural suitability. In this context, PCA classifies the capability of soil by harmonizing soil properties within each class. In addition, PCA provides a visual representation of the main clustering patterns for identifying similarities and differences among soil characteristics [48].

2.7. Mapping Soil Properties Using Inverse Distance Weighted (IDW)

The IDW tool in ArcGIS10.7 software was used to produce interpolation maps of chemical, fertility, physical, and environment parameters. This approach works by computing the grid note by considering neighboring locations within a user-defined search radius. The IDW is widely used in soil investigations because it is easy to implement [49–54]. The local impact of the measurement point decreases with distance, as illustrated in the following equation:

$$z_p = \frac{\sum_{i=1}^{n}\left(\frac{z_i}{d_i}\right)}{\sum_{i=1}^{n}\left(\frac{1}{d_i}\right)} \quad (4)$$

where z_p is the value predicted at point P, z_i is the z value at measured point i, and d_i is the distance between point 0 and point 'i'.

Based on SPSS results, the geometrical interval classification method was used to produce most of the interpolation maps, because these data were not distributed normally, whereas natural breaks classification (Jenks) was used for EC, ESP, and $CaCO_3$ maps, as the data used for these maps were normally distributed.

2.8. Determination of Land Suitability

The Almagra model defines soil suitability in five different clusters, namely, optimum (S1), high (S2), moderate (S3), marginal (S4), and unsuitable (S5), for five traditional annual crops, including wheat, maize, and potato, as well as for semiannual and perennial crops such as alfalfa and citrus, respectively. The model was implemented in Micro-LEIS and uses soil variables and favorable crop conditions to evaluate suitability [29,31,55]. The variable generalization levels were determined based on crop requirements for each soil parameter using the most limiting factor method to define soil suitability classes. In this study, the Almagra model was implemented to assess the CS of soil suitability for five crops that are predominantly cultivated within the study area. The OS was based on manageable soil parameters, such as EC, ESP, and $CaCO_3$, without considering the interaction between them. Other soil parameters such as texture and depth were not considered owing to the difficulty in their modification.

The suggested OS was calculated based on Equation (5) [30]:

$$OS = CS - UR_s \qquad (5)$$

where OS, CS, and URs represent the optimal scenario, the current scenario, and the units of reduction, respectively.

The reduction units were defined by assessing CS to meet the suggested fixed value of OS to raise the final soil suitability class. Notably, when the soil under CS was unsuitable (S5) or marginally suitable (S4), higher URs were required relative to those of moderate suitability (S3), which required lower URs to meet the fixed OS value for each soil variable. Under OS, EC classes were reduced and the values varied from slightly to highly saline, with a fixed value of 2 dSm^{-1}, which represents nonsaline soil. For ESP, the projected value of OS was 5%. Finally, OS decreased the $CaCO_3$ values from 9.04% to <2.

3. Results and Discussion

3.1. Geomorphology of the Study Area

The geomorphological units of the study area were determined using Sentinel-2 satellite imagery, DEM, and field truth points (Figure 4). The study area included flood plain, lacustrine plain, and marine plain as the three main landscape features. These features are very common in the north of the Nile Delta and the southern areas of lakes such as Idku in Egypt [1,56]. The flood plain (713 km^2) formed from deposits of the Nile before the high dam's construction. There are many landforms under this landscape, i.e., river terraces, overflow basins, decantation basins, river levees, and meandering belt. The lacustrine plain (40 km^2) is formed from Holocene-era lacustrine sediments. This landscape includes fish farms, dry and wet sabkha, and coastal sand dunes. The marine plain (40 km^2) is located in the north zone of the study area, and includes sand sheet landforms. Water bodies (Lakes) represent 4 km^2 of the total area.

3.2. Spatial Analysis and Soil Physiochemical Properties

3.2.1. Chemical and Physical Soil Capability Indicators

Chemical soil capability indicators (CSCI) are dynamic indicators that vary over time as a result of land management. The CSCI were chosen based on their sensitivity to disturbance and their ability to execute soil ecosystem functions. CSCI included EC, pH, ESP, $CaCO_3$, and CEC as well as physical indicators including depth, as represented in Figure S1.

The spatial trends of EC and ESP increased in the upper part of the northwest of the study area (around 12–20 dS/m and 18–25%), respectively. The high values of ECe in certain areas of the study area may have resulted from the high salinity of the water table and the effects of lake water and seawater. This agrees with the common pattern of the northern delta, where most of the soil is categorized by high soil salinity [15,57]. This high sodium percentage can negatively affect soil properties such as soil structure and hydrology, consequently reducing crop productivity [7]. The highest values of pH (approximately 8.6–8.9) were found in sites in the northeast and southeast of the study area. The highest values of $CaCO_3$ (roughly 6–9%) were found in the middle and southwest of the studied area due to shell fragments, which can lead to solid layer formations impermeable to crops of plants and water in addition to fixation of P fertilizer [7,58]. From the interpolation map, the highest value of CEC (around 37–42 cmolc/kg) was found in sites in southwest and middle of study area. The profile depth ranged from 80–150 cm.

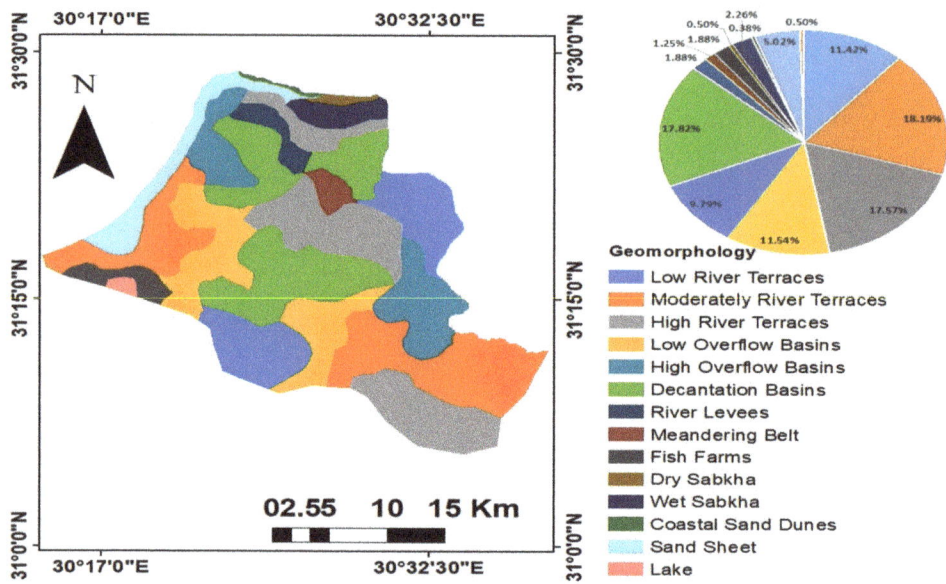

Figure 4. Geomorphological map illustrating the study area.

3.2.2. Fertility and Environmental Soil Capability Indicators

The spatial distribution map for available N, P, K, and SOM in Figure S2 shows that the trend of both N, with values ranging from 7.50 to 81 mg/kg, and P, with values ranging from 6.30 to 22.3 mg/kg, increased from north to south across the study area. On the other hand, the spatial trend of K, with values ranging from 9.30–457.1 mg/kg, increased in sites in the upper north and lower south of the study area. The highest values of SOM (0.9–1.22%) were found in the middle of the northeast and northwest of the study area. The IWQI map (Figure S3) is thought to be a useful tool in future agricultural management plans [18].

3.3. Multivariate Statistical Analysis

3.3.1. Descriptive Statistics of Soil Indicators

Fifteen soil characteristics were analyzed as prospective soil capability indicators. The descriptive statistics obtained based on the weighted mean of parameters of investigated soil profiles are provided in Table S3. The skewness and kurtosis of the tested soil properties revealed a normal distribution in EC, ESP, and $CaCO_3$, while other properties had skewed distribution. The normality test using the Anderson–Darling method obtained p values < 0.05 for all the tested soil properties.

3.3.2. Correlations of Soil Physicochemical Indicators and Principal Component Analysis

The Pearson correlation coefficient plot revealed both positive and negative coefficients at both $p < 0.01$ or $p < 0.05$ (Figure 5). A significant positive association was observed between depth and both EC and ESP, with $r = 0.38$ and 0.41, respectively. Similarly, significant positive coefficients of $r = 0.55, 0.56, 0.57, 0.5$, and 0.34 were observed between depth and AK, AN, AP, CEC, and OM, respectively. In addition, positive significant correlations of $r = 0.87$ between EC and ESP and 0.46 between EC and AP were detected. In contrast, negative correlation coefficients were observed between pH and other properties, except for ESP and $CaCO_3$. Notably, higher positive correlations were observed between OM and $CaCO_3$, AN, AP, AK, and CEC, with coefficients of $r = 0.48, 0.80, 0.71, 0.82$, and 0.89, respectively. In addition, a positive significant correlation between CEC and AN ($r = 0.94$), AP ($r = 0.84$), and AK ($r = 0.95$) was observed.

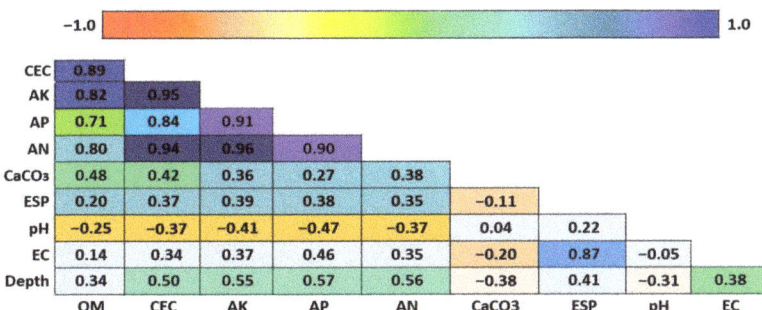

Figure 5. Correlation plot showing coefficients between soil properties. Note: $p < 0.01$ and/or $p < 0.05$. AK, AP, and AN represent available potassium, phosphorus, and nitrogen, respectively.

The factor loading results revealed the acceptable clustering of soil properties and confirmed the reliability of PCA for defining soil characteristics in different clusters [59]. PCA was used to assess land capability based on the variation in soil physicochemical properties and environmental conditions. The method uses eigenvalues, proportions of variance, and cumulative variance of PCs to estimate clusters based on soil characteristics. In this study, PCs with eigenvalues > 1 were retained, while those with values <1 were screened out. As a result, the first four groups with eigenvalues >1 were selected. The soil indicators and these four PCs are shown in Table 1. Notably, a cumulative variance of 91.24% for all the tested variables was observed, with PC1, PC2, PC3, and PC4 explaining about 51.12%, 18.37%, 12.48%, and 9.27% of the total variance, respectively. The factor loadings and component score coefficient outputs from the varimax method showed higher factor loads. The most representative physical and chemical indicators for PC1 based on their close correlation included AN, AP, AK, OM, CEC, and $CaCO_3$, which might be due to the association between natural conditions and the soil formation processes in the study area [60]. In contrast, PC2 was correlated with soil depth, pH, and IWQI. In addition, PC3 was linked with EC and ESP, while PC4 was attributed to ESP.

Table 1. Summary of PCA.

PC Parameters	PC1	PC2	PC3	PC4
Eigenvalue	5.62	2.02	1.37	1.02
Variability (%)	51.12	18.37	12.48	9.27
Cumulative (%)	51.12	69.49	81.96	91.24
	Component score coefficients			
Indicator	PC1	PC2	PC3	PC4
Depth	0.64	0.47	−0.44	0.04
EC (dSm^{-1})	0.49	0.70	0.36	−0.24
pH	−0.41	0.20	0.60	0.58
ESP	0.48	0.69	0.51	0.02
CaCO3	0.28	−0.76	0.52	0.06
AN	0.96	−0.15	0.02	−0.01
AP	0.93	−0.05	−0.00	−0.23
AK	0.97	−0.13	0.04	−0.07
CEC	0.96	−0.17	0.04	0.10
OM	0.83	−0.34	0.06	0.24
IWQI	0.42	0.20	−0.41	0.71

Table 2 shows the acceptable level of p values for the Bartlett sphericity and the KMO tests at $p = 0.05$. The Bartlett sphericity test revealed a p value of <0.0001, which confirmed the suitability of PCA for defining soil clusters based on their characteristics.

Table 2. The Kaiser–Meyer–Olkin (KMO) and Bartlett sphericity tests.

KMO and Bartlett Tests		
KMO Measure of Sampling Adequacy		0.692
Bartlett Test of Sphericity	Chi-square (approx. value)	138.160
	Degree of freedom (DF)	55
	p value	0.0001

The cluster analysis revealed two dissimilar clusters based on PC scores. A dendrogram showing hierarchical clustering of the four groups based on soil properties was obtained, with each group sharing soil profiles that contained a set of similar characteristics (Figure 6).

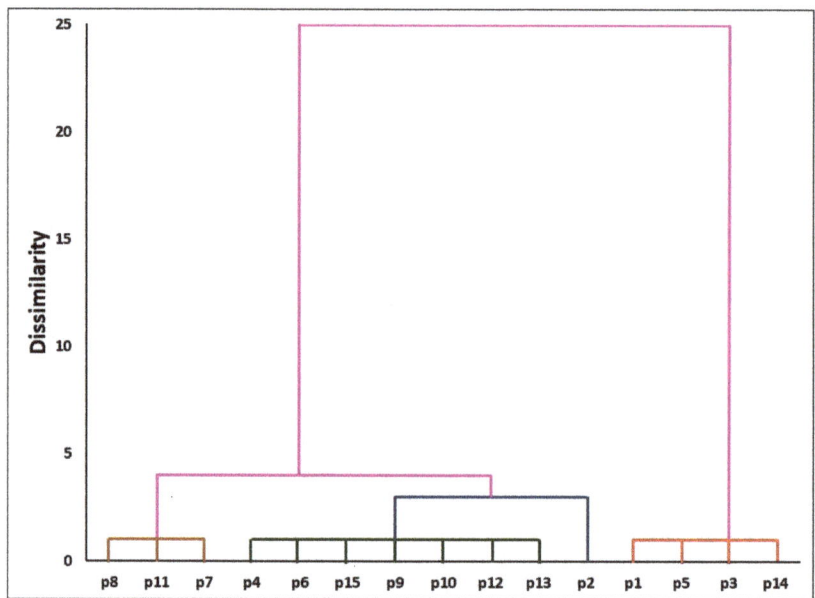

Figure 6. Agglomerative hierarchical clustering dendrogram showing clustering based on soil properties.

3.3.3. Assessment of Land Capability Based on PCA

Soil characteristics classification and its correlation with soil capability and crop suitability is an unprecedented soil analysis approach that can overcome the challenge of classifying soils into clusters based on similarities in their properties, which relies on the intricate determination of increasing and decreasing soil characteristics. The investigated area land capability map was constructed using PCA results; the map reflected the four previously identified groups (Figure 7). The statistical analysis of soil parameters for land capability clusters (C1–C4) are shown in (Table 3). The high capability cluster (C1) occupied 31.83% of the total investigated area, with the soils of this class being identified by moderate salinity, ESP, IWQI, and $CaCO_3$ values. The moderately high capability class (C2) accounted for 37.88% of the total study area. The limiting factors of this class were high $CaCO_3$ content of 9.04% and shallow soil depth of 80 cm. The moderate capability class (C3) accounted for 28.27% of the total study area, and the unit was characterized by a number of limitations, such as high pH and salinity values, which represented the major limiting factors for soil capability, and low SOM%. In addition, the soils of C3 showed a moderate ESP content of 14.01%. The low capability class (C4) represented a small area of 2.02% of the total study area. The soil chemical analysis of this class illustrated high salinity values and moderate ESP and SOM contents.

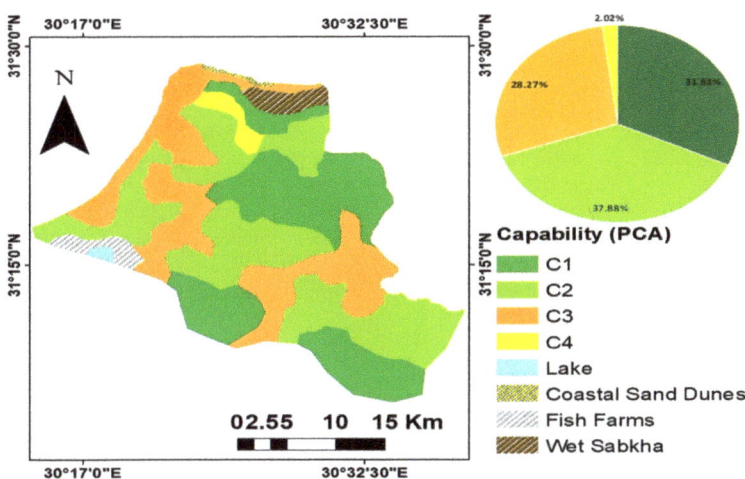

Figure 7. Land capability classes within study area.

Table 3. Statistical summary of soil properties in the four land capability clusters.

Classes	Depth	EC (dS/m)	pH	ESP	CaCO$_3$	AN	AP	AK	CEC	OM	IWQI
C1	108	2.22	8.58	7.01	2.80	13.55	8.43	14.18	9.34	0.38	30.67
C2	80	1.50	8.37	4.73	9.04	63.00	17.40	413.30	36.84	1.17	26.50
C3	123	5.32	8.67	12.76	3.97	43.60	12.53	272.13	32.26	0.93	37.40
C4	150	7.92	8.37	14.01	3.77	68.16	19.41	409.80	39.42	0.97	34.28

3.3.4. Soil Suitability

Soil profiles were evaluated based on their suitability for crop production by considering the specific soil property requirements of each crop to achieve maximum yield. The results showed that soil suitability of selected crops could be categorized into S2–S5 classes, with different limiting factors being identified in each class based on geomorphological units. The soil suitability was examined with five horticultural and field crops, namely, wheat, maize, alfalfa, potato, and citrus (Figure 8). Overall, cultivating field crops in the area demonstrated good potential for sustainable agricultural development (Figure S4); however, improved quality of irrigation water is highly necessary [18].

3.3.5. Soil Factors under Current and Optimal Scenarios

The key soil limiting factors in the study area were identified to be high salinity, increased sodium saturation, poor drainage, calcium carbonate, and rough soil texture (Figures S5 and S6). Reducing the manageable soil limiting severity of factors, such as EC, ESP, CaCO$_3$, and drainage, where possible, resulted in enhanced soil suitability for all selected crops under OS. In addition, under OS, soils in all suitability classes showed decreased salinity contents to 2 dSm^{-1}, which correspond to non-saline soil levels. No detectable change in salinity content was observed in nonsaline soil (<2 dSm^{-1}), while 10–18 reduction units were observed in highest-salinity soils with contents of 12–20 dSm^{-1}. Numerous soil management options have been proposed to decrease soil salinity, such as using low-salinity water to enhance the leaching of salts from the soil root zone [61]. The rate of plant growth under salt stress strongly varies among plant species [62,63]. Sodium saturation values can be reduced to low sodium levels of 5% with 4–20 reduction units. Previous studies have demonstrated that the addition of gypsum can lower high soil sodium saturation content [64–66] owing to its ability to absorb calcium instead of sodium in soil particles, directly leading to improved aggregation and decreased pH [65,67].

In addition, low values of SOM may affect the soil structure negatively [68]. Thus, it is recommended to raise the SOM level by adding organic amendments and residues of crops such as leguminous plants [69]. Similarly, about 1.6–7 reduction units are necessary to improve calcium carbonate content to the optimum <2% level. The best practice in the study area is to cultivate different seasonal crops and to avoid replanting the same plants in the same sites in order to maintain soil fertility and increase the SOM level [68–76]. This helps to maintain soil quality over the long run, which leads to an increased degree of crop yield and soil sustainability for different varieties of crops [68]. The spatial distribution of salinity, sodium saturation, and calcium carbonate under CS and their projected reduction units in each suitability class are shown in Figure 9. The status of the agricultural drainage system in the investigated area ranged from excessive to poor (Figure S6) and was predominantly poor under CS.

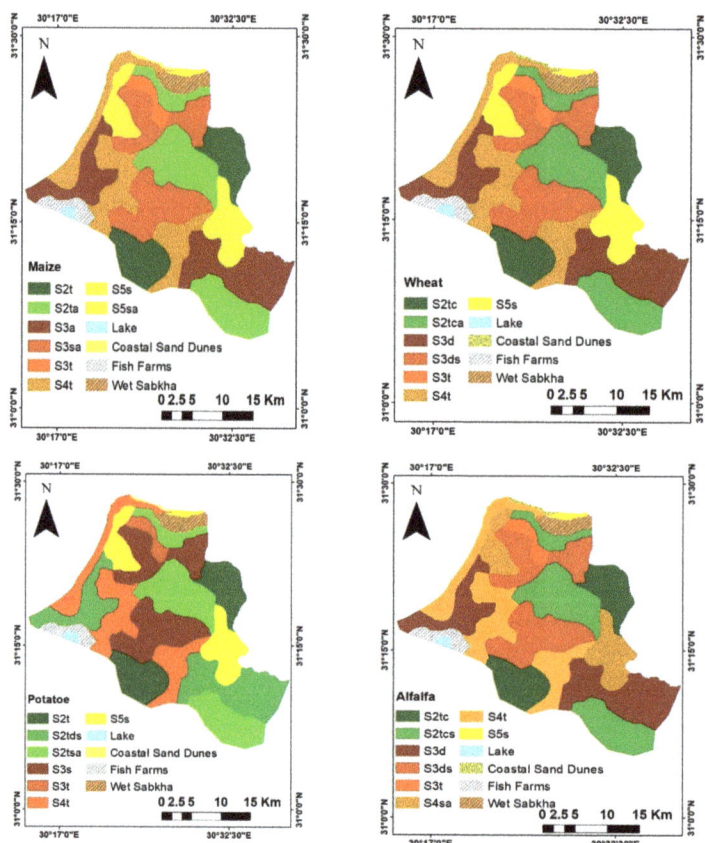

Figure 8. *Cont.*

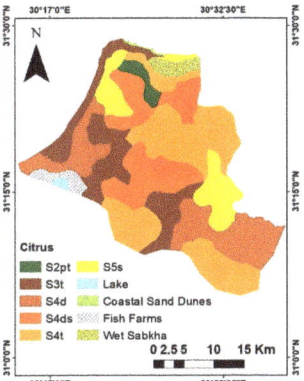

Figure 8. Maps showing soil suitability classes (S1–S5) for selected field and horticultural crops. Lowercase letters represent main soil limiting factors in each class; s, salinity; t, texture; a, sodium saturation; d, drainage; c, carbonate content; p, profile depth.

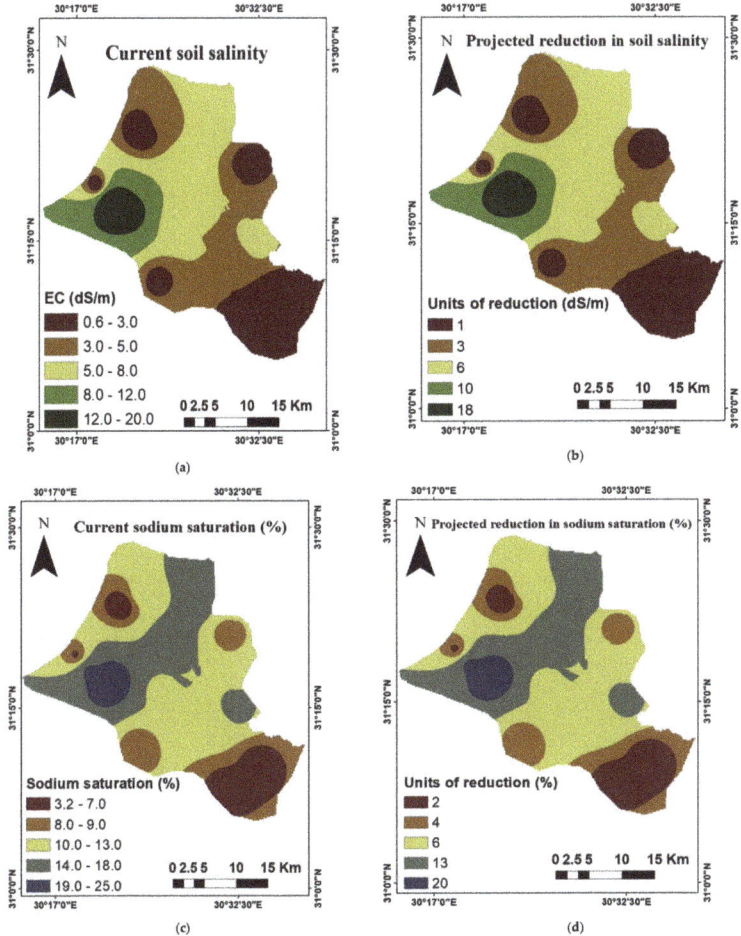

Figure 9. *Cont.*

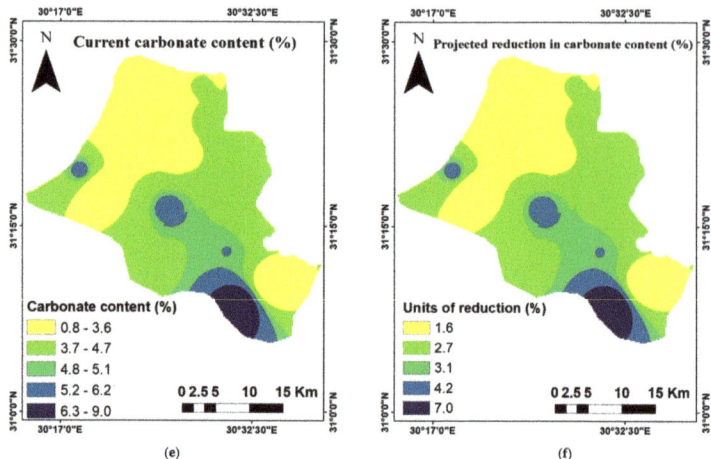

(e) (f)

Figure 9. Spatial distribution of soil factors under CS and their projected reduction units under OS: soil salinity (**a**,**b**), sodium saturation (**c**,**d**), and carbonate content (**e**,**f**).

3.3.6. Evaluation of Soil Suitability under CS and OS

Geomorphic features such as coastal sand dunes, wet sabkha, and fish farms, which cumulatively account for 4.52% of the total study area, were not considered in the suitability evaluation. In addition, water bodies (lakes), which account for 0.5% of the total study area, were not considered in the suitability evaluation. Under CS, subclasses 8–20 represented the main soil suitability subclasses, covering suitability classes S2, S3, S4, and S5 for most evaluated crops (Figure 8). With the application of OS, the moderate suitability class (S3) and the unsuitable class (S5) were enhanced to the suitable (S2) and marginally suitable (S4) classes, respectively, while the high suitability class (S1) showed increased area, which had significant effects on the suitability of maize crop (Table 4).

Table 4. The soil suitability for the five crops evaluated.

Class	Crops									
	Wheat		Maize		Potato		Alfalfa		Citrus	
	A	B	A	B	A	B	A	B	A	B
S1	–	–	–	39.9	–	–	–	–	–	–
S2	31.8	82.6	31.8	42.7	51.0	82.6	31.8	82.6	2.0	39.9
S3	39.9	–	39.9	–	20.7	–	39.9	–	17.4	17.4
S4	17.4	17.4	17.4	17.4	17.4	17.4	27.8	17.4	69.7	42.7
S5	10.9	–	10.9	–	10.9	–	0.5	–	10.9	–
Total	100	100	100	100	100	100	100	100	100	100

Note: (A) Current situation and (B) optimal scenario.

4. Conclusions

Integrated PCA and AHC analysis were used to classify soil capability within the study area, relying on the associations and interactions between soil characteristics. The study area could be classified into four classes relying on PCA. The main limiting factors within the study area included shallow depth, high salinity, and high $CaCO_3$ content in certain sites. Subsequently, multivariate analysis was used to assess soil capability based on its properties under different conditions. The observed crop suitability under CS can provide valuable information to decision-makers about key limiting factors. Moreover, evaluation of crop suitability under OS could potentially be used to predict the degree of improvements necessary to achieve agricultural sustainability. Similarly, remote sensing data are useful for extracting geomorphologic units, which are considered the base map for soil evaluation studies. GIS techniques are vital tools for mapping soil capability and crop suitability in order to achieve the best land use and food security in arid zones.

Supplementary Materials: The following supporting information can be downloaded at: https://www.mdpi.com/article/10.3390/land11071027/s1, Figure S1: Spatial distribution of chemical and physical soil properties: (a) electric conductivity (EC: dS/m), (b) soil reaction (pH), (c) exchangeable sodium percent (ESP), (d) calcium carbonate percentage ($CaCO_3$: %), (e) cation exchange capacity (CEC: cmolc/Kg), and (f) depth (cm); Figure S2; Spatial distribution of fertility soil properties: (a) (Available N: mg/kg), (b) (Available P: mg/kg), (c) (Available K: mg/kg), (d) Soil Organic Matter (SOM %); Figure S3:The IWQ index map of the study area Figure S4: Cultivated orchards (a) Mango and (b) Orange in the study area; Figure S5: Saline soils near the fish ponds south of Idku lake in the studied area; Figure S6: Very poorly drained soil in the study area; Table S1: Scores of all parameters; Table S2: Final SC range of study area; Table S3: Statistical characterization of the weighted mean of the studied soil profiles properties (n = 61).

Author Contributions: Conceptualization, R.A.E.B. and M.S.S. methodology, M.S.S.; software, R.A.E.B. and M.S.S.; validation R.A.E.B. and M.S.S.; formal analysis, R.A.E.B. and M.S.S.; investigation, M.S.S.; resources R.A.E.B. and M.S.S.; data curation, R.A.E.B., A.A.E.B., M.M.I. and M.S.S.; writing—original draft preparation, R.A.E.B.; writing—review and editing, A.A.E.B., M.M.I., E.S.M. and D.E.K.; visualization, A.A.E.B., M.M.I. and M.S.S.; supervision, A.A.E.B., M.M.I. and M.S.S.; project administration, A.A.E.B. and M.M.I.; funding acquisition, E.S.M. and D.E.K. All authors have read and agreed to the published version of the manuscript.

Funding: This research received no external funding.

Data Availability Statement: Not applicable.

Acknowledgments: This paper was supported by the RUDN University Strategic Academic Leadership Program.

Conflicts of Interest: The authors declare no conflict of interest.

References

1. Baroudy, A.A.; Ali, A.M.; Mohamed, E.S.; Moghanm, F.S.; Shokr, M.S.; Savin, I.; Poddubsky, A.; Ding, Z.; Kheir, A.M.S.; Aldosari, A.A.; et al. Modeling land suitability for rice crop using remote sensing and soil quality indicators: The case study of the Nile Delta. *Sustainability* **2020**, *12*, 9653. [CrossRef]
2. Tahmasebinia, F.; Tsumura, Y.; Wang, B.; Wen, Y.; Bao, C.; Sepasgozar, S.; Alonso-Marroquin, F. Floating Cities Bridge in 2050. In *Smart Cities and Construction Technologies*; IntechOpen: London, UK, 2020.
3. Debiagi, F.; Madeira, T.B.; Nixdorf, S.L.; Mali, S. Pretreatment eficiency using autoclave high-pressure steam and ultrasonication in sugar production from liquid hydrolysates and access to the residual solid fractions of wheat bran and oat hulls. *Appl. Biochem. Biotechnol.* **2020**, *190*, 166–181. [CrossRef] [PubMed]
4. Xiang, T.; Malik, T.H.; Nielsen, K. The impact of population pressure on global fertilizer use intensity, 1970–2011: An analysis of policy-induced mediation. *Technol. Forecast. Soc.* **2020**, *152*, 119895. [CrossRef]
5. Gerten, D.; Heck, V.; Jägermeyr, J.; Bodirsky, B.L.; Fetzer, I.; Jalava, M.; Kummu, M.; Lucht, W.; Rockström, J.; Schapho, S.; et al. Feeding ten billion people is possible within four terrestrial planetary boundaries. *Nat. Sustain.* **2020**, *3*, 200–208. [CrossRef]
6. Tir, J.; Diehl, P.F. Demographic Pressure and Interstate Conflict. In *Environmental Conflict*; Routledge: London, UK, 2018; pp. 58–83.
7. Shokr, M.S.; Abdellatif, M.A.; El Baroudy, A.A.; Elnashar, A.; Ali, E.F.; Belal, A.A.; Attia, W.; Ahmed, M.; Aldosari, A.A.; Szantoi, Z.; et al. Development of a Spatial Model for Soil Quality Assessment under Arid and Semi-Arid Conditions. *Sustainability* **2021**, *13*, 2893. [CrossRef]
8. Bakr, N.; Bahnassy, M.H. *Egyptian Natural Resources: In the Soils of Egypt*; Springer: Cham, Switzerland, 2019; pp. 33–49.
9. Satoh, M.; Aboulroos, S. *Irrigated Agriculture in Egypt: Past, Present and Future*; Springer: Cham, Switzerland, 2017.
10. Abd-Elmabod, S.K.; Fitch, A.C.; Zhang, Z.; Ali, R.R.; Jones, L. Rapid urbanisation threatens fertile agricultural land and soil carbon in the Nile Delta. *J. Environ. Manag.* **2019**, *252*, 109668. [CrossRef]
11. Saleh, A.M.; Belal, A.B.; Mohamed, E.S. Land resources assessment of El-Galaba basin, South Egypt for the potentiality of agriculture expansion using remote sensing and GIS techniques. *Egypt. J. Remote Sens. Space Sci.* **2015**, *18*, S19–S30. [CrossRef]
12. Food and Agriculture Organization of the United Nations. *Guidelines for Soil Profile Description*, 3rd ed.; Food and Agriculture Organization of the United Nations: Rome, Italy, 2006.
13. Mandal, S.; Choudhury, B.U.; Satpati, L. Soil site suitability analysis using geo-statistical and visualization techniques for selected winter crops in Sagar Island, India. *Appl. Geogr.* **2020**, *122*, 102249. [CrossRef]

14. Abdelrahman, M.A.E.; Shalaby, A.; Mohamed, E.S. Comparison of two soil quality indices using two methods based on geographic information system. *Egypt. J. Remote Sens. Space Sci.* **2019**, *22*, 127–136. [CrossRef]
15. Hammam, A.A.; Mohamed, E.S. Mapping soil salinity in the East Nile Delta using several methodological approaches of salinity assessment. *Egypt. J. Remote Sens. Space Sci.* **2020**, *23*, 125–131. [CrossRef]
16. Hassan, A.M.; Belal, A.A.; Hassan, M.A.; Farag, F.M.; Mohamed, E.S. Potential of thermal remote sensing techniques in monitoring waterlogged area based on surface soil moisture retrieval. *J. Afr. Earth Sci.* **2019**, *155*, 64–74. [CrossRef]
17. Elbeih, S.F.; El-Zeiny, A.M. Qualitative assessment of groundwater quality based on land use spectral retrieved indices: Case study Sohag Governorate, Egypt. *Remote Sens. Appl. Soc. Environ.* **2018**, *10*, 82–92. [CrossRef]
18. El Behairy, R.A.; El Baroudy, A.A.; Ibrahim, M.M.; Kheir, A.M.S.; Shokr, M.S. Modelling and assessment of irrigation water quality index using GIS in semi-arid region for sustainable agriculture. *Water Air Soil Pollut.* **2021**, 232–352. [CrossRef]
19. Mohamed, E.S.; Baroudy, A.A.E.; El-beshbeshy, T.; Emam, M.; Belal, A.A.; Elfadaly, A.; Aldosari, A.A.; Ali, A.M.; Lasaponara, R. Vis-NIR Spectroscopy and Satellite Landsat-8 OLI Data to Map Soil Nutrients in Arid Conditions: A Case Study of the Northwest Coast of Egypt. *Remote Sens.* **2020**, *12*, 3716. [CrossRef]
20. Burrough, P.A.; McDonnell, R.; McDonnell, R.A.; Lloyd, C.D. *Principles of Geographical Information Systems*; Oxford University Press: Oxford, UK, 2015.
21. Shokr, M.S.; El Baroudy, A.A.; Fullen, M.A.; El-beshbeshy, T.R.; Ali, R.R.; Elhalim, A.; Guerra, A.J.T.; Jorge, M.C.O. Mapping of heavy metal contamination in alluvial soils of the Middle Nile Delta of Egypt. *J. Environ. Eng. Landsc. Manag.* **2016**, *24*, 218–231. [CrossRef]
22. Abd-Elmabod, S.K.; Mansour, H.M.; Hussein, A.A.; Zhang, Z.; Anaya-Romero, M.; de la Rosa, D.; Jordán, A. Influence of Irrigation Water Quantity on the Land Capability Classification. *Plant Arch.* **2019**, *19*, 2253–2261.
23. Belal, A.A.; Mohamed, E.S.; Abu-Hashim, M.S.D. Land evaluation based on GIS-spatial multi-criteria evaluation (SMCE) for agricultural development in dry Wadi, Eastern Desert. *Egypt. Int. J. Soil Sci.* **2015**, *10*, 100–116. [CrossRef]
24. Mohamed, E.S.; Ali, A.; El-Shirbeny, M.; Abutaleb, K.; Shaddad, S.M. Mapping soil moisture and their correlation with crop pattern using remotely sensed data in arid region. *Egypt. J. Remote Sens. Space Sci.* **2019**, *23*, 347–353. [CrossRef]
25. Mohamed, E.S.; Saleh, A.M.; Belal, A.B.; Gad, A. Application of near-infrared reflectance for quantitative assessment of soil properties. *Egypt. J. Remote Sens. Space Sci.* **2018**, *21*, 1–14. [CrossRef]
26. Csomós, E.; Héberger, K.; Simon-Sarkadi, L. Principal component analysis of biogenic amines and polyphenols in Hungarian wines. *J. Agric. Food Chem.* **2002**, *50*, 3768–3774. [CrossRef]
27. Muñoz-Rojas, M.; Doro, L.; Ledda, L.; Francaviglia, R. Application of CarboSOIL model to predict the effects of climate change on soil organic carbon stocks in agro-silvo-pastoral Mediterranean management systems. *Agric. Ecosyst. Environ.* **2015**, *202*, 8–16. [CrossRef]
28. Muñoz-Rojas, M.; Abd-Elmabod, S.K.; Zavala, L.M.; De la Rosa, D.; Jordán, A. Climate change impacts on soil organic carbon stocks of Mediterranean agricultural areas: A case study in Northern Egypt. *Agric. Ecosyst. Environ.* **2017**, *238*, 142–152. [CrossRef]
29. De la Rosa, D.; Mayol, F.; Diaz-Pereira, E.; Fernandez, M. A land evaluation decision support system (MicroLEIS DSS) for agricultural soil protection. *Environ. Modell. Softw.* **2004**, *19*, 929–942. [CrossRef]
30. Abd-Elmabod, S.K.; Bakr, N.; Muñoz-Rojas, M.; Pereira, P.; Zhang, Z.; Cerdà, A.; Jordán, A.; Mansour, H.; De la Rosa, D.; Jones, L. Assessment of soil suitability for improvement of soil factors and agricultural management. *Sustainability* **2019**, *11*, 1588. [CrossRef]
31. Abd-Elmabod, S.K.; Jordán, A.; Fleskens, L.; Phillips, J.D.; Muñoz-Rojas, M.; Van der Ploeg, M.; Anaya-Romero, M.; De la Rosa, D. Modelling agricultural suitability along soil transects under current conditions and improved scenario of soil factors. In *Soil Mapping and Process Modeling for Sustainable Land Use Management*; Elsevier: Amsterdam, The Netherlands, 2017; pp. 193–219. [CrossRef]
32. Climatological Normal for Egypt. *The Normal for Beheira Governorate from 1960–2011*; Ministry of Civil Aviation, Meteorological Authority: Cairo, Egypt, 2011.
33. Soil Survey Staff. *Keys to Soil Taxonomy, USDA-NRCS*, 11th ed.; U.S. Government Print Office: Washington, DC, USA, 2014.
34. Dawoud, M.A.; Darwish, M.M.; El-Kady, M.M. GIS-based groundwater management model for Western Nile Delta. *Water Resour. Manag.* **2005**, *19*, 585–604. [CrossRef]
35. El Behairy, R.A. Using New Techniques for Studying Land Resources in Some Areas of North West Nile Delta, Egypt. Master's Thesis, Faculty of Agriculture, Tanta University, Cairo, Egypt, 2021.
36. Said, M.E.S.; Ali, A.M.; Borin, M.; Abd-Elmabod, S.K.; Aldosari, A.A.; Khalil, M.M.N.; Abdel-Fattah, M.K. On the use of multivariate analysis and land evaluation for potential agricultural development of the Northwestern Coast of Egypt. *Agronomy* **2020**, *10*, 1318. [CrossRef]
37. El Baroudy, A.A. Geomatics-based soil mapping and degradation risk assessment of Nile delta soils. *Pol. J. Environ. Stud.* **2010**, *1123*, 1131.
38. Antoniadis, V.; Shaheen, S.M.; Levizou, E.; Shahid, M.; Niazi, N.K.; Vithanage, M.; Ok, Y.S.; Bolan, N.; Rinklebe, J. A critical prospective analysis of the potential toxicity of trace element regulation limits in soils worldwide: Are they protective concerning health risk assessment?—A review. *Environ. Int.* **2019**, *127*, 819–847. [CrossRef] [PubMed]
39. Rhoades, J.D. Salinity: Electrical Conductivity and Total Dissolved Solids. In *Methods of Soil Analysis Part 3, Chemical Methods*; Sparks, D.L., Ed.; Soil Science Society of America Book Series, No. 5; Soil Science Society of America, American Society of Agronomy: Madison, WI, USA, 1996; pp. 417–435.

40. Thomas, G.W. Soil pH and Soil Acidity. In *Methods of Soil Analysis Part 3, Chemical Methods*; Sparks, D.L., Ed.; Soil Science Society of America Book Series, No. 5; Soil Science Society of America, American Society of Agronomy: Madison, WI, USA, 1996; pp. 475–490.
41. Summer, M.E.; Miller, W.P. Cation Exchange Capacity and Exchange Coefficients. In *Methods of Soil Analysis Part 3. Chemical Methods*; Sparks, D.L., Ed.; Soil Science Society of America Book Series, No. 5; Soil Science Society of America, American Society of Agronomy: Madison, WI, USA, 1996; pp. 1201–1229.
42. Lavkulich, L.M. *Methods Manual: Pedology Laboratory*; Department of Soil Science, University of British Columbia: Vancouver, BC, Canada, 1981.
43. Page, A.L.; Miller, R.H.; Keeney, D.R. *Methods of Soil Analysis (Part 2): Chemical and Microbiological Properties*, 2nd ed.; The American Society of Agronomy: Madison, WI, USA, 1982.
44. De La Mora-Orozco, C.; Flores-Lopez, H.; Rubio-Arias, H.; Chavez-Duran, A.; Ochoa-Rivero, J. Developing a water quality index (WQI) for an irrigation dam. *Int. J. Environ. Res. Public Health* **2017**, *14*, 439. [CrossRef] [PubMed]
45. Simsek, C.; Gunduz, O. IWQ index: A GIS-integrated technique to assess irrigation water quality. *Environ. Monit. Assess.* **2007**, *128*, 277–300. [CrossRef] [PubMed]
46. Jolliffe, I.T.; Cadima, J. Principal component analysis: A review and recent developments. *Philos. Trans. A Math. Phys. Eng. Sci.* **2016**, *374*, 20150202. [CrossRef]
47. Chen, Y.D.; Wang, H.Y.; Zhou, J.M.; Xing, L.; Zhu, B.S.; Zhao, Y.C.; Chen, X.Q. Minimum data set for assessing soil quality in farmland of Northeast China. *Pedosphere* **2013**, *23*, 564–576. [CrossRef]
48. Jagadamma, S.; Lal, R.; Hoeft, R.G.; Nafziger, E.D.; Adee, E.A. Nitrogen fertilization and cropping system impacts on soil properties and their relationship to crop yield in the central Corn Belt, USA. *Soil Till. Res.* **2008**, *98*, 120–129. [CrossRef]
49. Imperato, M.; Adamo, P.; Naimo, D.; Arienzo, M.; Stanzione, D.; Violante, P. Spatial distribution of heavy metals in urban soils of Naples city (Italy). *Environ. Pollut.* **2003**, *124*, 247–256. [CrossRef]
50. McGrath, D.; Zhang, C.; Carton, O.T. Geostatistical analyses and hazard assessment on soil lead in Silvermines area, Ireland. *Environ. Pollut.* **2004**, *127*, 239–248. [CrossRef]
51. Lee, C.S.L.; Li, X.; Shi, W.; Cheung, S.C.N.; Thornton, I. Metal contamination in urban, suburban, and country park soils of Hong Kong: A study based on GIS and multivariate statistics. *Sci. Total Environ.* **2006**, *356*, 45–61. [CrossRef]
52. Franzen, D.W.; Peck, T.R. Field soil sampling density for variable rate fertilization. *J. Prod. Agric.* **1995**, *8*, 568–574. [CrossRef]
53. Weisz, R.; Fleischer, S.; Smilowitz, Z. Map generation in highvalue horticultural integrated pest management: Appropriate interpolation methods for site-specific pest management of Colorado potato beetle (Coleoptera: Chrysomelidae). *J. Econ. Entomol.* **1995**, *88*, 1650–1657. [CrossRef]
54. Ali, R.R.; Moghanm, F.S. Variation of soil properties over the landforms around Idku lake, Egypt. *Egypt. J. Remote Sens. Space Sci.* **2013**, *16*, 91–101. [CrossRef]
55. De la Rosa, D.; Cardona, F.; Paneque, G. Evaluación de suelos para diferentesusosagrícolas. Un sistemadesarrollado para regionesmediterráneas. *An. Edafol. Agrobiol.* **1977**, *36*, 1100–1112.
56. Elbasiouny, H.; Abowaly, M.; Abu_Alkheir, A.; Gad, A.A. Spatial variation of soil carbon and nitrogen pools by using ordinary kriging method in an area of North Nile Delta, Egypt. *Catena* **2014**, *113*, 70–78. [CrossRef]
57. Abdel-Fattah, M.K.; Abd-Elmabod, S.K.; Aldosari, A.A.; Elrys, A.S.; Mohamed, E.S. Multivariate Analysis for Assessing Irrigation Water Quality: A Case Study of the Bahr Mouise Canal, Eastern Nile Delta. *Water* **2020**, *12*, 2537. [CrossRef]
58. Wandruszka, R.V. Phosphorus retention in calcareous soils and the effect of organic matter on its mobility. *Geochem. Trans.* **2006**, *7*, 6. [CrossRef]
59. Nehrani, S.H.; Askari, M.S.; Saadat, S.; Delavar, M.A.; Taheri, M.; Holden, N.M. Quantification of soil quality under semi-arid agriculture in the northwest of Iran. *Ecol. Indic.* **2020**, *108*, 105770. [CrossRef]
60. Mohamed, E.S.; Abu-Hashim, M.; Abdelrahman, M.A.; Schütt, B.; Lasaponara, R. Evaluating the effects of human activity over the last decades on the soil organic carbon pool using satellite imagery and GIS techniques in the Nile Delta Area, Egypt. *Sustainability* **2019**, *11*, 2644. [CrossRef]
61. Zalacáin, D.; Martínez-Pérez, S.; Bienes, R.; García-Díaz, A.; Sastre-Merlín, A. Salt accumulation in soils and plants under reclaimed water irrigation in urban parks of Madrid (Spain). *Agric. Water Manag.* **2019**, *213*, 468–476. [CrossRef]
62. Qadir, M.; Schubert, S. Degradation processes and nutrient constraints in sodic soils. *Land Degrad. Dev.* **2002**, *13*, 275–294. [CrossRef]
63. Jacobsen, S.-E.; Jensen, C.R.; Liu, F. Improving crop production in the arid Mediterranean climate. *Field Crops Res.* **2012**, *128*, 34–47. [CrossRef]
64. Food and Agriculture Organization. Salt-affected soils and their management. In *Soils Bulletin*; Food and Agriculture Organization: Rome, Italy, 1988; p. 39.
65. Chi, C.M.; Zhao, C.W.; Sun, X.J.; Wang, Z.C. Reclamation of saline-sodic soil properties and improvement of rice (*Oriza sativa* L.) growth and yield using desulfurized gypsum in the west of Songnen Plain, northeast China. *Geoderma* **2012**, *187*, 24–30. [CrossRef]
66. Rasouli, F.; Pouya, A.K.; Karimian, N. Wheat yield and physico-chemical properties of a sodic soil from semi-arid area of Iran as affected by applied gypsum. *Geoderma* **2013**, *193–194*, 246–255. [CrossRef]
67. Temiz, C.; Cayci, G. The effects of gypsum and mulch applications on reclamation parameters and physical properties of an alkali soil. *Environ. Monit. Assess.* **2018**, *190*, 347. [CrossRef] [PubMed]

68. Zakarya, Y.M.; Metwaly, M.M.; AbdelRahman, M.A.E.; Metwalli, M.R.; Koubouris, G. Optimized Land Use through Integrated Land Suitability and GIS Approach in West El-Minia Governorate, Upper Egypt. *Sustainability* **2021**, *13*, 12236. [CrossRef]
69. Michalopoulos, G.; Kasapi, K.A.; Koubouris, G.; Psarras, G.; Arampatzis, G.; Hatzigiannakis, E.; Kavvadias, V.; Xiloyannis, C.; Montanaro, G.; Malliaraki, S. Adaptation of Mediterranean olive groves to climate change through sustainable cultivation practices. *Climate* **2020**, *8*, 54. [CrossRef]
70. Leteinturier, B.; Herman, J.; Longueville, F.D.; Quintin, L.; Oger, R. Adaptation of a crop sequence indicator based on a land parcel management system. *Agric. Ecosyst. Environ.* **2006**, *112*, 324–334. [CrossRef]
71. Singha, C.; Swain, K.C. Land Suitability Evaluation Criteria for Agricultural crop selection: A Review. *Agric. Rev.* **2016**, *37*, 125–132. [CrossRef]
72. Lenz-Wiedemann, V.I.S.; Klar, C.W.; Schneider, K. Development and test of a crop growth model for application within a Global Change decision support system. *Ecol. Model.* **2010**, *221*, 314–329. [CrossRef]
73. Lorenz, M.; Fürst, C.; Thiel, E. A methodological approach for deriving regional crop rotations as basis for the assessment of the impact of agricultural strategies using soil erosion as example. *J. Environ. Manag.* **2013**, *127*, S37–S47. [CrossRef]
74. Abuzaid, A.S.; Jahin, H.S.; Asaad, A.A.; Fadl, M.E.; AbdelRahman, M.A.E.; Scopa, A. Accumulation of Potentially Toxic Metals in Egyptian Alluvial Soils, Berseem Clover (*Trifolium alexandrinum* L.), and Groundwater after Long-Term Wastewater Irrigation. *Agriculture* **2021**, *11*, 713. [CrossRef]
75. Abuzaid, A.S.; AbdelRahman, M.A.E.; Fadl, M.E.; Scopa, A. Land Degradation Vulnerability Mapping in a Newly-Reclaimed Desert Oasis in a Hyper-Arid Agro-Ecosystem Using AHP and Geospatial Techniques. *Agronomy* **2021**, *11*, 1426. [CrossRef]
76. AbdelRahman, M.A.E.; Rehab, H.H.; Yossif, T.M.H. Soil fertility assessment for optimal agricultural use using remote sensing and GIS technologies. *Appl. Geomat.* **2021**, *13*, 605–619. [CrossRef]

Article

Assessment of Soil Fertility Status under Soil Degradation Rate Using Geomatics in West Nile Delta

Mohamed A. E. AbdelRahman [1], Mohamed M. Metwaly [2], Ahmed A. Afifi [3], Paola D'Antonio [4] and Antonio Scopa [4,*]

[1] Division of Environmental Studies and Land Use, National Authority for Remote Sensing and Space Sciences (NARSS), Cairo 11769, Egypt
[2] Data Reception, Analysis and Receiving Station Affairs Division, National Authority for Remote Sensing and Space Sciences, Cairo 11769, Egypt
[3] Soils and Water Use Department, National Research Centre, Giza 12622, Egypt
[4] Scuola di Scienze Agrarie, Forestali, Alimentari ed Ambientali (SAFE), Università degli Studi della Basilicata, Viale dell'Ateneo Lucano, 10, 85100 Potenza, Italy
* Correspondence: antonio.scopa@unibas.it; Tel.: +39-0971-205240

Abstract: The presence of a noticeable rate of degradation in the land of the Nile Delta reduces the efficiency of crop production and hinders supply of the increasing demand of its growing population. For this purpose, knowledge of soil resources and their agricultural potential is important for determining their proper use and appropriate management. Thus, we investigated the state of soil fertility by understanding the effect of the physical and chemical properties of the soil and their impact on the state of land degradation for the years 1985, 2002 (ancillary data), and 2021 (our investigation). The study showed that there are clear changes in the degree of soil salinity as a result of agricultural management, water conditions, and climatic changes. The soil fertility is obtained in four classes: Class one (I) represents soils of a good fertility level with an area of about 39%. Class two (II) includes soils of an average fertility level, on an area of about 7%. Class three (III) includes soils with a poor level of fertility, with an area of about 17%. Class four (IV) includes soils of a very poor level of fertility with an area of about 37% of the total area. Principal component analysis (PCA) has revealed that the parameters that control fertility in the studied soils are: C/N, pH, Ca, CEC, OM, P, and Mg. Agro-pedo-ecological units are important units for making appropriate agricultural decisions in the long term, which contribute to improving soil quality and thus increasing the efficiency of soil fertility processes.

Keywords: El-Beheira–Nile Delta; fertility parameters; physical-chemical properties; soil fertility; soil degradation

Citation: AbdelRahman, M.A.E.; Metwaly, M.M.; Afifi, A.A.; D'Antonio, P.; Scopa, A. Assessment of Soil Fertility Status under Soil Degradation Rate Using Geomatics in West Nile Delta. *Land* **2022**, *11*, 1256. https://doi.org/10.3390/land11081256

Academic Editors: Chiara Piccini and Rosa Francaviglia

Received: 21 July 2022
Accepted: 4 August 2022
Published: 6 August 2022

Publisher's Note: MDPI stays neutral with regard to jurisdictional claims in published maps and institutional affiliations.

Copyright: © 2022 by the authors. Licensee MDPI, Basel, Switzerland. This article is an open access article distributed under the terms and conditions of the Creative Commons Attribution (CC BY) license (https://creativecommons.org/licenses/by/4.0/).

1. Introduction

Soil salinization, and physical and biological degradation are among the most prominent global environmental challenges because they have a negative impact on agricultural output and fair development. The agriculture of Nile Delta has many problems related to productivity [1], yet it remains the main economic activity and source of field crops for the population [2,3]. The assessment of soil fertility and degradation is therefore fundamental in order to help find the optimal conditions for plant growth [4]. Both organic and mineral components of soil create an interactive natural environment that fosters plant growth [5]. Its physical, chemical, and biological characteristics enable it to supply nutrients in adequate proportion and balance for plant growth [6–8]. This serves as the foundation for all input-based high-production systems. [9]. The demand for land for agricultural production has also increased due to Egypt's rising population. However, detailed information on the level of soil fertility/degradation in Egypt is still quite limited [10–13]. For the purpose of guiding agricultural management decisions, Egyptian farmers need the most recent information on soil quality [14]. The main economic activity is agriculture as a source of field crops for the population. Soil fertility and degradation are among the most important productivity problems facing agriculture in the Nile Delta. Therefore, it is still necessary to evaluate the state of the land and the dynamics of its changing properties to suggest optimum conditions for plant growth as one of the most important inputs to increase productivity [11–13].

Currently, technological advancements in the geospatial field have introduced greater ease for choice makers by presenting a list of alternatives for problem structuring [15,16] of actual global issues of a multidimensional nature, and evaluating alternatives by outlining the link between input and output maps [17]. The ever-growing availability of earth observation data and the well-established use of GIS leads to the development of automated workflows and toolboxes for environmental management [18–20]. They enable time savings, objective and non-biased judgments, and a systematic and spatially clear evaluation framework. [18,19]. With the fast development of GIS and computer technology, this approach has been widely employed in research programs. Currently, the most significant uses of GIS are in assessment, providing ecological capability maps, land management, and land planning [20]. This integrated strategy allows for the incorporation of expert judgments into geographical data. This integrated strategy has been employed in a number of research projects to deal with situations in which multi-criteria judgments were evaluated using geographical data [21]. Mapping soil deterioration is a time-consuming and labor-intensive process. Modeling degradation processes allows for the prediction of deterioration [22,23]. El-Beheira Governorate is one of the area's most vulnerable to land degradation [24].

Soil is the deposit factor for fertility, which is the fruit of the interaction between climate and vegetation, and since the soil cannot be restored to become rich in nutrients through time, knowing the causes of the actual degradation can accelerate the appropriate agricultural management related to improving soil fertility through good management of soil quality determinants. Thus, the identification of areas with low values of soil fertility indicates the long-term progression of degradation. Based on the foregoing, this study intends to assess fertility and degradation for sustainable land management.

2. Materials and Methods

2.1. Study Area

El-Beheira Governorate is located in the west of the Delta. It is bordered to the north by the Mediterranean Sea, to the east by the Rashid branch, to the west by the governorates of Alexandria and Matrouh, and to the south by the Giza governorate. According to Khalil et al. [24] the landscape of the newly cultivated soil in El-Beheira governorate is low to flat land and the maximum ground elevation is 30 m above sea level. Ground surface elevation is below sea level in Abu EL Matamir and Hush Isa counties, north of the governorate. Geomorphology characterization of the governorate is characterized by two major units: tablelands and alluvial plains. Tablelands are extended towards the southwest of the coastal plains and its surface is covered by sandy limestone on the western and southern sides. Tablelands are distinguished by three landforms which are ridges, depressions, and erg plain. The alluvial plains have a young alluvial plain and an old alluvial plain. The young alluvial plain represents a portion of the oldest cultivated land in the northwest of the Nile Delta. Its surface is covered with clay beds alternated with a thin band of silt where it lies between the Abu Mina Depression in the west and the Rosette branch of the Nile in the east, while the old alluvial plain lies in the south of the young alluvial plain toward the northern and eastern areas of Wadi El Natrun [25]. These plain slopes are toward the north and northeast and vary in elevation between 20 and 60 m. A sandy deposit has covered the northern part surface, whereas gravelly deposits dominate in the southern part, Figure 1.

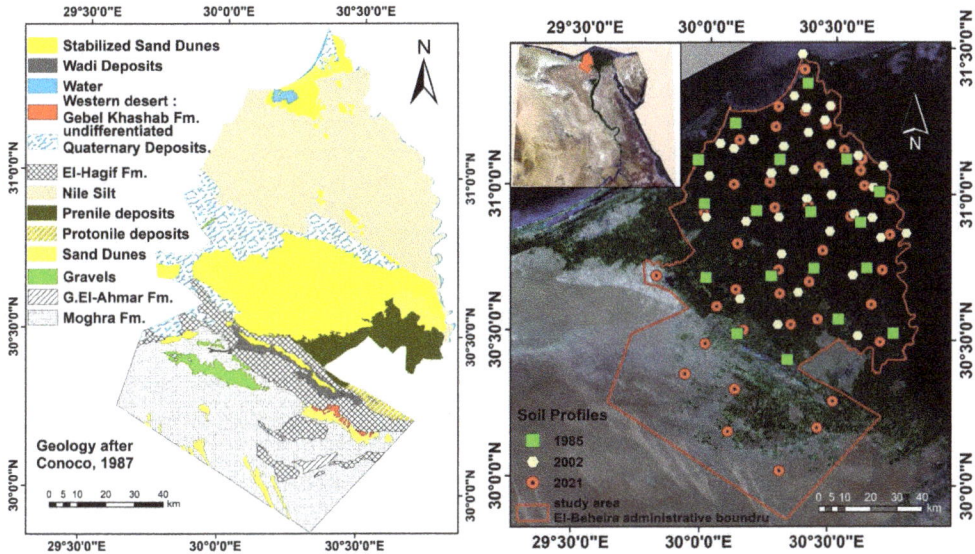

Figure 1. Location of the study area, Geology after Conoco [25].

Figure 2 gives indications of typical climate patterns and expected conditions (temperature, precipitation, sunshine, and wind). The diagram indicates that the soil moisture regime is torric and the soil temperature regime is thermic. The historical data (2004–2021) showed a mean yearly temperature of 21.2 °C and mean monthly temperatures ranging from 13.7 °C in January to 27.8 °C in August, while 34.5 °C in July and 7.8 °C in January were the highest and lowest temperatures, respectively, for the past ten years. Between 52% in May and 70% in November and December, the average monthly relative humidity varied. At around 2 p.m., the relative humidity reaches its lowest point (approximately 30%). In December and January, the wind blows at 11 km/h, while in May and June, it blows at 20 km h^{-1}. Last but not least, winds that are generally present are those that originate in the north, northwest, and northeast.

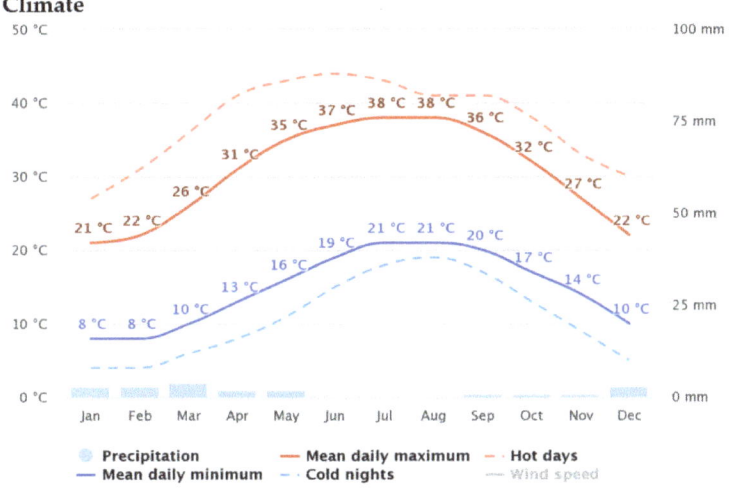

Figure 2. The climate diagrams are based on 30 years of hourly weather model simulations.

2.2. Soil Sampling and Analysis

Ancillary data for the years 1985 and 2002 were collected from [26,27]. The field campaign in 2021 consisted of a soil survey by digging profiles with a thickness ranging from about 40–150 cm, with a detailed study of the agricultural horizon. Then, soil samples were taken from different layers and horizons along the crop profile. These samples were taken from the different identified soil groups, taking into consideration the morphology of each soil group. A total of 121 soil samples were collected from 35 soil profiles distributed in the region to cover the existed landform units. Once the samples were taken, they were taken to a laboratory where physical and chemical analyses were performed. Chemical analyses included: soil organic carbon (OC), total nitrogen (N), available P, total P, exchangeable cations (Ca, Mg, K, Na), cation exchange capacity (CEC), and pH. OC was extracted by oxidation with potassium dichromate in a highly acidic solution and quantified using a TOC-5000A analyzer (Shimadzu, Japan). Total N was determined by the Kjeldahl method [28]. The available P and total P were determined by the Bray II method [29], and exchangeable cations were extracted by ammonium acetate ($C_2H_3O_2NH_4$) buffered at pH = 7, and quantified using an atomic absorption spectrophotometer (AAnova 350 Analytic Jena GmbH, Thuringia, Germany) [30]. CEC at pH 7 was determined using the ammonium acetate method [31]. The soil pH was determined in a 1:2.5 soil suspension with demineralized water. Physical analyses performed were bulk density (BD) and particle size distribution. The bulk density (BD) was obtained using the Koppeki cylinder method [32]. With regard to the particle size distribution was determined by the hydrometer method [33,34]. Binary and ternary diagrams models were used [35–37] to establish balances between soil textural, pH and nutrient concentrations These diagrams display various soil texture grades in connection to their agronomic significance. The poles of relative richness in a given cation in the equilibrium of the cationic balance (Ca/Mg/K) are shown in the triangular diagram of Dabin [36].

The SSI (structural stability index) is a physical parameter that determines the degree of degradability and erodibility of soil. It is defined by Pieri [38] according to the report:

$$SSI = \frac{1.724 \times OC}{L + A} \times 100; \quad 0 \leq ISS \leq \infty \tag{1}$$

with OM: soil organic matter content, A: clay content in the soil. L: Silt content in the soil. SSI > 9% indicates soils with a stable structure; 7% < SSI ≤ 9% indicates soils with a low risk of structural degradation; 5% < ISS ≤ 7% indicates soils with a high risk of structural degradation; 5% < ISS indicate soils with a degraded structure.

The beat index (BI) indicates the risk of erosion in compaction of one. Remy et al. [39] formula for estimating the risks of beating is written:

$$BI = \frac{(1.5 \times Lf) + (0.75 \times LG)}{(A - 10 \times OM)} - C \tag{2}$$

with C = 0.2 × (pH-7), LF = Fine silt, LG = coarse silt, A = clay, OM = organic matter. BI < 1.4% indicates non-beating soils; 1.4% < IB ≤ 1.6% indicates low-beating soils; 1.6% < BI ≤ 1.8% indicates beating soils; IB > 1.8% indicate very beating soils.

The Forestier index [35] indicates the reserve in bases exchangeable in the ground.

$$FI = \frac{S^2}{(A + Lf)} \tag{3}$$

When this index is above 1.5 (FI > 1.5) the reserve in exchangeable bases is good and when FI < 1.5 the reserve in the base is low.

The balances between certain physical-chemical properties have been established and reported on the diagrams according to the models used by other authors [27–30]. The different textural classes are given from the FAO textural diagram, thus characterizing the different groups of soils of north El-Beheira and its surroundings on the agro-pedo-ecological level. The Ca-Mg-K ternary fertility diagram after Dabin [36] highlights the thresholds of deficiency and a relative deficiency in a given cation in the equilibrium of the cationic balance. Dabin's [36] diagram on the N-pH equilibrium highlights the nitrogen contents carried on the abscissa and the pH carried in the ordinate. The limitations are defined by the pH values carried in ordinate, which define horizontal lines and show only the influence of the pH on the total nitrogen reserve. It defines four levels of chemical fertility (low, poor, medium, and good) of soils according to their degree of pH. Dabin's diagram [36] makes it possible to highlight this antagonism or synergy between K-Mg cations in the soil. The Ca/K diagram relating to the binary fertility diagram of Martin [37] establishes the balance between calcium and potassium in soils. The data obtained, after analysis of the samples in the laboratory, were processed statistically using SPSS (IBM SPSS Statistics for Windows, Version 25.0. Armonk, NY, USA: IBM Corp.). A descriptive statistical analysis of 15 variables was used to compare mean and standard deviations by soil group in the study area. Principal component analysis (PCA) determined the parameters that control fertility in the investigated soils.

According to FAO-ISRIC [23] guidelines, a quantitative assessment of soil degradation was performed with an emphasis on salinization and nutrient loss. The weighting of criteria and sub-criteria was carried out using the AHP method. In fact, AHP and the GIS in an integrated technique were used for mapping vulnerable areas to degrade. This process was done in three main steps. In the first step, the most important criteria and sub-criteria that affect vulnerability to degradation were determined. The initial criteria and sub-criteria were chosen based on the study area's conditions, expert comments, and a literature review. Five criteria were elected and weighted using AHP. Pairwise comparison was used to allocate weights to criteria rating (Table 1) and sub-criteria and the Consistency Ratio also was calculated to verify the coherence of the judgments [40].

Table 1. Pair wise comparison matrix of criteria in AHP.

	Clay	EC	ESP	OM	CaCO$_3$	Weightage
Clay	1	3	3	5	5	0.43
EC	0.33	1	3	4	4	0.26
ESP	0.33	0.33	1	3	3	0.15
OM	0.2	0.25	0.33	1	4	0.10
CaCO$_3$	0.20	0.25	0.33	0.25	1	0.05

Note: n = 5, λ_max = 5.45, RCI = 1.12, CI = 0.08; CR = 0.10. n: no. of parameters, λ_max: maximum eigenvalue, RCI: random consistency index, CI: consistency index, CR: consistency ratio.

For weighting (criteria and sub-criteria), a questionnaire was delivered, which was associated with a group of experts who were aware of the area and degradation perception. Experts used Saaty's scale of pairwise comparisons to evaluate the importance of criteria and sub-criteria [40]. Relative weights for the functions were calculated based on input from the experts.

The research indices' weight values, which are absolute integers between zero and 100%, indicate the priority given to them. This implies that the total weights applied to all parameters should equal 100%. Table 1 summarizes the degradation factors and how they were ranked in terms of their impact on degradation occurrences in the research region, Table 1. The weights assigned to each criterion/index are frequently based on a professional understanding of each parameter's relevance and are occasionally based on analytical techniques and literature.

The steps involved in soil assessment in the study area are shown sequentially in Figure 3. The figure shows the steps followed in the Model Builder in the GIS program. It started with building a database of various analyzes, then the rating and the creation of intermediate maps, then working and producing for all the factors, then getting the final results of the used indices and degradation rate which combined to produce agropedological zones.

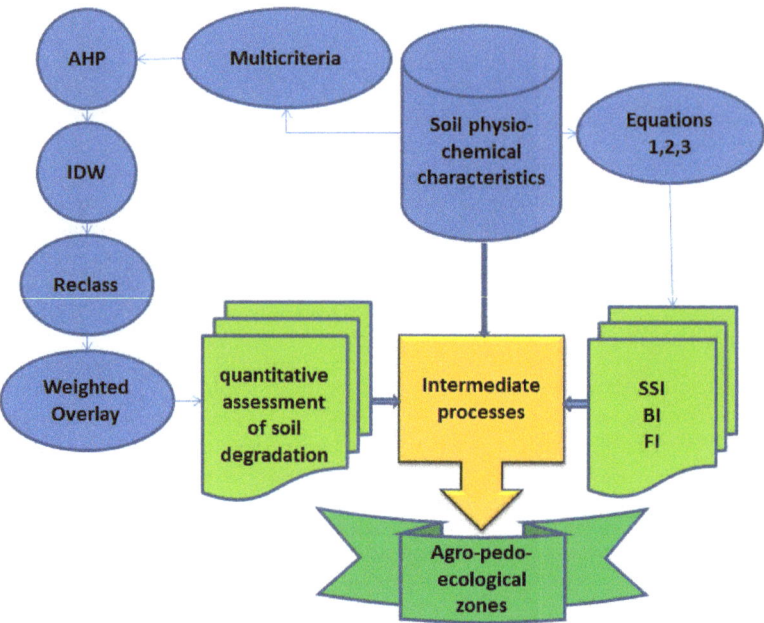

Figure 3. Methodology flowchart.

3. Results

Soil taxonomy contains 11 units, as shown in Figure 4. The alluvial soil texture is fine textured (clay loam to clay) and the water table is between 90 and 160 cm. The depth varies from the south of the alluvial soil, which is characterized by the depth of the soil profile being the deepest, and the depth decreases in the north of the alluvial soil, especially in the areas adjacent to Lake Edko, less than (70 cm) with low to moderate depth level. Overall, the majority of the alluvial areas have a high clay texture (44 to 65%). Concerning the coastal lands, with clay content (13 to 42%) with texture sandy to clay-loam. Soils are affected by salinity (EC > 4.0 dS m^{-1}) while the pH varies from being near neutral (pH = 7) to an alkaline value of 8.83.

3.1. Variation of Soils Physical Properties

The results show significant variability across different soils (Tables 2–4). The beating index (BI) shows that, with the exception of which has very flappy soils (BI > 1.8%), all old alluvial plain soils are non-beating (i.e., BI < 1.4), therefore, have a beating crust on their surface, thus increasing the cohesion of the soil and therefore its resistance to detachment. In theory, the beating crust should reduce the rate of erosion. However, even if it increases the soil's resistance to detachment, it greatly decreases the rate of water infiltration and increases the rate of runoff.

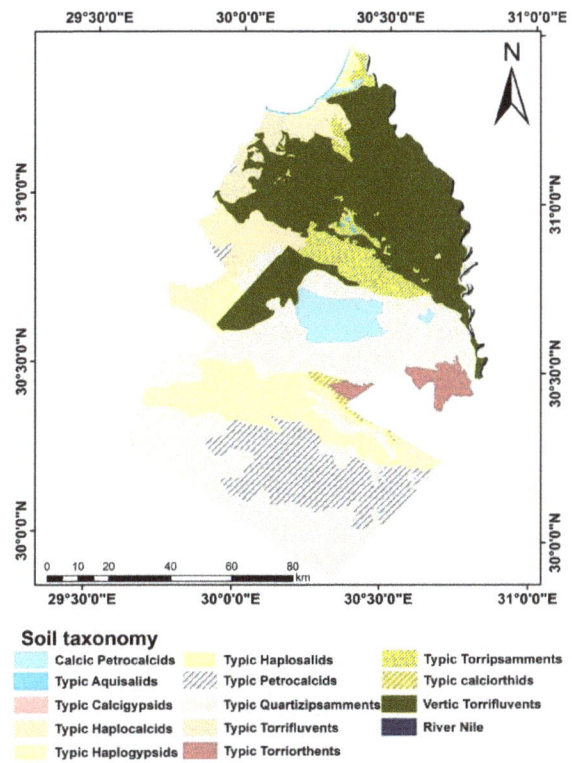

Figure 4. Soil taxonomy of the study area.

Table 2. Summary statistics of soil characteristics of study area.

Year 1985	Depth	Clay	CaCO$_3$	OM	EC	ESP	CEC	AN	AP	AK	pH
Count	18	18	18	18	18	18	18	18	18	18	18
Min	90	40	0.8	0.3	1.8	11.2	7.2	38.9	1.4	65.8	7.6
Max	130	52.3	12.81	1.9	42	20.5	46.2	87.35	5.9	92.3	8.37
Mean	105.6	45.1	4.3	1.2	18.9	14.4	28.2	58.7	3.1	79.4	8.1
Median	100	45.2	3.7	1.5	7.5	14.6	32.8	59.7	2.4	76.2	8.1
Stdev	13.4	3.5	3.1	0.6	17.2	2.3	13.2	17.2	1.3	9.0	0.2
Variance	179.1	12.4	9.6	0.4	297.3	5.5	174.0	296.1	1.8	80.9	0.1
Skewness	0.6	0.5	1.3	−0.6	0.3	0.9	−0.5	0.3	0.8	0.1	−0.7
Kurtosis	−0.8	−0.2	2.0	−1.3	−2.0	1.3	−1.3	−1.5	−0.5	−1.6	−0.3
Coefficient of variation	0.13	0.08	0.72	0.48	0.91	0.16	0.47	0.29	0.43	0.11	0.03

Table 3. Summary statistics of soil characteristics of study area.

Year (2002)	Depth	Clay	CaCO$_3$	OM	EC	ESP	CEC	AN	AP	AK	pH
count	35	35	35	35	35	35	35	35	35	35	35
Min	90	39.3	0.8	0.3	1.2	8.5	30.1	38.9	1.2	54.5	7.7
Max	150	62.4	5.9	2.3	4.6	15.4	53.6	91.0	8.6	112.3	8.4
Mean	119.4	48.1	3.0	1.7	2.5	12.6	40.3	68.8	3.8	89.2	8.0
Median	120	47.3	3.2	1.7	2.3	12.5	39.5	73.6	3.3	89.4	8.0
Stdev	17.5	4.7	1.2	0.4	0.9	1.8	6.0	15.5	2.2	15.5	0.2
Variance	305.5	22.4	1.5	0.2	0.9	3.2	35.4	239.8	4.6	240.6	0.0
Skewness	0.1	0.7	0.2	−1.2	0.6	−0.3	0.3	−0.6	0.9	−0.5	0.9
Kurtosis	−1.0	1.1	−0.4	2.4	−0.7	−0.7	−0.9	−1.0	−0.1	−0.4	0.3
Coefficient of variation	0.15	0.10	0.41	0.25	0.38	0.14	0.15	0.22	0.57	0.17	0.02

Table 4. Summary statistics of soil characteristics of study area.

Year (2021)	Depth	Clay	CaCO$_3$	OM	EC	ESP	CEC	AP	AK	pH
count	35	35	35	35	35	35	35	35	35	35
Min	90	3.5	1.2	0.3	0.8	8.5	30.1	1.2	54.5	6.8
Max	150	77.0	15.0	3.1	10.7	15.4	53.6	8.6	112.3	8.4
Mean	129.3	39.3	6.4	1.0	2.9	12.6	40.3	3.8	89.2	7.7
Median	130	43.4	5.8	0.9	2.5	12.5	39.5	3.3	89.4	7.7
Stdev	17.7	20.4	3.4	0.7	1.9	1.8	6.0	2.2	15.5	0.4
Variance	312.0	416.8	11.6	0.4	3.7	3.2	35.4	4.6	240.6	0.1
Skewness	−0.4	−0.5	0.7	1.8	2.4	−0.3	0.3	0.9	−0.5	−0.1
Kurtosis	−0.7	−0.8	−0.3	4.1	7.5	−0.7	−0.9	−0.1	−0.4	0.0
Coefficient of variation	0.14	0.52	0.54	0.63	0.67	0.14	0.15	0.57	0.17	0.05

The beating crust of its soils may exhibit mechanical resistance to root growth and stem expansion. It thus creates anaerobic conditions for the roots [41]. This beating crust is due to the significant presence of the clay fraction in the soil. Soils with the shoreline showed a high risk of structural degradation (5% < SSI < 7%) due to the lower organic matter content in these soils and a high clay content [42]. The rest of the soil groups show soils with a stable structure. Soils with a high risk of structural degradability have a high fertility probability of erodibility [43]. They reduce the rate of water infiltration, which determines the availability of water for plants, unlike soils with a stable structure that facilitates water infiltration. Soils rich in organic matter have physical phases favorable to plant development [42,43] because organic matter plays a physical role in the soil cohesion, structure, porosity, water retention or storage. All the soils in the study area have a neutral pH range than 6.8 to 8.4. The pH is a key element of the chemical composition of the soil and determines the availability of nutrients for plants and soil microorganisms [44].

3.2. Variation of Chemical Properties in Soil Groups

Chemically, except for south part soils which show the fertility index (FI) below 1.5 due to the low exchangeable base rates in their soils, the rest of the soil groups have an average IF above 1.5; this suggests that these soils have a good reserve of exchangeable bases and therefore good chemical fertility [45]. The CEC is a relative indicator of the fertility power of soil [46]. According to Chapman [47], CEC depends on the organic matter and clay content of the soil. Soils with a high CEC can retain more cations and has a high capacity to exchange them. This soil has a high CEC (39.47 ± 3.68 meq/100 g), this would be due to its very high clay content (77 ± 3.48%) and high organic matter rate (3.1 ± 0.49%). The presence of organic matter in soil significantly increases CEC in soil (e.g., 1% of OM contributes 2 meq/100 g of CEC matter to soils) [48]. Given the low presence of organic matter in the rest of the soil groups, the CEC remains average, and therefore they do not have a good ability to retain and exchange cations. This reflects a very high rate of exchangeable cations in these soils. For the rest of the soil groups, exchangeable cations are medium to high. The high level of exchangeable cations is linked to the heavy clay complex, which is rich in organic matter and therefore humus [49].

Biochemically, soils in the study area have medium to high levels of organic matter (OM) in soils. This justifies the richness of these soils' nutrients [50,51]. As clay and organic matter are the basis of the areas with heavy clay, their deficiency would largely contribute to the degradation of the fertility of these soils [51,52]. Nitrogen levels range from low to medium in all soils. The C/N mineralization rate is greater than 20 for all soils. This reflects a very low rate of mineralization caused by the low total nitrogen content. Mineralization is slow in this soil group and allows only a small amount of mineral nitrogen to the soil [53,54]. The P content is medium in young and old alluvial plains and low in all other soils. This reflects a rate not high enough to ensure proper nutrition of the plant [46]. Phosphorus deficiency in these soils is influenced by the high fixing power of soils due to the presence of iron oxides and hydroxides [55,56].

There is a correlation between the fertility indicators and the physicochemical parameters of the studied soil samples. The partial components (OM, C/N, pH, N, Mg, CEC, Ca, K) show the affinity between chemical characteristics of the soil ranges between 26 and 42%. These variables are positively correlated with the soil fertility components. The variables (exchangeable cations, Ca^{2+}, Mg^{2+}, K^+, P, total N, pH, CEC) are closely correlated, with a positive coordinate, while the variables C/N ratio and OM are very closely correlated.

Agronomically, the physical-chemical properties that control the fertility of alluvial soils are C/N, pH, Ca, CEC, MO, Mg, and P. Almost similar results for soil fertility indicators have been obtained by [54–57]. Organic restitutions through long-term fallows restore fertility soils depleted by several years of successive cropping [56–58].

According to the FAO textural diagram, the area soils are grouped into two categories:

1. Soils with a clay texture; poorly permeable and poorly aerated, preventing the smooth penetration of roots and soil micro-organisms. However, high clay contents in soils condition the fixation of the mineral elements of the adsorption complex [59–63].
2. Balanced textured soils; very suitable for development because they are very permeable and easy to work [64,64] and are therefore ideal for growing maize and rice [65]. Soils with balanced textures are excellent and suitable for most crops [57,66–68].

The calculation of the equilibrium of the cationic balance (Ca/Mg/K) shows that these soils are close to the optimal equilibrium. This testifies to a balance in absorption and good assimilation by the roots of plants [69,70] while the remaining soils show deficits in potassium and magnesium. This means that the texture complex is essentially dominated by calcium. This richness of the texture in calcium may explain the low pH of these soil groups [59].

This binary fertility diagram or N-pH diagram from [45] divides the soils of El-Beheira and its surroundings into two fertility classes: soils with poor fertility and soils with medium fertility. The limitation of the soils with poor fertility, and medium fertility is due to pH levels between 6.8 and 7.3. They are characterized by low to medium reserves in exchangeable bases and medium to high exchangeable base reserves.

The Dabin [36] K/Mg diagram shows that all soils in the study area are above the potassium and magnesium deficiency thresholds. For the most part, they have a good K-Mg balance (1 < Mg/K < 4). This reflects a nutritional balance between Mg and K. This reflects an excess of assimilation of Mg in the soil by the roots of plants, compared to K. There is an excess of assimilation of K by plants, compared to Mg. Too high a K/Mg ratio in light soils causes magnesium deficiency and therefore decreases yields, while in clay soils, too low a K^+/Mg^{2+} ratio slows down the rate of potassium uptake, thus limiting yields [57].

According to this Ca/Mg binary diagram, all alluvial soils and their surroundings are above the magnesium (Mg = 0.3 meq/100 g) and calcium (Ca = 1 meq/100 g) deficiency and deficiency thresholds. According to the work of [54,71], the decrease in calcium and magnesium in a nutrient solution would be due to the increase in potassium contents. The south-western part has a deficiency of Mg compared to Ca (Ca/Mg > 5) therefore a nutritional imbalance which indicates an excess of Ca in the soil compared to Mg. The rest of the soils in the study area have a perfect Ca-Mg (1 < Ca/Mg < 5) balance. This means that these soils are satisfactory and reflect a nutritional balance between Ca and Mg [72,73].

The balance between the saturation rate and pH makes it possible to highlight the influence of pH on the evolution of exchangeable bases in the soil. It was obvious in the soils with a good saturation rate (CEC > 50%). This reflects a low level of pH in these soils. Soils with an average saturation rate are located within the sure line and east and southern parts. This means that these soils are moderately (6.8 < pH < 8) with medium exchangeable cation contents. The pH values showing moderately pH soils are a limiting factor for plant nutrition [73–76].

3.3. Criteria for Assessing the Fertility of Soils

The statistical analysis of the fertility parameters as well as the balances between these parameters made it possible to assess the current fertility status (Table 5, Figure 5) of the soils. Then, they were grouped into fertility classes according to [61] modified by [36,45]. This made it possible to define four levels of fertility of soils.

Table 5. Criterion for assessing soil fertility classes [61] modified by [36,45] for the year 2021.

Characteristic	Level I (No Limitation)	Level II (Moderate Limitation)	Level III (Severe Limitation)	Level IV (Very Severe Limitation)
OM (%)	>2	1–2	0.5–1	<0.5
N (%)	>0.08	0.045–0.080	0.030–0.045	<0.03
P (ppm)	>20	10–20	5–10	<5
K (meq/100 g)	>0.4	0.2–0.4	0.1–0.2	<0.1
EC (dS m^{-1})	>4	4–8	8–15	<16
CEC (%)	>60	40–60	15–40	<15
CEC (meq/100 g)	>25	10–25	5–10	<5
pH	>5.5	5.1–5.5	4.75–5.1	<4.75
BI	≤1.4	1.6–1.4	1.8–1.6	≥1.8
FI	>1.5	-	-	<1.5
SSI	>9	7–9	5–7	<5
CaCO$_3$ (%)	<10	10–20	20–40	>40
ESP	<13	-	-	>13

Level I: soils with no or low limitations; Level II: soils with no more than three moderate limitations associated with low limitations; Level III: soils with more than three moderate limitations associated with severe limitation; Level IV: soils with more than one severe limitation.

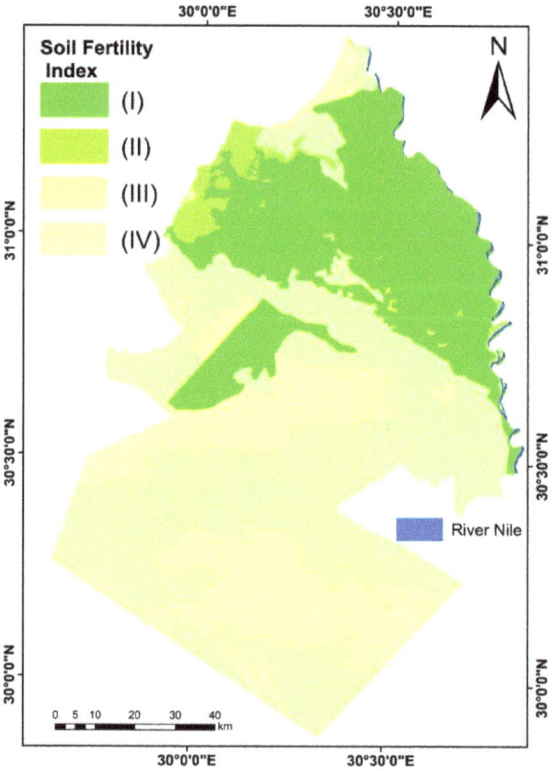

Figure 5. Soil fertility of the study area.

According to the soil fertility assessment (Table 5), four soil fertility classes can be differentiated. Class (I) include groups together soils with a good level of fertility such as *Vertic Torrifluvents*. They present medium limitations in pH and phosphorus. This means that these soils necessarily need the addition of CaO in order to be improved. Class (II includes soils with a medium level of fertility, such as *Typic Torrifluvents*. These soils have medium limitations in FI, pH, CEC, and moderate limitations in phosphorus. It is necessary to provide these soils with agricultural inputs rich in phosphate fertilizer, without forgetting the practice of fallowing for a long period of time in order to allow a reconstitution of the soil properties. Class (III), which includes soils with a poor level of fertility, such as *Typic Haplocalcids, Typic Calcigypsids, Typic Haplogypsids, Typic Quartizipsamments*. They present severe to very severe limitations in FI, SSI, BI, pH, CEC, and medium to severe limitations in P. This means that these soils need organic fertilizer inputs to repair the seemingly very poor physical properties in order to facilitate good soil aeration and sufficient retention of infiltration water. The practice of fallowing for a long period of time is essential for the good fertility of these soils. Lime is important to improve the acidity of these soils, which present high toxicity risks. A calcium amendment would favor the availability of P and Mg to the plant, which would facilitate the installation of the roots. Class (IV) groups together soils with a good level of fertility, such as *Typic Torripsamments*, rock escarpment (*Calcic petrocalcids*), and rock land *(Typic petrocalcids)*. They present severe limitations in nutrient depletions. This means that these soils necessarily need the addition of all fertilizers depending on the crop requirements in order to be improved.

3.4. Spatial Distribution of Fertility of El-Beheira Soils

The soil fertility spatial distribution map (Figure 5) shows that poorly fertile soils are located north (surrounding the Idku lake, costal area and west and south of the study area). They cover 57% of the study area. Average fertility soils are located to the west of the study area. They cover 7% of the study area. Soils with good fertility spread over the entire study area. They cover 66.65% of the study area or an area of about 39% of the total area.

3.5. Quantitative Assessment of Land Degradation

From Table 6 and Figures 6–8 soil quality degradation is one of the main causes of land degradation. Soil can be severely degraded due to compaction, salinization and sodification, and the use of chemical fertilizers that prevent the land from regenerating, and soil quality also declines as a result of chemical fertilizers for agriculture, increasing the pollution of water and land and thus reducing the value of the land.

Table 6. Area of different weighted values for the years 1985, 2002, and 2021.

	Weighted Value	Area (Hectares)	Area %
Year 1985	7	8427.24	1.04
	8	613,254.24	75.44
	9	191,219.31	23.52
Year 2002	7	1157.67	0.14
	8	15,566.94	1.91
	9	788,877.72	97.04
	10	7298.46	0.90
Year 2021	4	3144.51	0.39
	5	27,734.31	3.41
	6	121,162.14	14.90
	7	143,513.01	17.65
	8	257,577.75	31.69
	9	259,769.07	31.96

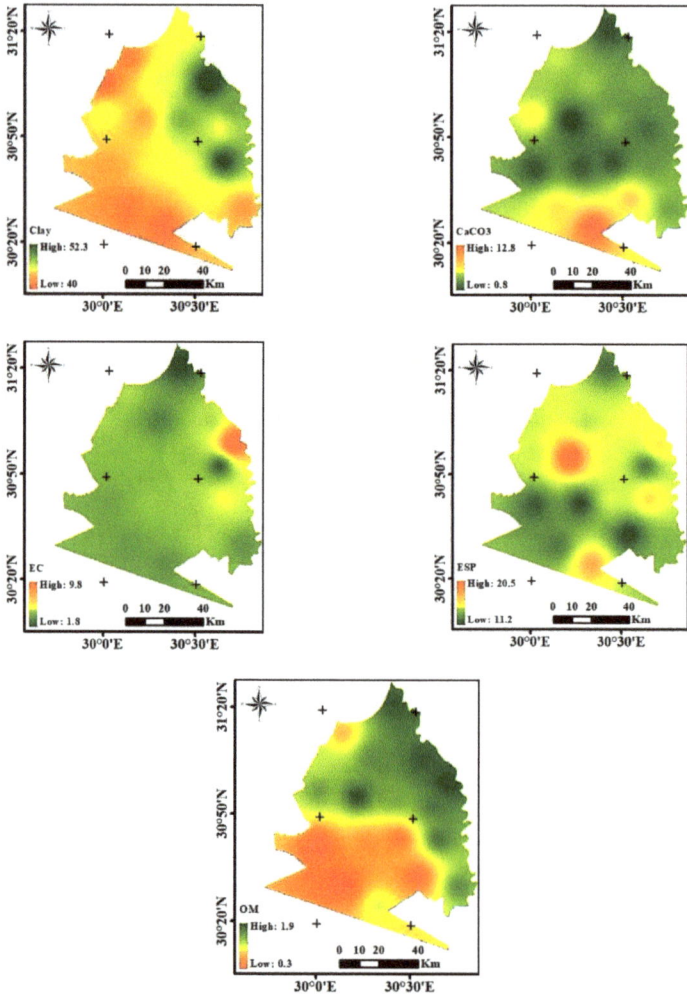

Figure 6. Outranking multi-criteria analyses (year 1985).

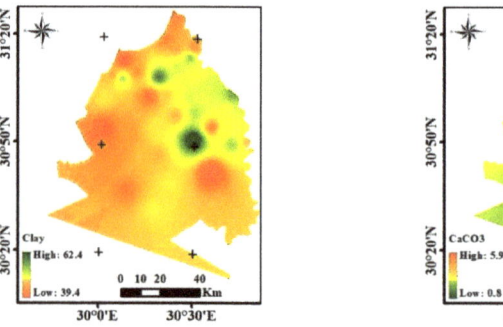

Figure 7. *Cont.*

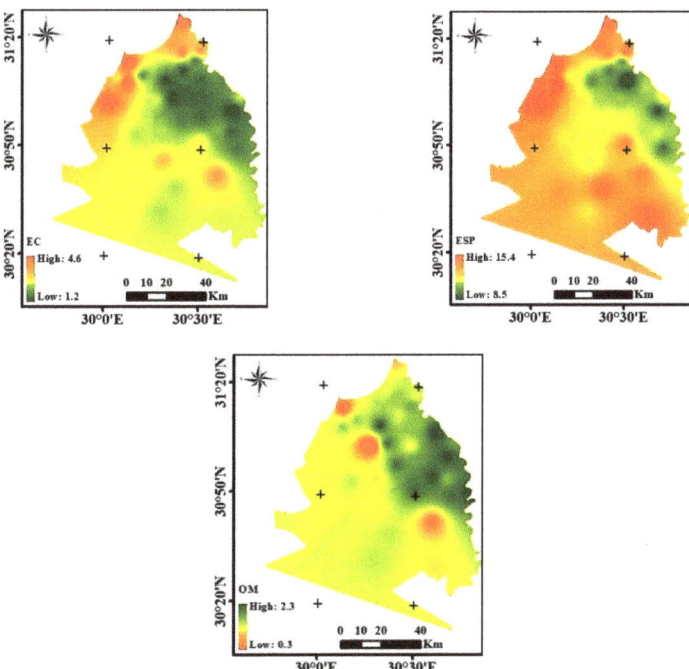

Figure 7. Outranking multi-criteria analyses (year 2002).

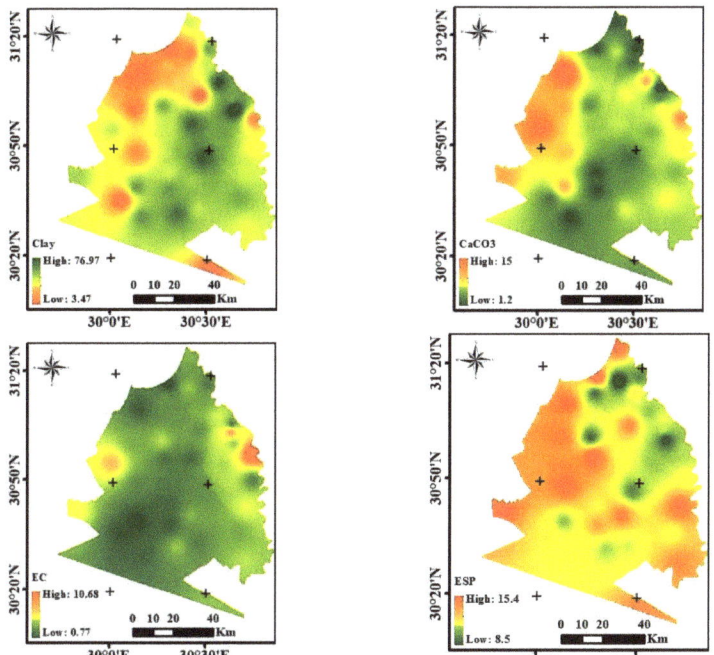

Figure 8. Cont.

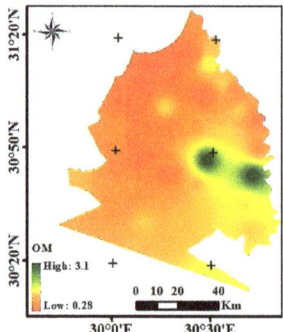

Figure 8. Outranking multi-criteria analyses (year 2021).

From Figure 9, the nature of the land covers changed as the result of changes in the environmental conditions between 1985 and 2021. These changes in the physiochemical properties lead to the deterioration of land areas and vegetation cover. The changes in all features are included, i.e., water bodies, cultivated lands, wetlands, dry lands, and bare soils. The study showed that the area of water bodies was not significantly affected; inverse with the area of cultivated lands and wetlands, and its effect is directly on the area of dry and bare soils. This was directly affected by anthropogenic activities.

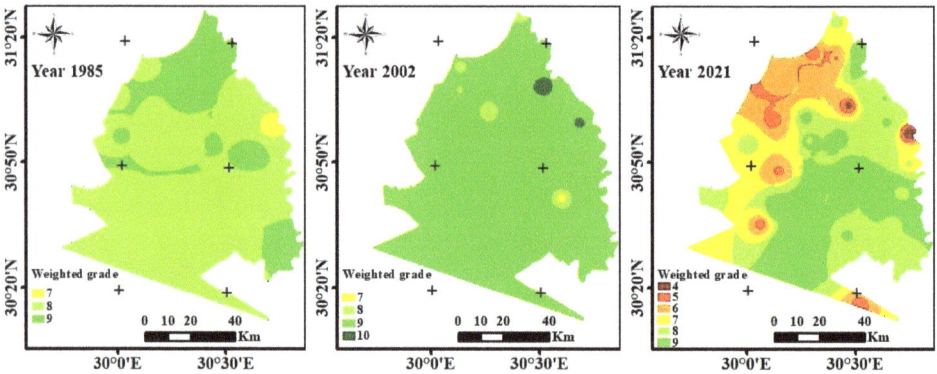

Figure 9. Outranking multi-criteria evaluation for quantitative land degradation in 1985, 2002, and 2021.

The estimations of degradation are made without taking the salinity and alkalinity of the groundwater into account and are based solely on the calculated climatic index. The study area is at a low to very high risk of salinization (Figure 9). Higher results for the current state and risk of degradation by salinization are a result of the soil, topography, and human activities. This emphasizes the value of good management and solid agricultural practices through effective drainage and irrigation methods. In the research area, sodification levels range from negligible to severe. Higher values are attributed to the soil, topography, and human activities for the current condition and risk of sodification degradation (Figure 10). Soil compaction risk, which is determined based on bulk density, is referred to as physical degradation risk (Figure 11). Biological degradation is calculated upon the organic matter deterioration (Figure 12).

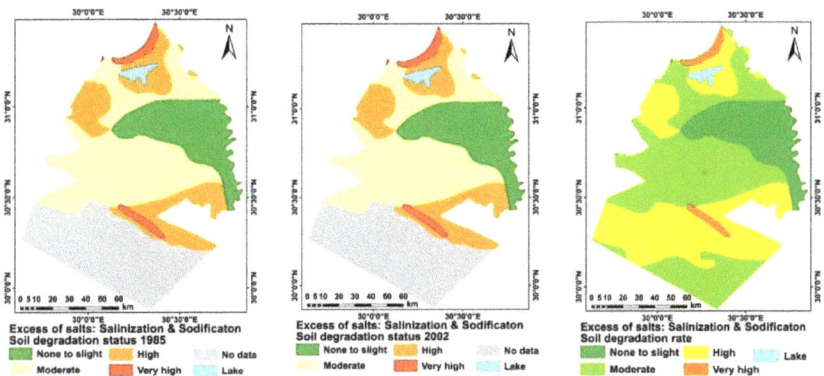

Figure 10. Assessment of excess of salts degradation.

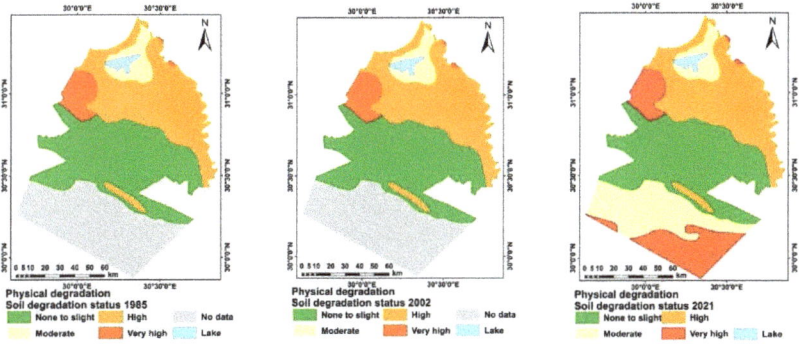

Figure 11. Assessment of physical degradation.

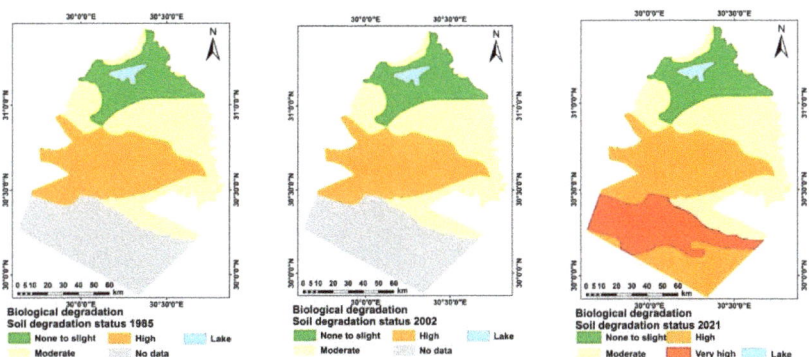

Figure 12. Assessment of biological degradation.

By combining the degree and relative extents of the various degradation classes, the Arc GIS spatial model generated the quantification of total land degradation (Figure 13). More than 50% of the units have deterioration rates that range from moderate to very high. Landforms and soil degradation severity levels are found to be related. Particularly in the north and clay-covered areas, it is at high levels, although it is only experiencing mild levels of degradation in sandy soils and river terrace soils wadi deposits, however, were a severely deteriorated area.

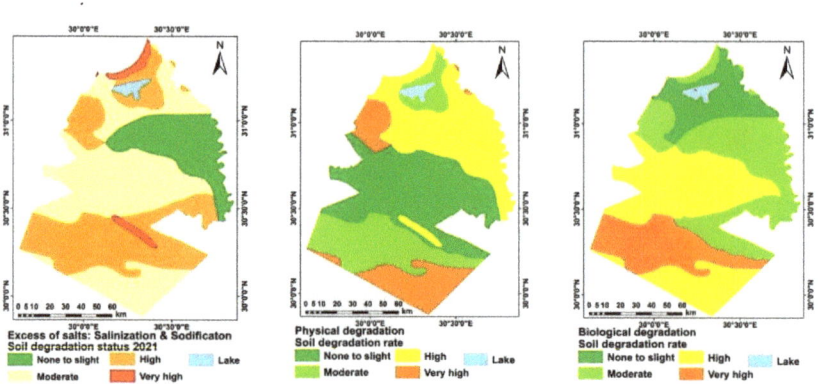

Figure 13. Assessment of overall land degradation rate.

To be successful in combating land degradation, a better awareness of its origins, effect, degree, and familiarity with climate, soil, water, land cover, and socioeconomic aspects is required. As a result, assessing land degradation is a fundamental aim of a decision support system for reversing deterioration. The study intends to use a neural network approach to measure the dynamic of land degradation in the north delta over a 50-year period. The research region is in Egypt's north delta (El-Beheira Governorate). Spatial models for overall qualitative land deterioration were constructed in 1985, 2002, and 2021 using the Model Builder tool in ArcGIS 10.3 (spatial analyst extension) for land degradation mapping. Land degradation variables (salinization, alkalinization, compaction, lime concentration, and water logging) were collected in a raster and each data set was rated on a scale of 1 to 5 (very low, low, moderate, high, and very high scale). The data sets were then weighted based on their impact on the overall model (more weight = greater effect). From 1985 to 2021, the total degradation change for highly degraded soil grew, whereas it dropped for very highly degraded soil. However, the low deteriorated soil grew over time, at the cost of the highly degraded soils, as a result of the reclamation process and soil management.

4. Discussions

The majority of the agricultural soils in these areas are saline and have low productivity due to the predominance of soluble salts because of the dry, arid climate [77]. In addition, the use of low-quality irrigation wastewater to meet the increasing irrigation water requirements is a result of the scarcity of surface water resources [78]. This leads, under harsh climatic conditions, to an increase in the concentration of salts in the upper horizon through evaporation, especially in clay soils.

The groundwater table depths of the old agricultural lands were typically between 120 and 130 cm, which is deep, with some parts having (80 cm), which had a moderate depth level. Most soils have a high clay texture overall (45 to 68%), and some additionally have a sandy clay loam texture (40–50%). According to reports, the former reclaimed agricultural area in the north of Egypt had a deep groundwater table that ranged in depth from 100 to 150 cm and was of fine texture (clay loam to clay). Going toward the Mediterranean Sea, these soils were stated to have more clay [79]. Furthermore, it has been noted that soils in arid and semi-arid areas have a higher build-up of calcium and magnesium carbonates and sulphates and have pH values that are somewhat alkaline (7.8 to 8.6) [80]. Due to the nature of the parent material from which these soils were created and the insufficient and limited leaching, it has been observed that soils along the northwestern coast are very calcareous. Additionally, due to the limited translocation of carbonates, the predominance of the arid climatic conditions may have aided in the creation of zones or horizons. These soils, which are equivalent to fluvio-marine plain soils in the area, contain low gypsum content (0.10–1.45%), $CaCO_3$ content (0.35–2.30%), and organic matter at the top layer that ranged between 1.11 and 2.55% [81]. The El-Beheira governorate's reclaimed soils, which had been irrigated with drainage waters since the 1960s, were reported to have a texture that ranged from sandy, silty loam to clay, a calcium carbonate content between 2 and 20%, and very little organic matter [82]. However, it was shown that these soils' acquired organic matter concentration was adequate for agricultural productivity despite the existing arid circumstances [82].

These old, reclaimed lands have a clay texture (30–60% clay) and a non-saline character (EC $\leq$ 4.0 dS m^{-1}), which may be related to the fact that the soil is from an old flood plain [12]. Due to poor drainage, these soils have a moderate level (100 cm) and rather deep-water table [83]. There is a second stratum of sandy clay loam in this soil as well. These soils had a pH that varied from 7.2 to 8.1, and possess a sandy clay loam area with a major clayey texture (50 to 69%). The pH was in the desired range (7–8). The maximum EC value was 21.13 dS m^{-1}, which indicates that salinity is only moderately affecting the EC range [83]. The pH of some reclaimed sandy soils west of the Nile delta and calcareous soils inside El-Beheira Governorate lands was alkaline (7.5 to 8.5), according to other studies [78–80,82–85].

As a result, the parent material from which some of these soils are generated may be to blame for the alkaline soil pH of these lower soil layers [83,85–87]. Old reclaimed agricultural areas may have EC values as high as 15 dS m^{-1} [83,86]. Additionally, fluvio-marine plain soils were said to have an EC varying from 2.24 to 25.09 dS m^{-1} to the north of Egypt [81]. These values were comparable to the EC of flood plains (clayey agricultural soils) to the north of the delta, where salinity ranged from 1.6 to 20.5 dS m^{-1} [88]. Overall, some soils' lower levels showed a noticeable rise in salt. The upward migration of salts from lower strata during dry periods, during irrigation intervals, and during the period without irrigation at the end of the season may explain this occurrence [87]. The loss of moisture from the top layers through evaporation and, in some regions, the capillary rise of salty fluids from the shallow water table may also be contributing factors to the increase in soil salinity values [87].

The majority of coastal lands were clayey (25–57%), with sandy clay loam everywhere. Salinity (EC 4.0 dS m^{-1}) affects coastal soils [13]. These soils' pH varies from the top layer to the lower level, with the top layer having a small area of near neutrality (pH = 7.5). However, the majority of the land was found to have an alkaline pH, with a value of 8.81 from 60 to 80 cm. This range is found naturally in reclaimed desert soil that had an alkaline (pH) throughout its layers and profile, and these results are consistent with other research [83,87] that claimed coastal region soils and recently reclaimed soils have a pH range of 7.9–8.5. Furthermore, according to [87], soils from recovered lakes had a pH that was slightly alkaline (8.0–8.31). However, it has been claimed that a soil pH above 8.70 indicates that $CaCO_3$ predominates and that $MgCO_3$ or Na_2CO_3 is present in these soils [89]. From another angle, the inflow of lake water at these soils may be the cause of this pH increase [87]. Some of these soils have increased salinity, which may be the result of poor drainage, seepage from low-quality lake water, or intrusion of seawater [81]. Overall, higher EC values were demonstrated, possibly as a result of the limited permeability of the clayey soils next to the lake [87]. Higher EC values for the top layer of these soils may be a sign of seawater intrusion causing waterlogging [79,87]. However, it was noted that the arid natural regions to the north of Egypt had soil salinities that varied from 6.5 to 31 dS m^{-1} [88]. Additionally, some recently recovered alluvial/marine soils near northern lakes may have EC values as high as 30 dS m^{-1} [83]. The prolonged exposure of these soils to lake seawater may be to blame for the rise in soil salinity. In addition, the lake is shallow, with a depth range of 10 to 90 cm, with the eastern and central portions of the lake having the deepest depths reported [90]. Furthermore, it has been noted that a shallow saline water table can cause salt to flow through the soil by capillary rise and evapotranspiration [78–96]. This can result in salt buildup on the surface of the soil. Additionally, the high groundwater table may be a contributing factor to the limited leaching capacity and salinity of these soils [79–87]. Furthermore, it was claimed that elevated Mg^{2+} concentrations indicated that these soils were primarily of marine origin [89].

ESP ranges between 6.3 and 10.5%, cation exchange capacity is rather low as it ranges between 2.4 to 6.8 meq/100 g soils, and the available nitrogen, phosphorous, and potassium are ranges from 0.3 to 0.7, 0.9 to 1.6, and 5.8 to 9.3 ppm, respectively. In the southern part of the study area, the soil texture is sandy, with profile depths ranging from 60 to 120 cm. The findings show that the aeolian plain (sand ripples of varied elevations and almost flat to undulating sand sheets), table land (almost flat and gradually slope areas), and old deltaic plain are the primary physiographic units in the examined area (sequence of old river terraces) [91,92,94].

Soil erosion is also a major threat for soil degradation that directly affects biodiversity [97,98]. Soil erosion occurs slowly in a discontinuous process in small parts of the northern region (coastal and lacustrine deposits) by precipitation, while wind erosion occurs at the same rate in the southern part of the region (aeolian and wadi deposits). Although the erosion process is not at an alarming rate (non to slight), it does lead to the loss of topsoil and ecological degradation.

The study area is classified into five agro-pedo-ecological zones (SQ1, SQ2, SQ3, SQ4, and SQ5) as shown in Figure 14. Qs1 (alluvial deposits) areas produce mainly rice, cotton and wheat, but water shortages and soil degradation have undermined agriculture in the region. The nature of rice cultivation, which consumes large amounts of water, helps to improve the quality of the land by ridding the soil of the salts in it. The study area is already experiencing a water crisis; given the existing environmental problems, water scarcity imposes limits on the economic development. The area is characterized by high land quality and soil fertility, despite the problems of deterioration in it, such as salinization, compaction, and waterlogging.

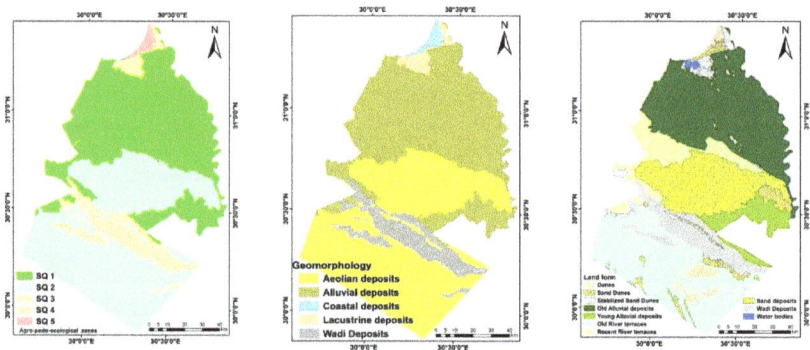

Figure 14. Agro-pedo-ecological zones of the study area, associated with geomorphology and landform.

SQ2 (aeolian deposits) would increase the cultivated area, especially since the types of lands there are good, the fertility rate and their quality are moderate, and there are all kinds of crops such as wheat, corn, vegetables, and fruits. There is an area 50–70 cm away from a mud flat, that was an historic passage for the Nile River that helped to develop agriculture in the area, resulting in a high quality of crops.

SQ3 (lacustrine deposits) and SQ4 (costal deposits); the most important problems inherent in these two areas are salinity, alkalinity, and soil waterlogging, and only salinity-tolerant plants such as rice are present in them. Sabkha is a morphological feature in the area along with salt crust on soil surface.

SQ5 (wadi deposits); lands are marginal, in addition to the exposure of the roots by the effect of the wind on the crops. It also has wetlands, where groundwater resides from the surface of the soil, impeding the growth of plants

Since landform is a key component of soil formation and a soil mapping criterion [94], it is practical to employ landform when arranging soil data. Geomorphology enhanced the results and enabled us to employ computer-assisted techniques by combining them with satellite data, field observations, and geomorphology. The ever-growing availability of earth observation data and the well-established use of GIS lead to the development of automated workflows and toolboxes for environmental management [99–101].

5. Conclusions

A considerable amount of the Nile Delta's northern region is degraded physically and excessively by salts. Additionally, the processes of salinization, sodification, and soil compaction range from extremely low to very high in various soil units. The objective of this study was to assess the soil fertility of the northern half of the El-Beheira Governorate based on physical-chemical properties and fertility parameters. The data obtained in the laboratory were processed by the statistical method using the SPSS software and it appears that soils have four fertility classes: the class of soils with good fertility with an area of about 39%; the class of soils with average fertility with an area of about 7% and the classes of soils with poor fertility (class III + IV) with an area of about 54%. Principal component analysis (PCA) has revealed that the parameters that control fertility in Foumban soils are: C/N, pH, Ca, CEC, MO, P, and Mg. The fertility assessment makes it possible to understand that the major problem of the soils in the study area is the risk of high acidity and phosphorus. To overcome this major problem, it would be wise for farmers to use lime (CaO) to reduce the risk of toxicity overall, and to use calcium fertilizers that would promote the availability of P and Mg to plants. Soils with medium and poor fertility have a poor physical phase. An amendment of these soils with organic fertilizer (fluent, compost, manure) would facilitate the formation of the clay-humic complex, thus allowing good retention of water in the soil. Finally, an amendment in mineral fertilizer would correct the CEC and would bring a high rate of exchangeable cations to the soils, thus increasing the sum of the bases. We recommend a detailed study for each agro-pedo-ecological zone, for better management planning to protect the soils from continuous deteriorations.

Author Contributions: Conceptualization, M.A.E.A., M.M.M., A.A.A., P.D. and A.S.; methodology, M.A.E.A., M.M.M., A.A.A., P.D. and A.S.; software, M.A.E.A., M.M.M., A.A.A., P.D. and A.S.; validation, M.A.E.A., M.M.M., A.A.A., P.D. and A.S.; formal analysis, M.A.E.A., M.M.M. and A.A.A.; investigation, M.A.E.A., M.M.M. and A.A.A.; resources, M.A.E.A., M.M.M. and A.A.A.; data curation, M.A.E.A., M.M.M. and A.S.; writing—original draft preparation, M.A.E.A., M.M.M., A.A.A., P.D. and A.S.; writing—review and editing, M.A.E.A., M.M.M., A.A.A., P.D. and A.S.; visualization, M.A.E.A.; supervision, M.A.E.A., M.M.M., A.A.A. and A.S.; project administration, M.A.E.A., M.M.M. and A.A.A.; funding acquisition, M.A.E.A., M.M.M., A.A.A., P.D. and A.S. All authors have read and agreed to the published version of the manuscript.

Funding: This research received no external funding.

Institutional Review Board Statement: Not applicable.

Informed Consent Statement: Not applicable.

Data Availability Statement: Not applicable.

Acknowledgments: The manuscript presents a participation between the scientific institutions in two countries (Egypt and Italy), and in particular, the authors are grateful for their support in carrying out the work to: (1) National Authority for Remote Sensing and Space Sciences (NARSS), Cairo 11769, Egypt. (2) National Research Centre, Giza 12622, Egypt. ((3) SAFE-Università degli Studi della Basilicata.

Conflicts of Interest: The authors declare no conflict of interest.

References

1. Alfiky, A.; Kaule, G.; Salheen, M. Agricultural Fragmentation of the Nile Delta; A Modeling Approach to Measuring Agricultural Land Deterioration in Egyptian Nile Delta. *Proced. Environ. Sci.* **2012**, *14*, 79–97. [CrossRef]
2. Zhao, X.; Liu, Y.; Thomas, I.; Salem, A.; Wang, Y.; Alassal, S.E.; Jiang, F.; Sun, Q.; Chen, J.; Finlayson, B.; et al. Herding then farming in the Nile Delta. *Commun. Earth Environ.* **2022**, *3*, 88. [CrossRef]

3. Fishar, M.R. Nile Delta (Egypt). In *The Wetland Book: II: Distribution, Description, and Conservation*; Finlayson, C.M., Milton, G.R., Prentice, R.C., Davidson, N.C., Eds.; Springer: Dordrecht, The Netherlands, 2018; pp. 1251–1260.
4. AbdelRahman, M.A.E.; Natarajan, A.; Srinivasamurty, C.A.; Hegde, R. Estimating soil fertility status in physically degraded land using GIS and remote sensing techniques in Chamarajanagar district, Karnataka, India. Egypt. *J. Remote Sens. Space Sci.* **2016**, *19*, 95–108. [CrossRef]
5. Schoonover, J.E.; Crim, J.F. An Introduction to Soil Concepts and the Role of Soils in Watershed Management. *J. Contemp. Water Res. Educ.* **2015**, *154*, 21–47. [CrossRef]
6. Ogura, T.; Date, Y.; Masukujane, M.; Coetzee, T.; Akashi, A.; Kikuchi, J. Improvement of physical, chemical and biological properties of aridisol from Botswana by the incorporation of torrefied biomass. *Sci. Rep.* **2016**, *6*, 28011. [CrossRef] [PubMed]
7. Onwuka, B.; Mang, B. Effects of soil temperature on some soil properties and plant growth. *Adv. Plants Agric. Res.* **2018**, *8*, 34–37. [CrossRef]
8. Page, K.L.; Dang, Y.P.; Dalal, R.C. The Ability of Conservation Agriculture to Conserve Soil Organic Carbon and the Subsequent Impact on Soil Physical, Chemical, and Biological Properties and Yield. *Front. Sustain. Food Syst.* **2020**, *4*, 21. [CrossRef]
9. Abuzaid, A.S.; El-Husseiny, A.M. Modeling crop suitability under micro irrigation using a hybrid AHP-GIS approach. *Arab J. Geosci.* **2022**, *15*, 1217. [CrossRef]
10. Ghanem, H.G.; El-Gabry, Y.A.; Okasha, E.M.; Ganzour, S.K. Improving Some Irrigation Efficiencies, Soil Fertility, Yield and Quality of Wheat under Deficit Irrigation by Integrated N-Fertilization. Egypt. *J. Chem.* **2021**, *64*, 2201–2212.
11. Aboelsoud, H.M.; AbdelRahman, M.A.E.; Kheir, A.M.S.; Eid, M.S.M.; Ammar, K.A.; Khalifa, T.H.; Scopa, A. Quantitative Estimation of Saline-Soil Amelioration Using Remote-Sensing Indices in Arid Land for Better Management. *Land* **2022**, *11*, 1041. [CrossRef]
12. AbdelRahman, M.A.E.; Afifi, A.A.; D'Antonio, P.; Gabr, S.S.; Scopa, A. Detecting and Mapping Salt-Affected Soil with Arid Integrated Indices in Feature Space using Multi-Temporal Landsat Imagery. *Remote Sens.* **2022**, *14*, 2599. [CrossRef]
13. AbdelRahman, M.A.E.; Afifi, A.A.; Scopa, A. A Time Series Investigation to Assess Climate Change and Anthropogenic Impacts on Quantitative Land Degradation in the North Delta, Egypt. *ISPRS Int. J. Geo-Inf.* **2022**, *11*, 30. [CrossRef]
14. Gaafar, A.A.; Morsy, I.M.; Yehia, H.M. Geospatial Analysis of Soil Characteristics and Sensitivity to Desertification of Some Alluvial Deposits, El Behira Governorate, Egypt. *Alex. Sci. Exch. J.* **2017**, *38*, 137–148. [CrossRef]
15. Memarbashi, E.; Azadi, H.; Barati, A.A.; Mohajeri, F.; Passel, S.V.; Witlox, F. Land-use suitability in Northeast Iran: Application of AHP-GIS hybrid model. *ISPRS Int. J. Geo-Inf.* **2017**, *6*, 396. [CrossRef]
16. Naz, A.; Rasheed, H. Modeling the rice land suitability using GIS and multi-criteria decision analysis approach in Sindh, Pakistan. *J. Basic Appl. Sci.* **2017**, *13*, 26–33. [CrossRef]
17. Erener, A.; Mutlu, A.; Düzgün, H.S. A comparative study for landslide susceptibility mapping using GIS-based multi-criteria decision analysis (MCDA), logistic regression (LR) and association rule mining (ARM). *Eng. Geol.* **2016**, *203*, 45–55. [CrossRef]
18. Ayoade, M.A. Suitability assessment and mapping of Oyo State, Nigeria, for rice cultivation using GIS. *Theor. Appl. Climatol.* **2017**, *129*, 1341–1354. [CrossRef]
19. Mokarram, M.; Hojati, M. Using ordered weight averaging (OWA) aggregation for multi-criteria soil fertility evaluation by GIS (case study: Southeast Iran). *Comput. Electron. Agric.* **2017**, *132*, 1–13. [CrossRef]
20. Maleknia, R.; Khezri, E.; Zeinivand, H.; Badehian, Z. Mapping Natural Resources Vulnerability to Droughts Using Multi-Criteria Decision Making and GIS (Case Study: Kashkan Basin Lorestan Province, Iran). *J. Range. Sci.* **2017**, *7*, 376–386.
21. Guarini, M.R.; Battisti, F.; Chiovitti, A. A Methodology for the Selection of Multi-Criteria Decision Analysis Methods in Real Estate and Land Management Processes. *Sustainability* **2018**, *10*, 507. [CrossRef]
22. Lal, R. Restoring Soil Quality to Mitigate Soil Degradation. *Sustainability* **2015**, *7*, 5875–5895. [CrossRef]
23. FAO-ISRIC. *Guiding Principles for the Quantitative Assessment of Soil Degradation with a Focus on Salinization, Nutrient Decline and Soil Pollution*; FAO: Rome, Italy, 2004.
24. Khalil, A.A.; Essa, Y.H.; Hassanein, M.K. Monitoring Agricultural Land Degradation in Egypt Using MODIS NDVI Satellite Images. *Nat. Sci.* **2014**, *12*, 15–21.
25. Egyptian Geological Survey and Mining Authority. *Geologic Map of Egypt: Egyptian General Authority for Petroleum (UNESCO Joint Map Project), 20 Sheets, Scale 1:50,000*; Egyptian Geological Survey and Mining Authority: Cairo, Egypt, 1987.
26. Ministry of Development of Egypt. New communities and Land Reclamation. In *Land Master Plan of Egypt*; Ministry of Development of Egypt: Giza, Egypt, 1986.
27. ASRT. *A Provisional Methodology for Digital Land Resource Data Base for Agricultural Use. (2000–2005)*; Final Report; Academy of Scientific Research and Technology (ASRT): Cairo, Egypt, 2005.
28. Kjeldahl, J. Neue Methode zur Bestimmung des Stickstoffs in organischen Körpern. *Fresenius, Zeitschrift f. anal. Chemie* **1883**, *22*, 366–382. [CrossRef]
29. Bray, R.H.; Kurtz, L.T. Determination of total, organic and available forms of phosphorus in soil. *Soil Sci.* **1945**, *59*, 39–46. [CrossRef]
30. Shuman, L.M.; Duncan, R.R. Soil exchangeable cations and aluminum measured by ammonium chloride, potassium chloride, and ammonium acetate. *Commun. Soil Sci. Plant Anal.* **2008**, *21*, 1217–1228. [CrossRef]

31. Ross, D.; Ketterings, Q. Recommended methods for determining soil cation exchange capacity. In *Recommended Soil Testing Procedures for the Northeastern United States*, 3rd ed.; Sims, T., Wolf, A., Eds.; The Northeast Coordinating Committee for Soil Testing (NECC-1312) Agricultural Experiment Stations of Connecticut: New Haven, CT, USA, 2011; pp. 75–86.
32. Blake, G.R. *Bulk Density, Soils Analysis*; America Society of Agronomy and Crop Science Society of America: Madison, WI, USA, 1982. [CrossRef]
33. Day, P.R. Particle fractionation and particle size analysis. In *Methods of Soil Analysis Part I. Agronomy n 9*; Black, C.A., Ed.; American Society of Agronomy: Madison, WI, USA, 1965.
34. Boverwijk, A. Particle size analysis of soils by means of the hydrometer method. *Sedim. Geol.* **1967**, *1*, 403–406. [CrossRef]
35. Forestier, J. Fertilité des sols des caféières en RCA. *Agron. Trop.* **1960**, *15*, 543–567.
36. Dabin, B. Les facteurs de fertilité des sols des régions tropicales en culture irriguée. *Bul. Assoc. Française D'etude Sol.* **1961**, 108–130.
37. Martin, D. Fertilité chimique des sols d'une ferme du Congo. *Cah. ORSTOM. Sér. Pédol.* **1979**, *17*, 47–64.
38. Pieri, C.J.M.G. *Fertility of Soils: A Future for Farming in the West African Savannah*; Springer: Berlin/Heidelberg, Germany, 1992; pp. 1–348.
39. Remy, J.C.; Marin-Laflèche, A. L'analyse de terre: Réalisation d'un programme d'interprétation automatique. *Ann. Agron.* **1974**, *25*, 607–632.
40. Saaty, T.L. Decision making with the Analytic Hierarchy Process. *Int. J. Serv. Sci.* **2008**, *1*, 83–98. [CrossRef]
41. Russell, R.S.; Goss, M.J. Physical aspects of soil fertility-The response of roots to mechanical impedance. *Neth. J. Agric. Sci.* **1974**, *22*, 305–318. [CrossRef]
42. Walsh, E.; McDonnell, K.P. The influence of added organic matter on soil physical, chemical, and biological properties: A small-scale and short-time experiment using straw. *Arch. Agron. Soil Sci.* **2012**, *58*, 201–205. [CrossRef]
43. Piaszczyk, W.; Lasota, J.; Błońska, E. Effect of Organic Matter Released from Deadwood at Different Decomposition Stages on Physical Properties of Forest Soil. *Forests* **2020**, *11*, 24. [CrossRef]
44. Borah, K.K.; Bhuyan, B.; Sarma, H.P. Lead, arsenic, fluoride, and iron contamination of drinking water in the tea garden belt of Darrang district, Assam, India. *Environ. Monit. Assess.* **2010**, *169*, 347–352. [CrossRef] [PubMed]
45. Nguemezi, C.; Tematio, P.; Yemefack, M.; Tsozue, D.; Silatsa, T.B.F. Soil quality and soil fertility status in major soil groups at the Tombel area, South-West Cameroon. *Heliyon* **2020**, *6*, e03432. [CrossRef] [PubMed]
46. Bedolla-Rivera, H.I.; Xochilt Negrete-Rodríguez, M.d.l.L.; Medina-Herrera, M.d.R.; Gámez-Vázquez, F.P.; Álvarez-Bernal, D.; Samaniego-Hernández, M.; Gámez-Vázquez, A.J.; Conde-Barajas, E. Development of a Soil Quality Index for Soils under Different Agricultural Management Conditions in the Central Lowlands of Mexico: Physicochemical, Biological and Ecophysiological Indicators. *Sustainability* **2020**, *12*, 9754. [CrossRef]
47. Chapman, H.D. Cation-Exchange Capacity. In *Agronomy Monographs*; Norman, A.G., Ed.; American Society of Agronomy, Soil Science Society of America: Madison, WI, USA, 2016; pp. 891–901. [CrossRef]
48. Kong, X.; Li, D.; Song, X.; Zhang, G. Quantitative Estimation of the Changes in Soil CEC after the Removal of Organic Matter and Iron Oxides. *Agric. Sci.* **2021**, *12*, 1244–1254. [CrossRef]
49. Minhal, F.; Ma'as, A.; Hanudin, E.; Sudira, P. Improvement of the chemical properties and buffering capacity of coastal sandy soil as affected by clays and organic by-product application. *Soil Water Res.* **2020**, *15*, 93–100. [CrossRef]
50. Loveland, P.; Webb, J. Is there a critical level of organic matter in the agricultural soils of temperate regions: A review. *Soil Tillage Res.* **2003**, *70*, 1–18. [CrossRef]
51. Rawat, J.; Saxena, J.; Sanwai, P. Biochar: A Sustainable Approach for Improving Plant Growth and Soil Properties. In *Biochar: An Imperative Amendment for Soil and the Environment*; Abrol, V., Sharma, P., Eds.; IntechOpen: London, UK, 2019. [CrossRef]
52. Kome, G.K.; Enang, R.K.; Tabi, F.O.; Yerima, B.P.K. Influence of Clay Minerals on Some Soil Fertility Attributes: A Review. *Open J. Soil Sci.* **2019**, *9*, 155–188. [CrossRef]
53. Smolander, A.; Henttonen, H.M.; Nöjd, P.; Soronen, P.; Mäkinen, H. Long-term response of soil and stem wood properties to repeated nitrogen fertilization in a N-limited Scots pine stand. *Eur. J. For. Res.* **2022**, *141*, 421–431. [CrossRef]
54. Ballot, C.S.A.; Mawussi, G.; Atakpama, W.; Moita-Nassy, M.; Yangakola, T.M.; Zinga, I.; Silla, S.; Kperkouma, W.; Dercon, G.; Komlan, B.; et al. Caractérisation physico- chimique des sols en vue de l'amélioration de la productivité du manioc (Manihot esculenta Crantz) dans la région de Damara au centre-sud de Centrafrique. *Agron. Afr.* **2016**, *28*, 9–23.
55. Troeh, F.R.; Thompson, L.M. *Soils and Soil Fertility*, 6th ed.; Blackwell: Ames, IA, USA, 2005; p. 489.
56. Dreschel, P.; Reck, B. Composted shrubprunings and other organic manures for smallholder farming systems in southern Rwanda. *Agrofor. Syst.* **1997**, *39*, 1–12. [CrossRef]
57. Pypers, P.; Sanginga, J.-M.; Kasereka, B.; Walangululu, M.; Vanlauwe, B. Increased productivity through integrated soil fertility management in cassava–legume intercropping systems in the highlands of Sud-Kivu, DR Congo. *Field Crops Res.* **2011**, *120*, 76–85. [CrossRef]
58. Diacono, M.; Montemurro, F. Long-term effects of organic amendments on soil fertility. A review. *Agron. Sustain. Dev.* **2010**, *30*, 401–422. [CrossRef]
59. Taalab, A.S.; Ageeb, G.W.; Siam, H.S.; Mahmoud, S.A. Some Characteristics of Calcareous soils. A review. *Middle East J. Agric. Res.* **2019**, *8*, 96–105.
60. Chatterjee, D.; Datta, S.C.; Manjaiah, K.M. Fractions, uptake and fixation capacity of phosphorus and potassium in three contrasting soil orders. *J. Soil Sci. Plant Nutr.* **2014**, *14*, 640–656. [CrossRef]

61. Quemada, M.; Cabrera, M.L. CERES-N model predictions of nitrogen mineralised from cover crop residues. *Soil Sci. Soc. Am. Div. S-3—Soil Biol. Biochem.* **1995**, *59*, 1059–1065. [CrossRef]
62. Lin, H.-Y.; Chuang, T.-J.; Yang, P.-T.; Guo, L.-Y.; Wang, S.-L. Adsorption and desorption of Thallium(I) in soils: The predominant contribution by clay minerals. *Appl. Clay Sci.* **2021**, *205*, 106063. [CrossRef]
63. Nieder, R.; Benbi, D.K.; Scherer, H.W. Fixation and defixation of ammonium in soils: A review. *Biol. Fertil. Soils* **2011**, *47*, 1–14. [CrossRef]
64. Radočaj, D.; Jurišić, M.; Zebec, V.; Plaščak, I. Delineation of Soil Texture Suitability Zones for Soybean Cultivation: A Case Study in Continental Croatia. *Agronomy* **2020**, *10*, 823. [CrossRef]
65. Epule, T.E.; Chehbouni, A.; Dhiba, D.; Etongo, D.; Driouech, F.; Brouziyne, Y.; Peng, C. Vulnerability of maize, millet, and rice yields to growing season precipitation and socio-economic proxies in Cameroon. *PLoS ONE* **2021**, *16*, e0252335. [CrossRef] [PubMed]
66. Almendro-Candel, M.B.; Lucas, I.G.; Navarro-Pedreño, J.; Zorpas, A.A. Physical Properties of Soils Affected by the Use of Agricultural Waste. In *Agricultural Waste and Residues*; Aladjadjiyan, A., Ed.; IntechOpen: London, UK, 2018.
67. Dou, F.; Soriano, J.; Tabien, R.E.; Chen, K. Soil Texture and Cultivar Effects on Rice (*Oryza sativa*, L.) Grain Yield, Yield Components and Water Productivity in Three Water Regimes. *PLoS ONE* **2016**, *11*, e0150549. [CrossRef]
68. Sonneveld, M.P.W.; Hack-ten Broeke, M.J.D.; van Diepen, C.A.; Boogaard, H.L. Thirty years of systematic land evaluation in the Netherlands. *Geoderma* **2010**, *156*, 84–92. [CrossRef]
69. Hussain, S.; Shaukat, M.; Ashraf, M.; Zhu, C.; Jin, Q.; Zhang, J. *Salinity Stress in Arid and Semi-Arid Climates: Effects and Management in Field Crops, in Climate Change and Agriculture*; IntechOpen: London, UK, 2019; Available online: https://www.intechopen.com/chapters/68075 (accessed on 1 June 2022).
70. El-Hadidi, E.M.; El-Dissoky, R.A.; AbdElhafez, A.A.H. Foliar Calcium and Magnesium Application Effect on Potato Crop Grown in Clay Loam Soils. *J. Soil Sci. Agric. Eng. Mansoura Univ.* **2017**, *8*, 1–8. [CrossRef]
71. Cole, J.C.; Smith, M.W.; Penn, C.J.; Cheary, B.S.; Conaghan, K.J. Nitrogen, phosphorus, calcium, and magnesium applied individually or as a slow release or controlled release fertilizer increase growth and yield and affect macronutrient and micronutrient concentration and content of field-grown tomato plants. *Sci. Hortic.* **2016**, *211*, 420–430. [CrossRef]
72. Gransee, A.; Führs, H. Magnesium mobility in soils as a challenge for soil and plant analysis, magnesium fertilization and root uptake under adverse growth conditions. *Plant Soil* **2013**, *368*, 5–21. [CrossRef]
73. Koulibaly, B.; Traoré, O.; Dakuo, D.; Lalsaga, R.; Lompo, F.; Zombre, N.P. Acidification des sols ferrugineux et ferrallitique dans les systèmes de production cotonnière au Burkina Faso. *Int. J. Biol. Chem. Sci.* **2014**, *8*, 2879–2890. [CrossRef]
74. Jakobsen, S.T. Nutritional Disorders between Potassium, Magnesium, Calcium, and Phosphorus in Soil. In *Plant and Soil*; JSTOR: New York, NY, USA, 1993; Volume 154, pp. 21–28. Available online: http://www.jstor.org/stable/42938994 (accessed on 20 July 2022).
75. Neina, D. The Role of Soil pH in Plant Nutrition and Soil Remediation. *Appl. Environ. Soil Sci.* **2019**, *2019*, 5794869. [CrossRef]
76. Gentili, R.; Ambrosini, R.; Montagnani, C.; Caronni, S.; Citterio, S. Effect of Soil pH on the Growth, Reproductive Investment and Pollen Allergenicity of *Ambrosia artemisiifolia* L. *Front. Plant. Sci.* **2018**, *9*, 1335. [CrossRef] [PubMed]
77. AbdelRahman, M.A.E.; Engel, B.; Eid, M.S.M.; Aboelsoud, H.M. A new index to assess soil sustainability based on Temporal Changes of Soil Measurements Using Geomatics–An example from El-Sharkia, Egypt. *All Earth* **2022**, *34*, 147–166. [CrossRef]
78. El-Dars, F.M.S.E.; Salem, W.A.; Fahim, M.M. Soil spatial variability in arable land south of Lake Idku, North- West Nile Delta, Egypt. *Environ. Sci. Indian J.* **2014**, *9*, 325–344.
79. Kotb, T.H.S.; Watanabe, T.; Ogino, Y.; Tanji, K.K. Soil salinization in the Nile Delta and related policy issues in Egypt. *Agric. Water Manag.* **2000**, *43*, 239–261. [CrossRef]
80. Bashour, I.I.; Sayegh, A.H. *Methods of Analysis for Soils of Arid and Semi-Arid Regions*; Food and Agriculture Organization of the United Nations (FAO): Rome, Italy, 2007; Volume 128.
81. Elewa, H.H.; El Nahry, A.H. Hydro-environmental status and soil management of the River Nile Delta, Egypt. *Environ. Geol.* **2009**, *57*, 759–774. [CrossRef]
82. Rhoades, J.D.; Kandiah, A.; Mashali, A.M. *The Use of Saline Waters for Crop Production, FAO Irrigation and Drainage Paper 48*; Food and Agriculture Organization of the United Nations (FAO): Rome, Italy, 1992.
83. Abdel-Hamid, M.A.; Shrestha, D.P. Soil Salinity Mapping in the Nile Delta, Egypt Using Remote Sensing Techniques. *Int. Soc. Photogramm. Remote Sens.* **1992**, *29*, 783–787.
84. Badr, M.A.; El-Tohamy, W.A.; Zaghloul, A.M. Yield and water use efficiency of potato grown under different irrigation and nitrogen levels in an arid region. *Agric. Water Manag.* **2012**, *110*, 9–15. [CrossRef]
85. Shehata, A.A.; Hamdy, M.A.; El Badry, D.D. Gypsum application and leaching of saline alkali soils in El-Beheira Governorate. *Egypt. J. Soil Sci.* **1983**, *23*, 63–73.
86. Abdelaty, E.F.; Aboukila, E.F. Detection of Soil Salinity for Bare and Cultivated Lands Using Landsat ETM+ Imagery Data: A Case Study from El-Beheira Governorate, Egypt. *Alex. Sci. Exch. J.* **2017**, *38*, 642–653.
87. Ali, R.R.; Moghanm, F.S. Variation of soil properties over the landforms around Idku lake, Egypt. *Egypt. J. Remote Sens. Space Sci.* **2013**, *16*, 91–101. [CrossRef]
88. Hamed, Y. Soil structure and salinity effects of fish farming as compared to traditional farming in north- eastern Egypt. *Land Use Policy* **2008**, *25*, 301–308. [CrossRef]

89. Sayed, A.S.A. Evaluation of the Land Resources for Agricultural Development—Case Study: El- Hammam Canal and Its Extension, NW Coast of Egypt. Ph.D. Thesis, Department of Geosciences, University of Hamburg, Hamburg, Germany, 2013.
90. Shakweer, L. Impacts of Drainage Water Discharge on the Water Chemistry of Lake Edku. *Egypt. J. Aquat. Res.* **2006**, *32*, 264–282.
91. Ismail, M.; Yacoub, R.K. Digital soil map using the capability of new technology in Sugar Beet area, Nubariya, Egypt. *Egypt. J. Remote Sens. Space Sci.* **2012**, *15*, 113–124. [CrossRef]
92. Ali, R.R.; El Baroudy, A.A. Use of GIS in Mapping the Environmental Sensitivity to Desertification in Wadi El Natrun Depression, Egypt. *Aust. J. Basic Appl. Sci.* **2008**, *2*, 157–164.
93. Abdel-Hamid, M.A.; Ismail, M.; Nasr, Y.A.; Kotb, Y. Assessment of Soils of Wadi El-Natrun Area, Egypt Using Remote Sensing and GIS Techniques. *J. Am. Sci.* **2010**, *10*, 195–206.
94. Ali, R.R. Digital Soil Mapping for Optimum Land Uses in some Newly Reclaimed Areas West of the Nile Delta, Egypt. *Aust. J. Basic Appl. Sci.* **2008**, *2*, 165–173.
95. Afifi, A.A.; Darwish, K.M. Detection and impact of land encroachment in El-Beheira governorate, Egypt. *Model. Earth Syst. Environ.* **2018**, *4*, 517–526. [CrossRef]
96. Aziz, A.M.; Shahin, S.A.; Essa, E.F.; Abd El-Hady, M. Sustainability of the Soils for Orchards Land Use of Al-Nagah Area, Beheira, Egypt. *Plant Arch.* **2019**, *19*, 738–744.
97. Stefanidis, S.; Alexandridis, V.; Ghosal, K. Assessment of Water-Induced Soil Erosion as a Threat to Natura 2000 Protected Areas in Crete Island, Greece. *Sustainability* **2022**, *14*, 2738. [CrossRef]
98. Orgiazzi, A.; Panagos, P. Soil biodiversity and soil erosion: It is time to get married: Adding an earthworm factor to soil erosion modelling. *Glob. Ecol. Biogeogr.* **2018**, *27*, 1155–1167. [CrossRef]
99. Stefanidis, S.; Chatzichristaki, C.; Stefanidis, P. An ArcGIS toolbox for estimation and mapping soil erosion. *J. Environ. Prot. Ecol.* **2021**, *22*, 689–696. Available online: https://scibulcom.net/en/article/RudumPyvp1Dc9aCFph69 (accessed on 1 June 2022).
100. Naghibi, S.A.; Hashemi, H.; Pradhan, B. APG: A novel python-based ArcGIS toolbox to generate absence-datasets for geospatial studies. *Geosci. Front.* **2021**, *12*, 101232. [CrossRef]
101. Roux, C.; Alber, A.; Bertrand, M.; Vaudor, L.; Piégay, H. FluvialCorridor: A new ArcGIS toolbox package for multiscale riverscape exploration. *Geomorphology* **2015**, *242*, 29–37. [CrossRef]

Article

Prediction of Soil Organic Carbon at Field Scale by Regression Kriging and Multivariate Adaptive Regression Splines Using Geophysical Covariates

Daniela De Benedetto [1], Emanuele Barca [2,*], Mirko Castellini [1], Stefano Popolizio [3], Giovanni Lacolla [4] and Anna Maria Stellacci [3]

1. Council for Agricultural Research and Economics-Agriculture and Environment Research Center (CREA-AA), 70126 Bari, Italy; daniela.debenedetto@crea.gov.it (D.D.B.); mirko.castellini@crea.gov.it (M.C.)
2. Water Research Institute (IRSA)—National Research Council (CNR), 70185 Bari, Italy
3. Department of Soil, Plant and Food Sciences, University of Bari "A. Moro", 70126 Bari, Italy; stefano.popolizio@uniba.it (S.P.); annamaria.stellacci@uniba.it (A.M.S.)
4. Department of Agricultural and Environmental Science, University of Bari "A. Moro", 70126 Bari, Italy; giovanni.lacolla@uniba.it
* Correspondence: emanuele.barca@ba.irsa.cnr.it

Citation: De Benedetto, D.; Barca, E.; Castellini, M.; Popolizio, S.; Lacolla, G.; Stellacci, A.M. Prediction of Soil Organic Carbon at Field Scale by Regression Kriging and Multivariate Adaptive Regression Splines Using Geophysical Covariates. *Land* **2022**, *11*, 381. https://doi.org/10.3390/land11030381

Academic Editor: Manuel López-Vicente

Received: 18 January 2022
Accepted: 2 March 2022
Published: 4 March 2022

Publisher's Note: MDPI stays neutral with regard to jurisdictional claims in published maps and institutional affiliations.

Copyright: © 2022 by the authors. Licensee MDPI, Basel, Switzerland. This article is an open access article distributed under the terms and conditions of the Creative Commons Attribution (CC BY) license (https:// creativecommons.org/licenses/by/ 4.0/).

Abstract: Knowledge of the spatial distribution of soil organic carbon (SOC) is of crucial importance for improving crop productivity and assessing the effect of agronomic management strategies on crop response and soil quality. Incorporating secondary variables correlated to SOC allows using information often available at finer spatial resolution, such as proximal and remote sensing data, and improving prediction accuracy. In this study, two nonstationary interpolation methods were used to predict SOC, namely, regression kriging (RK) and multivariate adaptive regression splines (MARS), using as secondary variables electromagnetic induction (EMI) and ground-penetrating radar (GPR) data. Two GPR covariates, representing two soil layers at different depths, and X geographical coordinates were selected by both methods with similar variable importance. Unlike the linear model of RK, the MARS model also selected one EMI covariate. This result can be attributed to the intrinsic capability of MARS to intercept the interactions among variables and highlight nonlinear features underlying the data. The results indicated a larger contribution of GPR than of EMI data due to the different resolution of EMI from that of GPR. Thus, MARS coupled with geophysical data is recommended for prediction of SOC, pointing out the need to improve soil management to guarantee agricultural land sustainability.

Keywords: SOC spatial distribution; regression kriging (RK); multivariate adaptive regression splines (MARS); secondary variables; electromagnetic induction technique (EMI); ground-penetrating radar (GPR)

1. Introduction

Soil organic carbon (SOC) is one of the most important indicators for assessing soil quality and overall soil health [1]. SOC plays a key role in unveiling soil structure development, nutrient turnover and stability, soil water retention, regulation of greenhouse gases, and susceptibility or resilience to land degradation [2]. SOC stock is thus a main factor in soil health, fertility, quality, and productivity [3] and supports important soil-derived ecosystem services (ESs) including water filtration and erosion control, soil strength and stability, nutrient conservation, and climate change adaptation and mitigation by sequestration of atmospheric CO_2 [4]. By selecting key soil indicators under different land use and management practices, Shukla et al. [5] concluded that SOC was the main soil quality indicator and suggested using SOC to monitor soil quality changes [6].

SOC distribution is influenced by many factors, including climate variables (temperature and rainfall), topographical features, soil texture, parent material, vegetation, land-use types, and human management at different spatial scales [7].

Agronomic management strategies, with particular regard to fertilization, soil tillage, and irrigation, may significantly modify SOC content and its labile fractions, mainly in the shallower soil layers [6,8–10]. Because of the interaction of the factors described, SOC spatial variation is often wide and complex, and the knowledge of its spatial distribution is the key information in agricultural productivity to improve food security, enhance crop production [11], and predict the effects of different agronomic management strategies. Among these strategies, irrigation with treated municipal wastewater can be considered important for saving limited freshwater resources and protecting the environment, but its effects should be monitored to avoid soil fertility decline in the medium to long term [9].

Conventional laboratory methods for quantifying this soil variable are destructive, time consuming, expensive, and hazardous for the environment. In addition, because of the associated costs, soil is sampled at relatively few spatial locations, which are often irregularly distributed over the study area. The small sample size does not allow meeting the criteria for soil quality assessment for precision farming or for using statistical methods taking into account residual autocorrelation [12]. Making a short review, a number of samples ranging from 50 [13] to 100 [14] is considered well suited for an accurate spatial analysis.

A strategy to enhance the quality of the estimation of SOC content and to reduce the spatial sampling intensity consists of incorporating secondary information correlated to the primary variable [15,16]. This multivariate approach allows utilization of secondary information, such as that derived from proximally and remotely sensed data, that is often much more abundant than information deriving from the primary target variable [17,18].

Proximal sensing data could provide strong support for characterizing the spatial variability at the field or even regional scale. These data are very attractive because of their high resolution, their noninvasive nature, the relatively low cost of data acquisition, the possibility for a mobile survey configuration, and their three-dimensional (3D) information, although their outcome is not a direct measurement of soil properties [19].

Among the geophysical methods, electromagnetic induction (EMI) and ground-penetrating radar (GPR) have been widely applied. EMI methods measure apparent electrical conductivity (EC_a), an integrated value of soil physical, chemical, and biological properties [20] that can capture soil spatial variability and characterize soil organic carbon distribution [21,22]. However, since soil properties vary in both the horizontal and vertical domains, soil needs to be described in three dimensions, and EMI sensors may have limitations when highly contrasting horizons are present [23]. Ground-penetrating radar (GPR) technology allows overcoming this limitation by measuring large volumes of soil (about cubic decimetres to cubic meters). Thus, GPR is suggested for field-scale determinations rather than for pointwise measurements, provides higher resolution of subsurface features, and is particularly suited to visualizing soil in two or three dimensions [24]. One of the most useful presentations of GPR data is to display horizontal maps of recorded reflection amplitudes, called "time slice" (or depth slice) maps [25]. There have been several studies involving GPR to determine thickness and characterize depths of organic soil materials [26,27], but few studies have been devoted so far to the potentiality of GPR to study the spatial variation of soil organic carbon.

The use of geophysical proximal sensor data as auxiliary information to effectively support an irregularly sampled target variable is not free from practical difficulties and experimental limits. This is because proximal sensing data are often massive, need to be collected on different spatial and temporal scales, and use different measurement supports. Several statistical methods are able to incorporate secondary information; for example, a multivariate extension of kriging, known as cokriging, is used for improving the prediction of a primary variable by using secondary information [28,29]. This technique assumes intrinsic stationarity, both of the target variables and of more intensively measured secondary variables, supposing a strong correlation between primary and secondary information [30].

These conditions are not always verified. Another way of taking into account the secondary variable is by checking for a spatial trend in the primary variable with respect to the secondary variable(s) and combining the deterministic part and the stochastic component, as in "hybrid methods" [29], or by adopting complex multivariate nonlinear approaches. In recent times, a number of hybrid interpolation techniques, which combine kriging with methods that use auxiliary information (covariates), have been developed and applied. Several authors have compared some of the techniques to incorporate trends and account for nonstationarity [31,32]. Two possible methods of nonstationary interpolation are regression kriging (RK) [28,33] and multivariate adaptive regression splines (MARS) [34]. In many cases, these techniques have been proven superior to common geostatistical methods, yielding more detailed results and higher accuracy of prediction, because they take advantage of being linear hybrid (RK) or nonlinear (MARS) [35]. MARS is a nonparametric predictive method that intrinsically models nonlinearities and interactions between variables, suitably managing local nonstationarity [34]. This method has been successfully applied in various fields, such as estimating the collapse potential for compacted soil, underground gas storage in bedded salt formations, and lateral spreading induced by earthquakes [36,37].

The regression kriging (RK) method is of straightforward use and often performs better than cokriging [38–40].

In this study, we compared the performance of RK and MARS to achieve the following objectives: (i) to prove that there are preferential nonlinear relationships between SOC and geophysical measurements, and (ii) to compare the performance of two nonstationary interpolation methods to effectively model SOC at the field scale. Machine learning techniques may open new perspectives to modelling SOC spatial distribution at the field and regional scales. The study was performed on a dataset deriving from a field experiment in which water of different qualities was used for irrigation.

To the best of authors' knowledge, no comparison between these methods has been presented before; therefore, it can be considered a novelty.

2. Materials and Methods

2.1. Study Area

Soil data were derived from a field experiment carried out in an olive grove located in Fasano (Apulia region, Southern Italy). The climate of the study area is "accentuated thermo-Mediterranean", as classified by UNESCO FAO [41,42], characterized by rather mild and rainy winters and warm and dry summer months. The soil of the experimental site is classified as loam (USDA classification), with an average content of silt, clay, and sand fractions of 35.28%, 21.74%, and 42.98%, respectively.

Olive trees were irrigated with treated municipal wastewater (TWW), and the following treatments were applied: irrigation with fresh water and full fertilization supply (FW); irrigation with TWW and full fertilization supply (R1); and irrigation with TWW and fertilizer supply reduced by the amount provided by TWW (R2) [10]. Treatments were arranged in a randomized complete block design (RCBD) with four replicates (Figure 1). Unit plot size was 108 m^2, with 3 plants per plot and a plant spacing of 6 m × 6 m; field size was 1296 m^2 (whole experimental area was 1728 m^2).

2.2. Soil Sampling and Soil Analysis

Soil samples with absolute coordinates were collected on a regular grid (April 2017) at 6 locations (subreplicates) per plot at a 0–0.20 m depth for a total of 72 observations (Figure 1); only 71 were used in this study. Soil organic carbon (SOC) was quantified on air-dried and sieved samples through dry combustion [43]. Further details about the experimental trial were reported by Barca et al. [44] and Stellacci et al. [10].

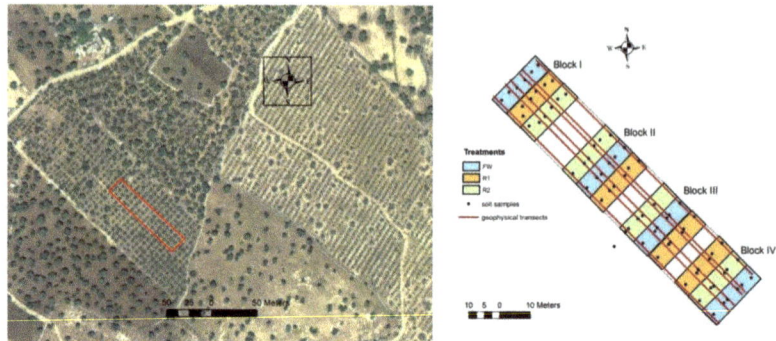

Figure 1. Location of the field experiment (Google Earth Pro, 2021) the soil sampling locations (black dots), and the electromagnetic induction (EMI) and ground-penetrating radar (GPR) acquisitions along transects (red lines).

2.3. Acquisition and Preprocessing of Auxiliary Information

A geophysical survey was carried out using an EMI sensor (EM38DD, Geonics Limited, Mississauga, ON, Canada) and a Georadar (RIS 2k-MF Multifrequency Array Radar-System, manufactured by IDS SpA, Italy) connected to the DGPS along 6 parallel transects by sliding the sensors on the surface (Figure 1) on the same day as soil sampling.

EMI soil survey is based on the principle that a transmitter coil in contact with the soil surface produces a time-varying primary magnetic field in the subsoil. The eddy currents induced in the soil generate a secondary magnetic field, which is recorded by a receiver coil in the EM unit. The apparent conductivity near the receiver is determined by the ratio of the magnitude of the secondary magnetic field to that of the primary magnetic field [22]. The EMI sensor used herein consisted of two perpendicularly superposed EM38 sensors that simultaneously measured apparent electrical conductivity (EC_a, expressed in mSm^{-1}) near the soil surface (0–0.75 m depth) with the horizontal mode (EC_a-H) and up to 1.5 m depth with the vertical mode (EC_a-V) [22]. Before operation, the instrument was set to zero at a height of 1.5 m, according to the manufacturer's instructions, and at the end of the survey, the zeroing was checked to detect possible drift. The survey was performed using a nonmetallic platform with wood cover, and the sensor was towed behind a tractor/The EC_a was recorded every second, with spatial resolution of 0.5 m, on average, along each transect.

Immediately after the EMI survey, the GPR survey was carried out by sliding the sensor along the surface. GPR data were collected with the common offset reflection method, using a monostatic system (the transmitting and receiving antenna placed in the same box) with two central frequencies of 600 and 1600 MHz (IDS Ing-manufactured, RIS 2k-MF Multifrequency Array Radar-System). The GPR worked with a time window of 60 ns and a temporal sampling interval of 0.05 ns; successive traces were collected every 0.024 m. GPR used electromagnetic pulse energy in the frequency range of 10 MHz to 1000 MHz. The transmitter component of the GPR system allowed the passage of generated pulse energy, which propagated through the subsurface materials, and the interactions with the material were sensed by the receiver component. Traditional surveys employ reflections of electromagnetic waves from boundaries between environments of different electromagnetic properties [45]. Theoretical aspects and working principles of radar components can be found in detail in Davis and Annan [46].

Both the data quality check and cleaning procedure characterized the preliminary data analysis. For EMI data, the points at which the instrument was stationary and any negative values were removed.

Processing the raw GPR data consisted of extracting quantifiable variables, such as attenuation, and displaying GPR data in horizontal maps at a specified time (or depth), called amplitude maps or time slices. The preprocessing of GPR signal amplitude data included the application of a set of filters [47] and the extraction of quantifiable variables.

The enveloped amplitude maps (time slices) were built by averaging the amplitude (or the square amplitude) of the radar signal, expressed in digital number (DN), within overlapping time windows of width Δt equal to the order of the dominant period of each antenna (2 and 1 ns for the 600 and 1600 MHz antennas, respectively). The total time interval was of 10 ns for the 600 MHz antenna because this time was comparable with the depth of the soil, and it was 6 ns for 1600 MHz because of the attenuation of radar signal. The time slices were then transformed in depth slices using the velocity of the radar waves determined through the analysis of hyperbolae [48]. Data preprocessing was performed with ReflexW Software [49].

In order to estimate the geophysical covariates at the same locations as the SOC measurements, geostatistical procedures were separately applied to EMI and GPR data by using a multivariate approach and fitting a linear model of coregionalization (LMC) to the experimental variograms. Each group of geophysical data was interpolated with ordinary cokriging (ck) on a 0.5 m × 0.5 m grid. The estimated covariates, migrated at the sample locations, were: the EC_a in horizontal (EC_aH) and vertical (EC_aV) modes; the amplitude for the 600 MHz antenna at ten depths from 0.05 m to 0.50 m with a step of 0.05 m (Amp600MHz_0.05 m-Amp600MHz_0.50 m); and the amplitude for 1600 MHz frequency antenna at eleven depths from 0.025 m to 0.275 m with a step 0.025 m (Amp1600MHz_0.025 m-Amp1600MHz_0.275 m).

Finally, 25 covariates were considered, namely, the 23 geophysical covariates plus the (two) geographical coordinates expressed in the WGS84 coordinate system.

2.4. Regression Kriging (Residual Kriging)

In the present paper, kriging combined with linear regression (RK), a hybrid interpolation technique, was applied [35,39] (see Figure 2). In mathematical terms, RK can be described as the sum of a deterministic (regression) component and kriging as shown in the following equation:

$$\hat{z}(s_0) = \hat{m}(s_0) + \hat{e}(s_0) = \sum_{k=1}^{p} \hat{\beta}_k \cdot q_k(s_0) + \sum_{i=1}^{N} \lambda_i \cdot e(s_i) \qquad (1)$$

where s_0 is the spatial location associated with the desired prediction, $\hat{m}(s_0)$ is the trend, $\hat{e}(s_0)$ is the interpolated residual, $\hat{\beta}_k$ are the estimated regressive coefficients, q_k are the covariates, p is the number of coviariates, λ_i are kriging weights, N is the number of observations, and $e(s_i)$ is the residual (i.e., the difference between the regression estimation minus the observation) at the generic observational location s_i.

From a practical standpoint, once the trend component has been estimated, the residual can be interpolated with kriging and then added to the previously estimated component. The prediction of the residual is a very critical step, because in principle, only the autocorrelated components should be estimated, neglecting the purely random component. Unfortunately, it is very difficult to separate the overall residual into the autocorrelated and the noncorrelated components. There are many different opinions about the best way to accomplish this issue [50,51]. In the present paper, the variography directly performed on the residuals provided results that did not depart much from those obtained with more sophisticated statistical methods; in other words, this approach did not significantly bias the final predictions. Therefore, the more straightforward approach, which brutally separates observations from trend values to obtain residuals, was preferred [29,52]. The validation of the RK method is usually carried out by means of the cross-validation procedure, and specifically the leave-one-out method [53]. Cross-validation is structured as a two-stage procedure. In the first stage, a leave-one-out method is applied, which consists of dropping an observation from the dataset and predicting this omitted value using the remaining data. Leave-one-out is iterated for each value in the dataset, and each time, a residual is computed as the difference between the observed and predicted values. The second stage of the cross-validation consists of making inferences about the residuals' distribution [54,55]. The R library [56] used to perform the aforementioned analysis was {Automap version 1.0–14}.

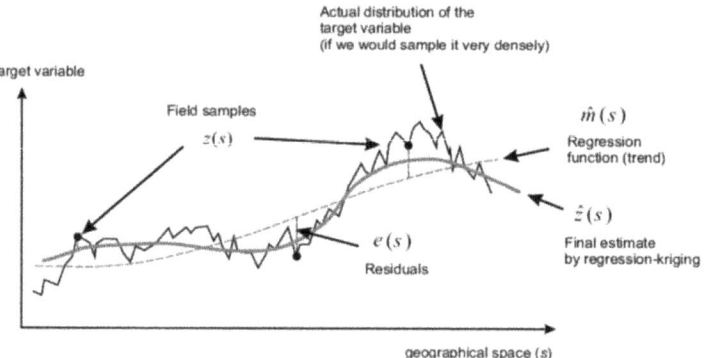

Figure 2. An example of the regression-kriging approach shown by means of a cross-section of the spatial random field (after Hengl, [35]).

2.5. Multivariate Adaptive Regression Splines (MARS)

MARS is a nonparametric and nonlinear predictive method that automatically models nonlinearities and interactions between variables managing suitably local nonstationarity [34]. Datasets are split into piecewise curves (splines) of differing slopes. Splines consist of two branches, i.e., left-sided (Equation (2)) and right-sided (Equation (3)) truncated functions, separated by a point called the *knot* [57].

$$b_q^-(x-t) = [-(x-t)]_+^q = \begin{cases} (t-x)^q & \text{if } x < t \\ 0 & \text{otherwise} \end{cases} \quad (2)$$

$$b_q^+(x-t) = [+(x-t)]_+^q = \begin{cases} (x-t)^q & \text{if } x > t \\ 0 & \text{otherwise} \end{cases} \quad (3)$$

$b_q^-(x-t)$ and $b_q^+(x-t)$ are splines describing the regions on the right and left sides of the knot (t), respectively, and q is the degree of the polynomial. The subscript "+" indicates that the result of the function is 0 outside the local definition domain. For each of the covariate variables, MARS selects the couple of splines and the knot location more in accordance with the response variable. In a next stage, the different splines are added up in a single multivariate model, which describes the response as a function of the covariates. The result is a nonlinear model assuming the form:

$$\hat{y} = a_0 \sum_{m=1}^{M} a_m B_m(x) \quad (4)$$

where $\hat{y}$ is the prediction of the response variable; a_0 is the known term; M is the number of basic splines; and B_m and a_m are the m-th basic spline and its coefficient, respectively [58].

Overall, a MARS analysis consists of three stages. Specifically, (i) the variable that best describes the response by means of the splines in terms of R^2 is selected. Afterwards, (ii) other covariates are added stepwise, always using splines, to build a multivariate model (i.e., the global MARS model). The aim of this addition is the improvement of model in terms of performance (R^2). The performance is computed on the training set. Since the global MARS model is usually affected by *overfitting*, it needs to be "pruned" in a further stage, for which iterations of the generalized cross-validations (GCV) alternated with 10-fold cross-validation are used [59]. The GCV index is a sum of squared errors (observations minus predictions) adjusted by embodying a penalty for reducing the model complexity. This criterion is used to prevent overfitting derived from an excessively accurate model with respect to the training set:

$$GCV = \frac{\frac{1}{n}\sum_{m=1}^{n}\left(y_i - \hat{f}_m(x_i)\right)^2}{(1 - C(M)/n)^2} \quad (5)$$

where C(M) is a parameter that penalizes models involving a large number of splines, defined as follows:

$$C(M) = (M+1) + dM \quad (6)$$

where M is the number of nonconstant splines (i.e., all terms of Equation (4) except a_0) in the MARS model and d is a user-defined penalty value for each spline optimization. Increases in the cost d cause the exclusion of splines. Substantially, d is increased during the pruning step in order to obtain smaller models. Besides its use during the pruning phase, GCV index is essential to rank covariates based on their importance in the model. The definition of the final model is reached in a third phase. This phase (iii) is performed by cross-validation or a new independent test set. The R library used to perform the aforementioned analysis herein is {earth} [59].

3. Results

3.1. Exploratory Data Analysis

Descriptive statistics showed that SOC data were normally distributed as confirmed by skewness and kurtosis values (Table 1) and by Shapiro–Wilk test (p = 0.656); for this reason, they were not subjected to a normal transform. The reported bubble plot (Figure 3) shows the spatial distribution of the SOC observations, evidencing some clusters of similar values.

Table 1. Summary statistics for SOC (g 100 g^{-1}).

Variable	N	Mean	Std	Min	Max	Skewness	Kurtosis
SOC	71	1.85	0.28	1.19	2.43	−0.21	−0.29

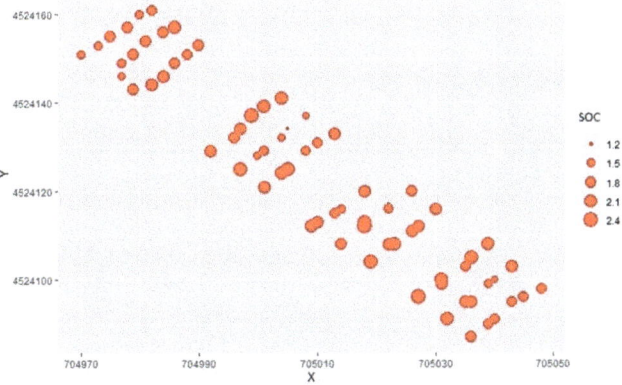

Figure 3. Bubble plot of spatial distribution of SOC values (g 100 g^{-1}).

The global Moran index provided an assessment of the spatial autocorrelation strength over the study area and is reported in Table 2. The result (I = 0.42) indicated a significant spatial autocorrelation (p = 0.00034). In addition to the global Moran index, the peak of the Moran index (local Moran index) was estimated by means of the computation of the mean of nearest neighbours. Afterwards, a lagged scatterplot provided the Moran computation at such distance lag. For the considered case, the mean of nearest neighbours was 2.63 m, and Figure 4 shows the Moran value corresponding to that distance, indicating a greater spatial correlation at short range (r = 0.75).

Table 2. Assessment of the global Moran index.

Spatial Autocorrelation Analysis (Original Data)			
Moran I	Variance	Expectation	p-Value
0.42	0.017	−0.014	0.00034

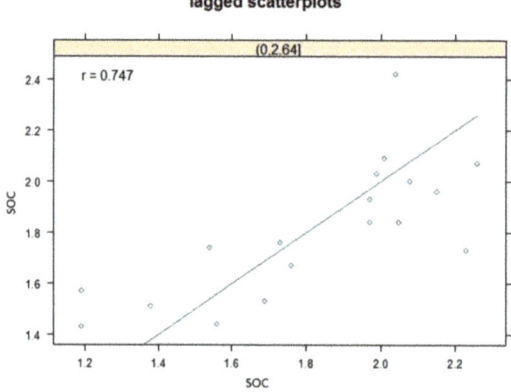

Figure 4. h-scatterplot for assessing local Moran I.

3.2. Linear Model Outcomes

The correlation matrix between SOC and the 25 covariates (23 geophysical variables plus the geographical coordinates) was first computed, and different sets of highly correlated covariates were derived and used to fit SOC data.

The following equation shows the first attempt to model SOC with the most correlated variables:

$$SOC \sim ckAmp0.05m\_600MHz + ckAmp0.1m\_600MHz + ckAmp0.4m\_600MHz$$

The five-point summary statistics and the coefficients of the linear model are reported in Tables 3 and 4. The outcomes seemed to indicate a larger contribution of the GPR data than of the EMI sensor data. The covariates related to the higher frequency antennae (1600MHz frequency) were therefore excluded.

Table 3. Five-point table of the linear model's residuals.

Min	1Q	Median	3Q	Max
−0.47	−0.16	0.01	0.15	0.63

In particular, the GPR data representations for both frequencies showed a first discontinuity in the radar signal at 0.1 m depth, a high level of spatial continuity along the soil profile at least to 0.30 m, and a second discontinuity after 0.30 m depth. Therefore, the selected covariates were representative of information derived by two different layers.

Table 4. Coefficients of the linear model.

	Estimate	Std_Error	t_Value	Pr(>|t|)
(Intercept)	−7.056e−01	1.14e+00	−0.62	0.54
ckAmp0.05m_600MHz	4.63e−06	7.06e−05	0.07	0.95
ckAmp0.1m_600MHz	2.07e−04	8.12e−05	2.55	0.013 *
ckAmp0.4m_600MHz	−6.34e−04	1.51e−03	−0.42	0.68

Signif. codes: 0.01, "*"; 0.05, ".".

The model was significant (F-statistic: 4.80 on 3 and 67 DF, p-value: 0.004) and showed a residual standard error of 0.26 with 67 degrees of freedom; multiple R-squared and adjusted R-squared were 0.177 and 0.14, respectively. Analysing Table 4, it was evident that there was a unique significant covariate, ckAmp0.1m_600MHz. The result showed the distribution of SOC to be significantly affected by the shallower layer, probably because it was comparable with the portion of sampled soil.

After many other attempts (not reported), a model was developed with the following optimal arrangement of the covariates:

$$SOC \sim X + Y + ckAmp0.35m\_600MHz$$

This model included the geographical coordinates and a unique geophysical covariate, ckAmp0.35m_600MHz (see Tables 5 and 6). This model was better that the aforementioned one, with all the covariates significant, a better value of R-squared (multiple R-squared: 0.26, adjusted R-squared: 0.22), and a more significant F-statistic p-value (F = 7.9 on 3 and 67 DF, p-value: 0.00018). Residual standard error was 0.24 with 67 degrees of freedom.

Table 5. Five-point table of the second linear model's residuals.

Min	1Q	Median	3Q	Max
−0.47	−0.16	−0.02	0.13	0.66

Table 6. The second linear model's coefficients with related statistics.

Coefficients	Estimate	Std. Error	t Value	Pr(>|t|)
(Intercept)	7.3e+04	2.1e+04	3.4	0.00153 **
X	−1.0e−02	3.9e−03	−2.7	0.01270 *
Y	−1.4e−02	4.1e−03	−3.5	0.00118 **
ckAmp0.35m_600MHz	−2.4e−03	6.0e−04	−4.0	0.00018 ***

Signif. codes: 0, "***"; 0.001, "**"; 0.01, "*"; 0.05, ".".

The model's residuals were then analysed. The Shapiro–Wilk Gaussianity test showed a nonsignificant departure from the normal distribution (W = 0.98567, p-value = 0.598); as a consequence, the Gaussian hypothesis was accepted. Afterwards, spatial autocorrelation analysis was performed to check at what extent the linear model filtered out the autocorrelation present in the raw data.

From Table 7, it was evident that in the linear model's residuals, there was still a significant quantity of spatial autocorrelation (p-value = 0.0012). Therefore, it made sense to apply regression kriging (RK) to exploit the residual autocorrelation with the aim of improving the goodness of fit.

Table 7. Linear model coefficients with related statistics.

Spatial Autocorrelation Analysis (Linear Model's Residuals)			
Moran I	Variance	Expectation	p-Value
0.29	0.01	−0.014	0.0012

3.3. Regression Kriging (RK)

Geostatistical analysis was then applied to the linear model's residuals with the aim of finding in them a structure that could represent their spatial variability.

The goodness of fit between the selected variogram model and the empirical variogram was evaluated by means of the SSErr index, which provides a value that helps user to judge the quality of the final model. For the case at hand, the value was SSErr = 0.00050, which appeared to be a satisfactory result. Moreover, by analysing the variogram parameters, reported in Table 8, it was possible to figure out the strength of the model by computing the nugget-to-sill ratio index [60], also called the spatial dependence index (SDI; [61]). For the case at hand, the observed value was 0.075, indicating high descriptive capability for the variogram model.

Table 8. Variogram model and parameters.

Model	Psill *	Range
Nugget	0.0042	0.0
Spherical	0.056	8.64

* Psill = Partial sill.

In Figure 5, the experimental variogram and the fitted nested model (nugget + spherical) are reported.

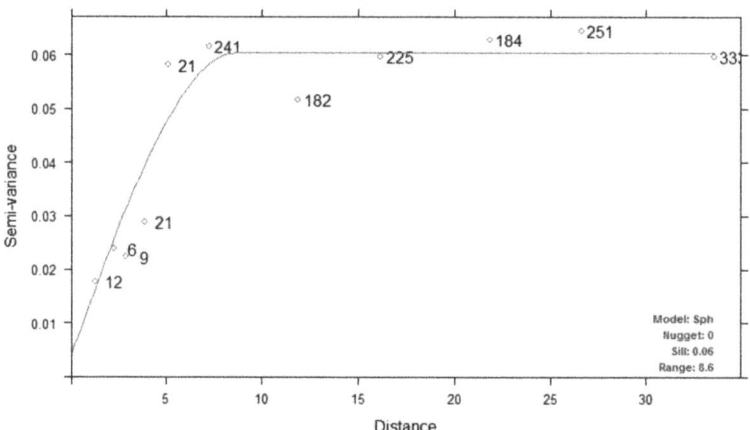

Figure 5. Experimental variogram and fitted variogram model.

Cross-validation statistics showed an MAE to RMSE ratio of 0.76, indicating a very good outcome. Mathematically, RMSE is always larger than MAE, because large errors are magnified by the square contained in the formula; therefore, the ratio between MAE and RMSE is always less than 1. However, the closer to 1 the ratio is, the fewer large errors made are by the model. This positive result was confirmed by a MAPE value far lower than 10% (Table 9). Computing the Lin coefficient (CCC) between observations and predictions, the outcomes were 0.65 for overall CCC, 0.68, for overall precision, and 0.95 for overall accuracy. The scatterplot of predicted versus observed values qualitatively showed the adequacy between the two data series (Figure 6).

Table 9. Accuracy metrics to assess the goodness of fit of the RK model.

Metric	MBE	MAE	RMSE	MAE/RMSE	MAPE	MIN	MAX
value	0.0013	0.15	0.20	0.76	8.47%	−0.49	0.42

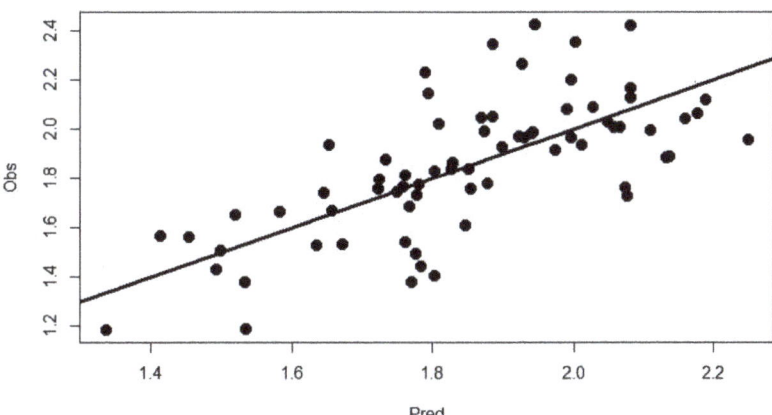

Figure 6. Scatterplot of predicted (RK model) vs. observed values.

The spatial distribution of SOC obtained through RK is reported in Figure 7.

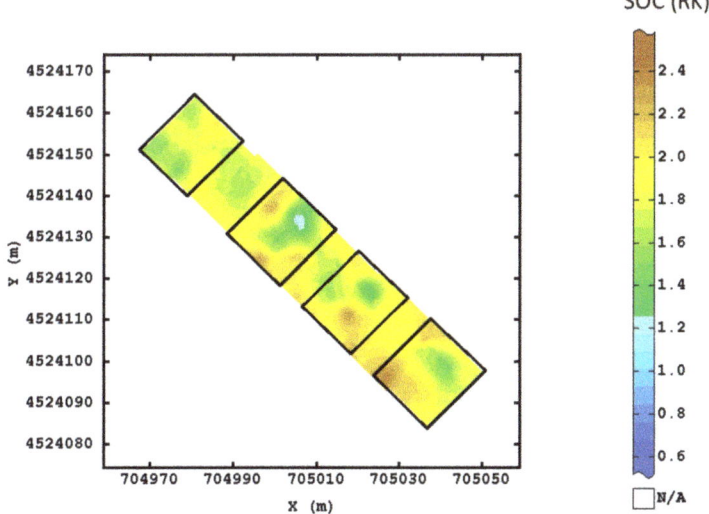

Figure 7. Map of SOC obtained with regression kriging. The black polygons indicate the four blocks in the RCB experimental design.

3.4. MARS Model Assessment

The original dataset was split into two complementary subsets, namely, training and test, corresponding to 80% and 20% of the original data, respectively.

Since the model is calibrated by means of the training dataset with the aim to predict the test data, the two subsets should be (statistically) similar at some extent. For this reason, after the splitting, subsets were subjected to the t-test for mean homogeneity and the Levene test for variance homogeneity. In addition, a univariate cluster analysis, carried out to assess the presence of clusters among data, showed that observations could be split into four groups. This represents another constraint about the splitting that has to be taken into account, i.e., the training and test subsets should be formed by a balanced quantity of elements extracted from all the clusters. Subsets were both checked for Gaussianity by means of Shapiro–Wilk test; results showed for both subsets a nonsignificant departure from normal distribution (W = 0.99, *p*-value = 0.90, for the training set; W = 0.97, *p*-value = 0.81, for the test set).

A Welch two-sample t-test showed that the means of the two subsets were not statistically different (t = −0.25, df = 20.36, *p*-value = 0.81). In addition, a boxplot confirmed the equality of the two means of the SOC variable subsets (Figure 8).

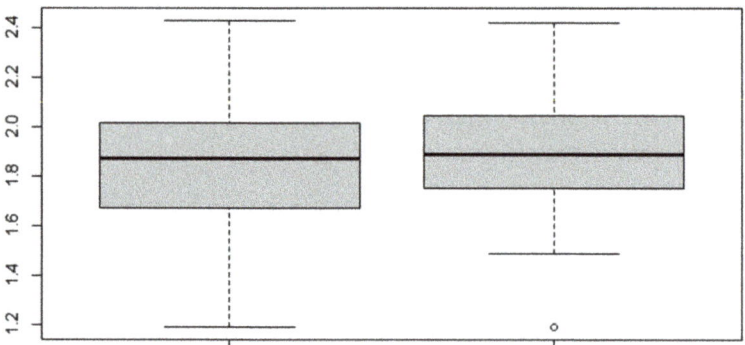

Figure 8. Boxplot for SOC comparison between training and test sets.

A Levene test, based on the absolute deviations from the median with a modified structural zero removal method and correction factor, showed the homogeneity of the group variances (test statistic = 0.059, *p*-value = 0.81). In Figure 9, the placement of the observations for the training (red points) and the test (green points) sets is reported.

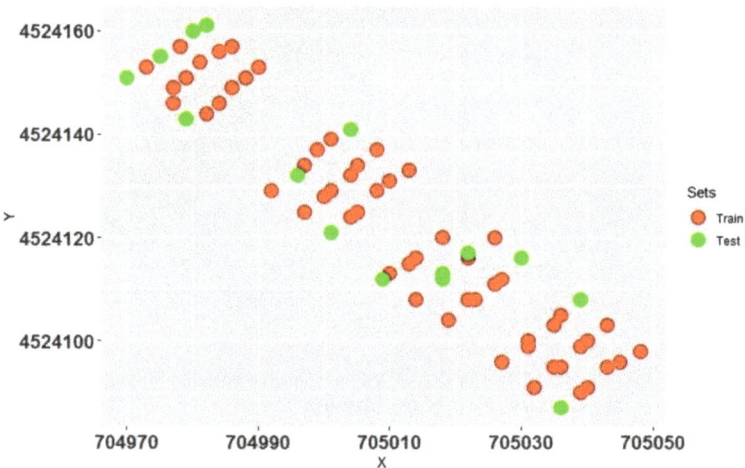

Figure 9. Spatial distribution of training and test sets points.

In summary, the two subsets could be considered similar according to the distribution, mean value, and variance comparisons. Therefore, the training set seemed to be appropriate to calibrate the model and the test set to check for overfitting.

The MARS model selected only 4 out of 25 predictors, namely, ckAmp0.35m_600MHz, X, ckEC$_a$Ver, and ckAmp0.1m_600MHz.

The model included the main GPR covariates selected previously. Regarding EMI data, only the apparent electrical conductivity measured in vertical polarization was selected because the two electrical conductivity variables were strongly correlated and therefore redundant. In addition, the sensor in vertical polarization had a maximum sensitivity approximately at a depth of 0.40 m, which was comparable with the time slices of GPR repeatedly selected (0.35 m).

From Table 10, it can be drawn that the MARS model was formed by four terms; apart from the intercept, the first was linear, and the remaining two were interactions between couples of covariates. After importance analysis was applied, by using the GCV and raw residual sum of squares (Rss) indices, the selected predictors were ranked accordingly (Table 11).

Table 10. MARS model structure.

MARS Terms	Coefficients
(Intercept)	2.0
h(ckAmp0.35m_600MHz-408)	$-1.12e-02$
h(13011-ckAmp0.1m_600MHz)*h(408-ckAmp0.35m_600MHz)	$-5.87e-06$
h(704990-X)*ckECaVer	$-2.68e-03$

Table 11. Covariates of the MARS model listed according to their importance rank with respect to GCV (generalized cross-validation) and Rss (raw residual sum of squares).

	GCV	Rss
ckAmp0.35m_600MHz	100.0	100.0
X	63.4	66.5
ckECaVer	63.4	66.5
ckAmp0.1m_600MHz	48.2	47.9

As first step, the Gaussianity of the residuals after the training was tested using the Shapiro–Wilk test; the residuals distribution could be considered Gaussian with a distribution $\sim N(0.0, 0.036)$ (W = 0.98, p-value = 0.50).

By applying a blind cross-validation with k-fold = 10, the resulting R^2 was 0.51, but it should be borne in mind that this was a pessimistic result, as the extractions of blocks of 10 elements (k-fold with k = 10) from the original dataset was performed 200 times in a purely random fashion, neglecting similar subsets. Moreover, the original dataset was relatively small and represents a not-very-homogeneous reality. Finally, the results in terms of goodness of fit were averaged.

The first step consisted of checking the correlation between predicted and observed values for the training set; the results showed a certain agreement (r = 0.72, p-value ≈ 0.0). In addition, correlation between residuals and predicted values of training subset was checked and was close to zero, as expected.

Afterwards, the MARS model calibrated on the training set was applied to predict SOC data from the test set, which was independent from the model calibration (training) set.

As a first step, the correlation between observations and (test set) predictions was analysed. This resulted in a highly significant correlation (r = 0.87, p-value ≈ 0.0). The value gained after the validation step surprisingly outperformed that of the training set, which is a rare event. The correlation between residuals and (test-set) predicted was not significant.

The residuals, according to the Shapiro–Wilk test, were Gaussian, with a distribution $\sim N(0.027, 0.025)$ (W = 0.93, p-value = 0.23).

Computing the Lin coefficient (CCC) between observations and predictions, the outcomes showed very good agreement (overall CCC, 0.81; overall precision, 0.88; overall accuracy, 0.93).

Since the observations were available, it was possible to compute the error metrics, which are reported in Table 12.

Table 12. Accuracy metrics to assess the goodness of fit of the MARS model.

Metric	MBE	MAE	RMSE	MAE/RMSE	MAPE	MIN	MAX
value	−0.03	0.13	0.16	0.80	6.7%	−0.42	0.19

The error indices were good overall; in particular, MAPE was below 10%, which value has been indicated in literature as a critical threshold. Another very interesting result concerned the ratio between MAE and RMSE, which was larger than that obtained with regression kriging (0.8 vs. 0.76). In conclusion, the MARS model could be considered effective whenever the coefficients of the covariates were not constant over the study domain and the covariates were intertwined in more complex ways than additively.

By comparing the error indices and Lin's coefficients of both methods, it became evident that MARS performed better than RK. The two methods were linear (RK) and nonlinear (MARS), respectively. The main difference concerns the interaction terms, since the MARS model has one linear term and two multiplicative terms (interactions) that represent the added value that allowed improving the predictive capability of MARS with respect to that of RK.

In Figure 10, the map of SOC predictions obtained with the MARS model is reported. Comparing the RK and MARS maps, they showed overall agreement, with a cluster of lower values in the northern part of the study area, a central part with the lowest values, and finally, a southern part with two clusters of larger values and a cluster of lower values.

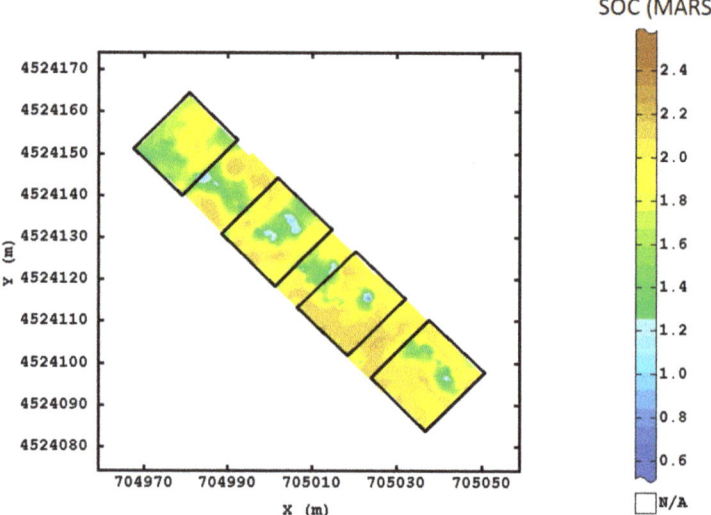

Figure 10. Map of SOC obtained with MARS model. The black polygons indicate the four blocks in the RCB experimental design.

Finally, to quantitatively compare the maps obtained by the two methods, a cross-correlogram was computed. The result was a value of 0.67 at the distance 0. Therefore, the map gained from RK can be considered a first approximation of that from MARS. This result underlines the reliability of the SOC spatial distribution predicted by the MARS model.

4. Discussion

Spatial prediction of SOC is critical for assessing the effect of agronomic management strategies on soil quality and crop productivity. In this scope, the sample size is a value that plays a key role in SOC prediction. Thus, it needs to be balanced between economic and predictive constraints. In fact, increasing the sample size may allow the application of statistical methods that take residual autocorrelation into account and thereby reduce the probability of inflation of the type I error rate [12], but at large cost. Regression kriging and MARS, incorporating covariate information often available at a finer resolution than primary data, such as proximal and remote sensor data, may improve the quality of SOC estimation without increasing the sampling size of the primary variable [62,63].

Outcomes obtained from linear models seem to highlight a larger informative contribution of GPR than of EMI data. From a physical standpoint, this result can be explained by the different nature of sensors' outcomes. In fact, GPR information results are more sensitive to near-surface effects than EMI data, which are integrated values over all soil layers [15]. However, unexpectedly [64], the covariates related to the higher-frequency antenna (1600 MHz frequency) were excluded, probably because they did not add further information or were redundant in this study case.

Two GPR covariates, namely, ckAmp0.35m_600MHz and ckAmp0.1m_600MHz, were selected by the MARS model. The same variables were also chosen by the final RK model (ckAmp0.35m_600MHz) and the preliminary RK model (ckAmp0.1m_600MHz). Similar importance was also assigned to the selected variables by both statistical methods, as shown by the ranking defined by GCV and Rss in MARS model, suggesting that their significance was physically based. In fact, the selected covariates were representative of information derived by two soil layers with different physical properties influencing radar signal and soil organic carbon distribution. The two methods also had the X geographical coordinate in common, indicating a larger continuity along this direction.

The main difference between the two approaches concerned the selection of the EMI covariate in vertical polarization performed only by the MARS model, indicating the different explanatory power of information brought by the two sensors. This result was tied to the intrinsic capability of the MARS model to intercept the interactions among variables and highlight nonlinear features underlying the data [34]. In addition, the coefficients of the MARS model were not constant but piecewise linear (splines), and therefore, their gradient varied over the studied domain [57]. This explains the larger descriptive capability of the MARS model and its ability to select hidden features with respect to regression kriging. Although MARS is not explicitly a spatial method, its capability of modelling covariate coefficients by means of flexible functions allows, when the geographic variables are included in the analysis, filtering out the spatial autocorrelation contained in the data, which makes it substantially a spatial method [65]. A confirmation of this was the statistical nonsignificance of the Moran I index obtained from the MARS residuals.

Studies on the spatial variability of SOC in agricultural soils remain a central theme in assessing the environmental sustainability of agricultural systems [66], because agronomic inputs could be rationalized in order to not impoverish the soil's fertility. Therefore, our results represent a knowledge contribution for future studies aimed at detecting the spatial distribution of soil organic carbon at the field scale. Geophysical methods show new applicative potentialities for environmental sciences (see, among others, [67]) and can represent support for research in this field. However, because of the complex interactions with soil properties, the use of geophysical measurements as covariates needs to be investigated in more detail to draw more precise conclusions. A limit of the present work could be its potential site-specificity, which could not be quantified in advance. Therefore, further experiments in different study areas and agroenvironments should be performed to test the performance of the methods under different conditions.

5. Conclusions

The results of our investigation showed that MARS outperformed RK in predicting SOC spatial distribution. The nonlinearity of MARS evidenced the contribution of EMI variables neglected by linear approaches. That result would have to be deepened in future works in consideration of the fact that EMI measures are more easily achievable than GPR ones.

The accuracy reached in mapping SOC with the support of MARS was remarkable and opens interesting perspectives in applying other, more powerful machine learning methods (e.g., deep learning) to even better exploit proximally sensed data. In the future, it is hoped that these machine learning methods will be successfully associated with mapping procedures and then applied at the regional and national level.

The use of relatively easy, accurate, and inexpensive geophysical methods for SOC estimation, together with application of advanced statistical techniques for SOC spatialization, can represent a viable solution to investigate agroecosystem sustainability.

Author Contributions: Conceptualization, A.M.S., E.B. and D.D.B.; methodology, E.B.; software, E.B.; validation, E.B., D.D.B., A.M.S., M.C. and G.L.; formal analysis, E.B., A.M.S. and D.D.B.; investigation, D.D.B.; resources, M.C. and S.P.; data curation, D.D.B.; writing—original draft preparation, E.B.; writing—review and editing, D.D.B., A.M.S., E.B., M.C. and G.L.; visualization, D.D.B.; supervision, A.M.S. and E.B.; project administration and funding acquisition, A.M.S. All authors have read and agreed to the published version of the manuscript.

Funding: The authors would like to thank the EU and MIUR for funding the present methodological contribution under the framework of the collaborative international consortium DESERT (ID-217) financed under the ERA-NET Cofund WaterWorks2014 Call. This ERA-NET is an integral part of the 2015 Joint Activities developed by the Water Challenges for a Changing World Joint Programme Initiative (Water JPI).

Institutional Review Board Statement: The study was conducted in accordance with the Declaration of Helsinki, and approved by the Institutional Review Board.

Informed Consent Statement: Not applicable.

Data Availability Statement: Not applicable.

Conflicts of Interest: The authors declare no conflict of interest.

References

1. Gregorich, E.G.; Carter, M.R.; Angers, D.A.; Monreal, C.M.; Ellert, B.H. Towards a minimum data set to assess soil organic matter quality in agricultural soils. *Can. J. Soil Sci.* **1994**, *74*, 367–385. [CrossRef]
2. Johnston, C.A.; Groffman, P.; Breshears, D.D.; Cardon, Z.G.; Currie, W.; Emanuel, W.; Gaudinski, J.; Jackson, R.B.; Lajtha, K.; Nadelhoffer, K.; et al. Carbon cycling in soil. *Front. Ecol. Environ.* **2004**, *2*, 522–528. [CrossRef]
3. Lorenz, K.; Lal, R.; Ehlers, K. Soil organic carbon stock as an indicator for monitoring land and soil degradation in relation to United Nations' Sustainable Development Goals. *Land Degrad. Dev.* **2019**, *30*, 824–838. [CrossRef]
4. Adhikari, K.; Hartemink, A.E. Linking soils to ecosystem services—A global review. *Geoderma* **2016**, *262*, 101–111. [CrossRef]
5. Shukla, M.K.; Lal, R.; Ebinger, M. Determining soil quality indicators by factor analysis. *Soil Till. Res.* **2006**, *87*, 194–204. [CrossRef]
6. Stellacci, A.; Castellini, M.; Diacono, M.; Rossi, R.; Gattullo, C. Assessment of Soil Quality under Different Soil Management Strategies: Combined Use of Statistical Approaches to Select the Most Informative Soil Physico-Chemical Indicators. *Appl. Sci.* **2021**, *11*, 5099. [CrossRef]
7. Fang, X.; Xue, Z.; Li, B.; An, S. Soil organic carbon distribution in relation to land use and its storage in a small watershed of the Loess Plateau, China. *CATENA* **2012**, *88*, 6–13. [CrossRef]
8. Ferrara, R.M.; Mazza, G.; Muschitiello, C.; Castellini, M.; Stellacci, A.M.; Navarro, A.; Lagomarsino, A.; Vitti, C.; Rossi, R.; Rana, G. Short-term effects of conversion to no-tillage on respiration and chemical-physical properties of the soil: A case study in a wheat cropping system in semi-dry environment. *Ital. J. Agrometeorol.* **2017**, *1*, 47–58.
9. Leogrande, R.; Stellacci, A.M.; Vitti, C.; Lacolla, G.; Moscelli, S.; Mastrangelo, M.; Vivaldi, G.A. Soil properties as affected by irrigation with treated municipal wastewater. In Proceedings of the XLVII Conference of Italian Society for Agronomy, Marsala, Italy, 12–14 September 2018.
10. Stellacci, A.M.; De Benedetto, D.; Leogrande, R.; Vitti, C.; Castellini, M.; Barca, E. Use of Mixed Effects Models accounting for residual spatial correlation to analyze soil properties variation in a field irrigated with treated municipal wastewater. In Proceedings of the XLVII Conference of Italian Society for Agronomy, Marsala, Italy, 12–14 September 2018.
11. Stevenson, F.J.; Cole, M.A. *Cycles of Soil*, 2nd ed.; Wiley: New York, NY, USA, 1999.
12. Littell, R.C.; Milliken, G.A.; Stroup, W.W.; Wolfinger, R.D.; Schabenberger, O. *SAS for Mixed Models*, 2nd ed.; SAS Institute Inc.: Cary, NC, USA, 2006.
13. Journel, A.G.; Huijbregts, C.J. *Mining Geostatistics*; Academic Press: Waltham, MA, USA, 1978.
14. Webster, R.; Oliver, M.A. How large a sample is needed to estimate the regional variogram adequately? In *Geostatistics Tróia '92*; Springer: Berlin, Germany, 1993; pp. 155–166.

15. Barca, E.; De Benedetto, D.; Stellacci, A.M. Contribution of EMI and GPR proximal sensing data in soil water content assessment by using linear mixed effects models and geostatistical approaches. *Geoderma* **2019**, *343*, 280–293. [CrossRef]
16. Piccini, C.; Marchetti, A.; Francaviglia, R. Estimation of soil organic matter by geostatistical methods: Use of auxiliary infor-mation in agricultural and environmental assessment. *Ecol. Ind.* **2014**, *36*, 301–314. [CrossRef]
17. Stevens, A.; Udelhoven, T.; Denis, A.; Tychon, B.; Lioy, R.; Hoffmann, L.; van Wesemael, B. Measuring soil organic carbon in croplands at regional scale using airborne imaging spectroscopy. *Geoderma* **2010**, *158*, 32–45. [CrossRef]
18. Nawar, S.; Mouazen, A.M. Predictive performance of mobile vis-near infrared spectroscopy for key soil properties at different geographical scales by using spiking and data mining techniques. *CATENA* **2017**, *151*, 118–129. [CrossRef]
19. Rossel, R.V.; Adamchuk, V.I.; Sudduth, K.A.; McKenzie, N.J.; Lobsey, C. Proximal Soil Sensing: An Effective Approach for Soil Measurements in Space and Time. *Adv. Agron.* **2011**, *113*, 243–291.
20. Heil, K.; Schmidhalter, U. The Application of EM38: Determination of Soil Parameters, Selection of Soil Sampling Points and Use in Agriculture and Archaeology. *Sensors* **2017**, *17*, 2540. [CrossRef] [PubMed]
21. Martinez, G.; Vanderlinden, K.; Ordóñez, R.; Muriel, J.L. Can Apparent Electrical Conductivity Improve the Spatial Characteriza-tion of Soil Organic Carbon? *Vadose Zone J.* **2009**, *8*, 586–593. [CrossRef]
22. McNeill, J.D. *Electromagnetic Terrain Conductivity Measurement at Low Induction Numbers*; Technical Note TN 6; Geonics Ltd.: Mississauga, ON, Canada, 1980.
23. Sudduth, K.A.; Kitchen, N.R.; Bollero, G.A.; Bullock, D.G.; Wiebold, W.J. Comparison of Electromagnetic Induction and Direct Sensing of Soil Electrical Conductivity. *Agron. J.* **2003**, *95*, 472–482. [CrossRef]
24. Grote, K.; Anger, C.; Kelly, B.; Hubbard, S.; Rubin, Y. Characterization of Soil Water Content Variability and Soil Texture using GPR Groundwave Techniques. *J. Environ. Eng. Geophys.* **2010**, *15*, 93–110. [CrossRef]
25. Conyers, L.B.; Goodman, D. *Ground Penetrating Radar: An Introduction for Archaeologists*; Altamira Press: London, UK, 1997.
26. Collins, M.; Schellentrager, G.; Doolittle, J.; Shih, S. Using ground-penetrating radar to study changes in soil map unit compo-sition in selected Histosols. *Soil Sci. Soc. Am. J.* **1986**, *50*, 408–412. [CrossRef]
27. Winkelbauer, J.; Völkel, J.; Leopold, M.; Bernt, N. Methods of surveying the thickness of humous horizons using ground pen-etrating radar (GPR): An example from the Garmisch-Partenkirchen area of the Northern Alps. *Eur. J. For. Res.* **2011**, *130*, 799–812. [CrossRef]
28. Goovaerts, P. Geostatistical approaches for incorporating elevation into the spatial interpolation of rainfall. *J. Hydrol.* **2000**, *228*, 113–129. [CrossRef]
29. Hengl, T.; Heuvelink, G.B.; Stein, A. A generic framework for spatial prediction of soil variables based on regression-kriging. *Geoderma* **2004**, *120*, 75–93. [CrossRef]
30. Webster, R.; Oliver, M.A. *Geostatistics for Environmental Scientists*; John Wiley & Sons Ltd.: Chichester, UK, 2001.
31. Bourennane, H.; King, D. Using multiple external drifts to estimate a soil variable. *Geoderma* **2003**, *114*, 1–18. [CrossRef]
32. Castrignanò, A.; Costantini, E.A.; Barbetti, R.; Sollitto, D. Accounting for extensive topographic and pedologic secondary information to improve soil mapping. *CATENA* **2009**, *77*, 28–38. [CrossRef]
33. Odeh, I.; McBratney, A.; Chittleborough, D. Further results on prediction of soil properties from terrain attributes: Heterotopic cokriging and regression-kriging. *Geoderma* **1995**, *67*, 215–226. [CrossRef]
34. Friedman, J.H. Multivariate adaptive regression splines. *Ann. Stat.* **1991**, *19*, 67. [CrossRef]
35. Hengl, T.A. *Practical Guide to Geostatistical Mapping*; Office for Official Publications of the European Communities: Luxembourg, 2009.
36. Zhang, W.; Goh, A. Multivariate adaptive regression splines for analysis of geotechnical engineering systems. *Comput. Geotech.* **2013**, *48*, 82–95. [CrossRef]
37. Garg, A.; Garg, A.; Tai, K. A multi-gene genetic programming model for estimating stress-dependent soil water retention curves. *Comput. Geosci.* **2014**, *18*, 45–56. [CrossRef]
38. Zhang, S.; Huang, Y.; Shen, C.; Ye, H.; Du, Y. Spatial prediction of soil organic matter using terrain indices and categorical variables as auxiliary information. *Geoderma* **2012**, *171–172*, 35–43. [CrossRef]
39. Minasny, B.; McBratney, A. Spatial prediction of soil properties using EBLUP with the Matérn covariance function. *Geoderma* **2007**, *140*, 324–336. [CrossRef]
40. Eldeiry, A.A.; Garcia, L.A. Comparison of Ordinary Kriging, Regression Kriging, and Cokriging Techniques to Estimate Soil Salinity Using LANDSAT Images. *J. Irrig. Drain. Eng.* **2010**, *136*, 355–364. [CrossRef]
41. United Nations Educational, Scientific and Cultural Organization-Food and Agriculture Organization of the United Nations (UNESCO-FAO). *Bioclimatic Map of the Mediterranean Zone*; UNESCO: Paris, France; FAO: Rome, Italy, 1963; 60p.
42. Leogrande, R.; Vitti, C.; Castellini, M.; Mastrangelo, M.; Pedrero, F.; Vivaldi, G.; Stellacci, A. Comparison of Two Methods for Total Inorganic Carbon Estimation in Three Soil Types in Mediterranean Area. *Land* **2021**, *10*, 409. [CrossRef]
43. Vitti, C.; Stellacci, A.M.; Leogrande, R.; Mastrangelo, M.; Cazzato, E.; Ventrella, D. Assessment of organic carbon in soils: A comparison between the Springer-Klee wet digestion and the dry combustion methods in Mediterranean soils (Southern Italy). *Catena* **2016**, *137*, 113–119. [CrossRef]
44. Barca, E.; Stellacci, A.M.; De Benedetto, D. Optimization of Sampling Design for Total Organic Carbon Assessment using Spatial Simulated Annealing: Comparison of Different Variogram Models Performances. In Proceedings of the XLVIII Conference of Italian Society for Agronomy, Perugia, Italy, 18–20 September 2019; pp. 223–224.

45. Annan, A.P. Electromagnetic principles of ground penetrating radar. In *Ground Penetrating Radar: Theory and Applications*; Jol, H.M., Ed.; Elsevier: Amsterdam, The Netherlands, 2009; pp. 3–40. ISBN 978-0-444-53348-7. [CrossRef]
46. Davis, J.L.; Annan, A.P. Ground-penetrating radar for high-resolution mapping of soil and rock stratigraphy. *Geophys. Prospect.* **1989**, *37*, 531–551. [CrossRef]
47. De Benedetto, D.; Quarto, R.; Castrignanò, A.; Palumbo, D.A. Impact of Data Processing and Antenna Frequency on Spatial Structure Modelling of GPR Data. *Sensors* **2015**, *15*, 16430–16447. [CrossRef] [PubMed]
48. Daniels, D.J. *Ground Penetrating Radar*, 2nd ed.; The Institution of Engineering and Technology: London, UK, 2004.
49. *User's Manual Online Version, Sandmeier Scientific Software*, Reflexw, v.6.1.1. Program for Processing and Interpretation of Reflection and Transmission Data. Reflexw: Karlruhe, Germany, 2012.
50. Kitanidis, P.K. Generalized covariance functions in estimation. *Math. Geol.* **1993**, *25*, 525–540. [CrossRef]
51. Lark, R.M.; Cullis, B.R.; Welham, S.J. On spatial prediction of soil properties in the presence of a spatial trend: The empirical best linear unbiased predictor (E-BLUP) with REML. *Eur. J. Soil Sci.* **2005**, *57*, 787–799. [CrossRef]
52. Hengl, T.; Heuvelink, G.B.; Rossiter, D.G. About regression-kriging: From equations to case studies. *Comput. Geosci.* **2007**, *33*, 1301–1315. [CrossRef]
53. Barca, E.; Porcu, E.; Bruno, D.; Passarella, G. An automated decision support system for aided assessment of variogram models. *Environ. Model. Softw.* **2017**, *87*, 72–83. [CrossRef]
54. Myers, J.C. *Geostatistical Error Management: Quantifying Uncertainty for Environmental Sampling and Mapping*; John Wiley and Sons: Hoboken, NJ, USA, 1997.
55. Chiles, J.P.; Delfiner, P. *Geostatistics, Modeling Spatial Uncertainty*; Wiley: Hoboken, NJ, USA, 1999.
56. R Core Team. *R: A Language and Environment for Statistical Computing*; R Foundation for Statistical Computing: Vienna, Austria, 2020. Available online: https://www.R-project.org/ (accessed on 17 January 2022).
57. Ghasemi, J.B.; Zolfonoun, E. Application of principal component analysis–multivariate adaptive regression splines for the simultaneous spectrofluorimetric determination of dialkyltins in micellar media. *Spectrochim. Acta Part A Mol. Biomol. Spectrosc.* **2013**, *115*, 357–363. [CrossRef]
58. Hiemstra, P.H.; Pebesma, E.; Twenhöfel, C.J.; Heuvelink, G.B. Real-time automatic interpolation of ambient gamma dose rates from the Dutch radioactivity monitoring network. *Comput. Geosci.* **2008**, *35*, 1711–1721. [CrossRef]
59. Milborrow, S. Derived from Mda: MARS by T. Hastie and R. Tibshirani. Earth: Multivariate Adaptive Regression Splines. R Package. 2011. Available online: http://www.milbo.users.sonic.net/earth/citing-earth.html (accessed on 17 January 2022).
60. Cambardella, C.A.; Elliott, E.T. Carbon and Nitrogen Dynamics of Soil Organic Matter Fractions from Cultivated Grassland Soils. *Soil Sci. Soc. Am. J.* **1994**, *58*, 123–130. [CrossRef]
61. Pasini, M.P.B.; Dal'Col Lúcio, A.; Cargnelutti, A.F. Semivariogram models for estimating fig fly population density throughout the year. *Pesqui. Agropecu. Bras.* **2014**, *49*, 493–505. [CrossRef]
62. Xie, X.-L.; Li, A.-B.; Mouazen, A. Improving spatial estimation of soil organic matter in a subtropical hilly area using covariate derived from vis-NIR spectroscopy. *Biosyst. Eng.* **2016**, *152*, 126–137. [CrossRef]
63. Mirzaee, S.; Ghorbani-Dashtaki, S.; Mohammadi, J.; Asadi, H.; Asadzadeh, F. Spatial variability of soil organic matter using remote sensing data. *CATENA* **2016**, *145*, 118–127. [CrossRef]
64. De Benedetto, D.; Castrignanò, A.; Sollitto, D.; Modugno, F.; Buttafuoco, G.; Lo Papa, G. Integrating geophysical and geosta-tistical techniques to map the spatial variation of clay. *Geoderma* **2012**, *171–172*, 53–63. [CrossRef]
65. Wang, X.; Yang, C.; Zhou, M. Partial Least Squares Improved Multivariate Adaptive Regression Splines for Visible and Near-Infrared-Based Soil Organic Matter Estimation Considering Spatial Heterogeneity. *Appl. Sci.* **2021**, *11*, 566. [CrossRef]
66. Castellini, M.; Stellacci, A.M.; Tomaiuolo, M.; Barca, E. Spatial variability of soil physical and hydraulic properties in a durum wheat field: An assessment by the BEST-Procedure. *Water* **2019**, *11*, 1434. [CrossRef]
67. Di Prima, S.; Winiarski, T.; Angulo-Jaramillo, R.; Stewart, R.D.; Castellini, M.; Najm, M.R.A.; Ventrella, D.; Pirastru, M.; Giadrossich, F.; Capello, G.; et al. Detecting infiltrated water and preferential flow pathways through time-lapse ground-penetrating radar surveys. *Sci. Total Environ.* **2020**, *726*, 138511. [CrossRef]

Article

Reducing Nitrogen Fertilization in Olive Growing by the Use of Natural Chabazite-Zeolitite as Soil Improver

Valeria Medoro [1], Giacomo Ferretti [2], Giulio Galamini [1], Annalisa Rotondi [3], Lucia Morrone [3], Barbara Faccini [1] and Massimo Coltorti [1,*]

1. Department of Physics and Earth Science, University of Ferrara, Via Giuseppe Saragat 1, 44122 Ferrara, Italy
2. Department of Chemical, Pharmaceutical and Agricultural Sciences, Via Luigi Borsari 46, 44121 Ferrara, Italy
3. Institute of Bioeconomy, National Research Council, Via Piero Gobetti 101, 40129 Bologna, Italy
* Correspondence: massimo.coltorti@unife.it

Abstract: In order to improve the sustainability and productivity of modern agriculture, it is mandatory to enhance the efficiency of Nitrogen (N) fertilizers with low-impact and natural strategies, without impairing crop yield and plant health. To achieve these goals, the ZeOliva project conducted an experiment using a zeolite-rich tuff as a soil amendment to improve the efficiency of the N fertilizers and allow a reduction of their inputs. The results of three years of experimentation performed in three different fields in the Emilia-Romagna region (Italy) are presented. In each field, young olive trees grown on zeolite-amended soil (−50% of N-input) were compared to trees grown on unamended soil (100% N-input). Soils and leaves were collected three times every year in each area and analyzed to monitor the efficiency of the zeolite treatment compared to the control. Vegetative measurements were performed along with analysis of pH, Soil Organic Matter and soluble anions in soil samples, whereas total C and N, C discrimination factor and N isotopic signature were investigated for both soils and leaves. Besides some fluctuations of nitrogen species due to the sampling time (Pre-Fert, Post-Fertilization and Harvest), the Total Nitrogen of leaves did not highlight any difference between treatments, which suggest that plant N uptake was not affected by lower N input in the zeolite treatment. Results, including vegetative measurements, showed no significant differences between the two treatments in all the observed variables, although the control received twice the N-input from fertilization. Based on these results, it is proposed that zeolite minerals increased the N retention time in the soil, allowing a better exploitation by plants which led to the same N uptake of the control notwithstanding the reduction in the N inputs. The use of zeolite-rich tuff in olive growing thus allows a reduction in the amount of fertilizer by up to 50% and improves the N use efficiency with many environmental and economic benefits.

Keywords: sustainable agriculture; soil; natural zeolite; chabazite; soil amendment; olive; nitrogen

1. Introduction

The low Fertilizer Use Efficiency (FUE) is one of the main causes of the altered equilibrium of agro-ecosystems [1] and it is responsible for relevant economic losses for farmers [2,3]. The role of N-based fertilizers is to provide an adequate amount of N to the plants and grant a good yield. However, after the addition of fertilizers to the soil, N is generally not efficiently uptaken by the plant, but it is lost in the surrounding environment through several pathways, causing the degradation of the soil, water and atmospheric compartments [4,5]. As pointed out by Drechsel et al. (2015) and Chien et al. (2016) [2,6], the apparent recovery efficiency (RE) of N by crops is lower than 55%. For this reason, to guarantee a crop yield able to sustain the future demands in terms of food for the population, there is an urgent need to: (1) improve the efficiency of agricultural practices, (2) reduce the N losses in the environment as harmful greenhouse gasses or leacheates, and (3) reduce the use of N based fertilizers [7–10]. Moreover, reducing the amount of fertilizers, especially those produced by synthetic processes such as urea [11], represents a great saving in terms of energy and exploitation of non-renewable resources. Improving FUE would also have great value for organic farming, which is known to have a limited set of products with low-N content available for fertilization purposes. The reduction of chemical fertilizers and pesticides is one of the biggest issues on which the EU council (Green Deal plan) is working. New strategies are being studied to decrease, by 2030, the amount of soil for crops, increase the biodiversity, grow up organic farming by +25% and preserve soil, water and human health. Bremmer et al. (2021) [12] reported that if the Green Deal objectives are not reached, the future scenario will be characterized by lower production, price increases, fewer European exports and more imports of agricultural products from outside Europe.

Thereby, the development of eco-friendly practices to reduce the use of fertilizers while improving their efficiency is necessary to increase the production in terms of quality and quantity and to guarantee human and environmental health, accordingly to the UE directions (Water Framework Directive 2000/60/CE, Directive 2009/128/CE for the pesticide use and Nitrates Directive 91/676/EEC). To reduce the leaching losses and increase the efficient use of the N-fertilizers, the N retention in the soil represents the key to limit the amount of N lost in the environment by giving "more time" to the plants to exploit the N reservoir.

Zeolite minerals are aluminosilicate with an open 3D structure formed by linked tetrahedra of $[SiO_4]^{4-}$ and $[AlO_4]^{5-}$ (the framework) and open cavities in the form of channels and cages, which are generally occupied by weakly bounded exchangeable cations and H_2O molecules. These highly reactive minerals have unique properties such as high cation exchange capacity (CEC), reversible dehydration and molecular sieve, which makes them very useful for many purposes, including agriculture [10,13–16]. Natural zeolites can be constituents of volcanic tuffs [17], and, from a geological point of view, a rock can be defined as "zeolitite" when it is constituted by more than 50% of zeolite minerals. When used as a soil amendment, zeolitites are useful for improving the capacity of the soil to retain nutrients and water, improving plant growth [18–24]. With this method, plants can uptake nitrogen more efficiently and the nitrogen losses in the surrounding environment can be significantly reduced [25–28]. In this context, their use as an inorganic amendment is becoming popular in many crops, such as maize, apple trees, sorghum, bean, aloe vera, corn and soy to cite some examples [26,29–37].

Many works have been conducted about nitrogen management in olive growing and its effects on plant growth [38–42], although only a few of them deal with zeolitite application [19,43]. Excessive dosing of mineral fertilizers is often observed as claimed by Fernández-Escobar (2011) [42] who reported that up to 200 kg-N/ha can be applied to adult olive trees. This quantity can satisfy their N demand for years, thus N fertilizers reduction in olive growing is an issue that needs to be deeply investigated.

This work aims at testing the use of zeolitite in olive-growing as a soil amendment for granting lower inputs of N-based fertilizers. It is expected that the zeolite minerals may influence the N dynamics in the soil, promoting a prolonged permanence of this nutrient and reducing the losses in the surrounding environment. This should be reflected in a more efficient uptake by plants and therefore in the possibility to significantly reduce the N inputs while maintaining crop quality and yield.

In this framework, the results of three years of experimentation in three different experimental sites are presented. An Italian chabazite-rich zeolitite was used as a soil amendment in olive growing to reduce the fertilizer N input by 50% with respect to common practices. During the experimentation, vegetative measurements were performed, and samples of soil and leaves were collected three times every year in order to measure a series of chemical parameters (including soil basic parameters, inorganic anions, nitrogen speciation and N-C stable isotopes), to account for differences between treatments and to evaluate the efficiency of this practice.

2. Materials and Methods

2.1. Zeolitite

The zeolitite (NZ) used in this experiment is a volcanic tuff quarried in Sorano (42°41′20.65″ N; 11°44′26.29″ E, Grosseto, Italy). This specific zeolitite has been widely studied in open-field and laboratory tests [26,31,44–46]. The NZ was composed of nearly 70% of zeolite minerals, mainly K-rich chabazite, which gives this NZ a very high CEC (Table 1). The NZ was employed in a granular form, with a particle size ranging between 3 and 6 mm. The main characteristics of the NZ are reported in Table 1.

Table 1. Apparent density (DA), water retention (WR) and cation exchange capacity (CEC) of the zeolitite used in the project; Quantitative Phase Analysis of the zeolitite. TZC refers to "total zeolitic content", i.e., the total content of zeolite minerals (chabazite, phillipsite and analcime). Data from Malferrari et al. (2013) [47].

		Phase	%	St.dev
DA (g cm^{-3})	0.56	chabazite	68.5	0.9
HR (%)	34.2	phillipsite	1.8	0.4
		analcime	0.6	0.3
CEC (meq g^{-1})		TZC	70.9	
Ca^{2+}	1.46	mica	5.3	0.6
Mg^{2+}	0.04	K-feldspar	9.7	0.7
Na$^+$	0.07	plagioclase	-	
K$^+$	0.6	pyroxene	2.9	0.4
Total	2.17	calcite	-	
		volcanic glass	11.2	1.0

2.2. Experimental Set-Up

To evaluate the effects of the NZ in increasing the efficiency of fertilizers and allowing a reduction in fertilizer input, two treatments were compared in 5-year old olive trees:

(1) CNT: 100% fertilizer N input and unamended soil (common practice);
(2) ZEO: 50% fertilizer N input and addition of natural zeolitite as soil amendment (500 g added to each plant at planting phase in 2016–2017 at a depth of 30–40 cm).

The fertilization reduction was performed according to the fertilization plan adopted at each field by the owner company. Different fertilizers were used in each field as well as slightly different amounts (see detailed description for each site). The experimentation started in February 2019. The monitoring lasted three years and was replicated in three different experimental fields located in various provinces of the Emilia-Romagna region, suited to olive growing (Figure 1).

At each site, three olive trees were selected randomly per each treatment (ZEO and CNT) to serve as replicates. Soil and leaf samples were collected three times each year (2019, 2020 and 2021): 1st before the fertilization (Pre-Fert) during the vegetative rest, 2nd after the fertilization (Post-Fert) during the vegetative recovery and 3rd at the olive harvest (Harvest) at each site (Figure 2). Soil samples were collected from the 0–30 cm soil layer and about 10 cm from the plant stem with an Eijkelkamp (Ø 30 mm × 500 mm) auger. Three subsamples were collected for each tree and mixed to form a single representative composite sample. For each tree, more than 20 leaves were randomly collected at each sampling. The total number of samples processed every year was 108 (considering 2 treatments, 3 experimental sites, 3 time points, resulting in 54 soil and 54 leaf samples); over the 3 years, a total of 324 samples was processed.

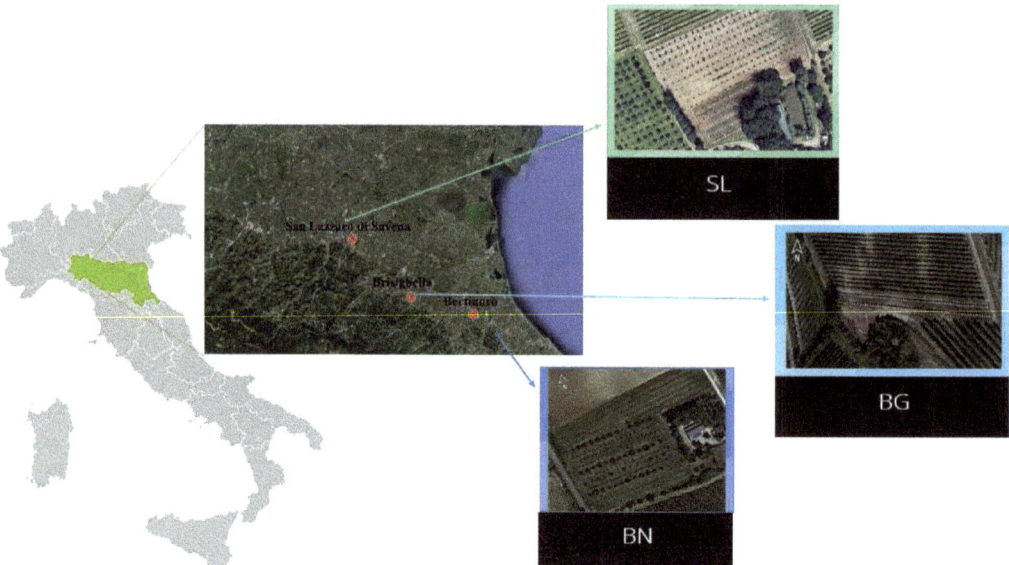

Figure 1. Geographical location of San Lazzaro (SL), Brisighella (BG) and Bertinoro (BN) experimental fields in Emilia-Romagna Region (Italy).

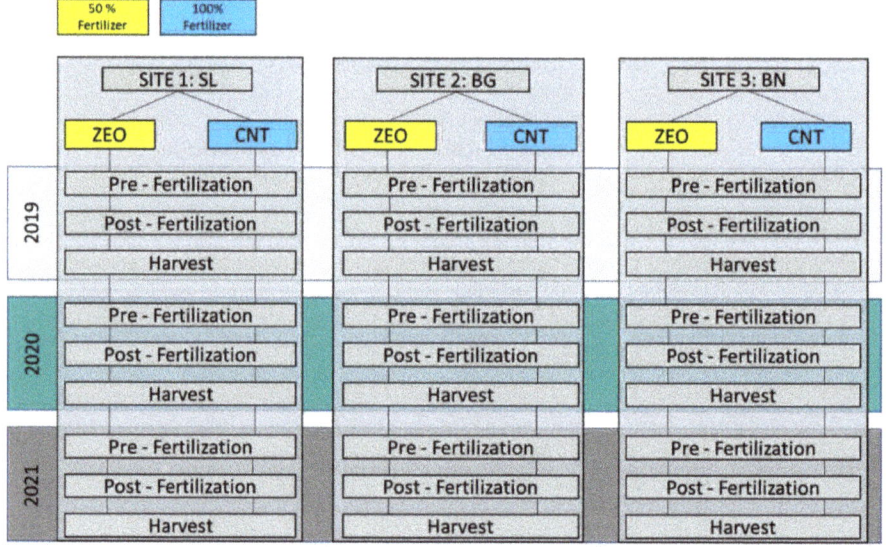

Figure 2. Experimental set-up of the experimental site of San Lazzaro (SL), Brisighella (BG) and Bertinoro (BN). At each site ZEO and CNT treatments were tested. Samples were sampled three times per year (Pre-Fertilization, Post-Fertilization and Harvest).

2.2.1. Site 1: San Lazzaro di Savena (SL)

The "SL" experimental field is located within the Bologna province and belongs to the "Azienda Agricola Bonazza" (organic regime). According to the soil map of the Emilia Romagna Region (GeoViewer—Geoportale) [48], the soil belongs to the unit CDV1 that is mainly represented by Hypocalcic Vertic Calcisol soils according to the World Reference

Base for Soil Resources (2022) [49,50]. The orchard consists of 6 rows of olive trees: 3 rows of Cv *Montecapra, Montebudello* and *Farneto* whose soil was treated with zeolitite before planting (ZEO) and 3 rows of the same Cv whose soil was left untreated (CNT). Then, 500 g of NZ were added to the ZEO treatment at transplanting (in March 2017) assuring contact with plant roots. Since 2019, organic fertilization has been halved (50% of fertilizer/year) only in the zeolite thesis (ZEO), whereas 100% of fertilizer was applied in the CNT.

In 2019, the fertilization was completed with an NP organic fertilizer (Phoenix NP, N 6%, C 2%) followed by a manure application in June for a global input of approximately 40 kg N/ha which corresponds to 118 g of N per tree in the CNT. Half of these dosages were used in the ZEO treatment.

In May 2020, Biouniversal fertilizer (N 11% and C 40%) was applied at a dosage of 37 kg N/ha in the CNT, corresponding to 55 g of N per tree, whereas half of the dosage was applied in the ZEO treatment.

In March 2021, Agriazoto11 (N 11% and C 39%) was applied in the same quantity as 2020. The olive grove was rainfed. The mean temperature for the overall period (2019–2021) was 15.3 °C and precipitation was approximately 635 mm, with the maximum rainfall recorded in 2019 (866 mm) and 570 mm and 468 mm for 2020 and 2021, respectively [51–53]. Three plants per treatment were randomly selected for soil and leaf sampling.

2.2.2. Site 2: Brisighella (BG)

The "BG" experimental field is located within the Ravenna province and belongs to "Azienda Agricola Giorgia". BG soil belongs to the cartographic unit BAN3/SOG according to the Emilia Romagna Region soil map that is mainly represented by Haplic Regosols (World Reference Base for Soil Resources (2022)) [49,50]. The olive grove consists of two olive rows of Cv *Nostrana di Brisighella* and three plants of both CNT and three ZEO treatments were selected for the sampling of soils and leaves. As in the SL area, 500 g of zeolitite per olive tree were added to the soil of the northern row (in May 2016) to create the ZEO treatment. Since the transplant, chemical fertilization has been halved (50% of fertilizer/year) in the ZEO treatment, whereas CNT received 100% of fertilizer.

In March 2019, the fertilization was performed using an organic-mineral fertilizer (Cosmo N 13%) using 100 kg N/ha in the CNT which corresponds to 185 g of N per tree and half of the dosage in the ZEO treatment. In June 2020, 50 kg N/ha of NH_4NO_3 (N, 34%) were applied to the CNT corresponding to 93 g of N per tree while half of the dosage was applied in the ZEO treatment. In March 2021, 37 kg N/ha of Urea (N, 46%) per tree were used in the CNT (corresponding to 69 g of N per tree) while half of the dosage was applied in the ZEO treatment. The orchard was irrigated with no differences between CNT and ZEO. The BG site showed the highest precipitation in 2019 (1072 mm), whereas during 2020 and 2021, precipitations were between 600 and 650 mm. The average temperature was 13.7 °C (2019–2021) but in July 2020 and August 2021, peaks of 40 °C were reached, surpassing the average temperature for that period in the last decade [51–53].

2.2.3. Site 3: Bertinoro (BN)

The "BN" experimental field is located within the Forlì-Cesena province and belongs to the "Azienda Agricola Tenute Unite". The soil belongs to the cartographic unit DEM/BAN3/DOG0 that is mainly represented by Haplic Cambisol according to the World Reference Base for Soil Resources (2022) [49,50]. The orchard is made up of different olive cultivars among which *Colombina, Correggiolo Pennita* and *Capolga di Romagna* were chosen to conduct the experiment. The set-up was similar to SL and BG sites: three plants were selected for CNT and three for ZEO treatments for soil and leaf sampling; in November 2016 the soil was amended with 500 g of zeolitite (ZEO treatment).

The BN site was managed with a considerably lower N input with respect to the other 2 sites. In 2019, Dermazoto (N 11% o, C 80%) was applied in March. The second fertilization was completed in June 2019 under the same conditions for a total of 7.5 kg N/ha (corresponding to 11 g per tree) while half of the dosage was used in the ZEO treatment.

The same fertilizer was applied also in May 2020 and March 2021, respectively at dosages of 5.6 and 6.7 kg N/ha, corresponding to 16.5 g and 19.8 g of N per tree, whereas half of the dosage was used in the ZEO treatment. The orchard was rainfed. BN recorded a mean temperature of 14.6 °C, aligned with the average temperature of the previous years. The mean precipitation for 2019–2021 was 595 mm, with the highest values recorded in 2019 (823 mm) and slightly lower than 500 mm in 2020 and 2021 [51–53].

2.3. Textural Analysis

Particle size analyses of four samples per area were conducted to characterize the soil texture. Samples were manually divided into quarters and opposite quarters were chosen for the analyses. To remove the organic matter, soils were treated with H_2O_2 and left to settle for 24 h. The sandy fraction was separated from the silty-loam fraction by a 63 μm sieving. The coarser fraction was dried at 105 °C for 24 h and weighted while the finest fraction was quantified with an X-ray sedigraph (Micromeritics 5100) at standard conditions, a dimensional range from 0.0884 mm to 0.00049 mm. A standard density value of 2.7 g/cm^3. 0.5 L of Sodium Esamexaphosphate with a low concentration (0.5%) was added to the finest fractions to simplify the grain scatter. All data obtained from the textural analyses were used for the USDA classification by Sedimcol software.

2.4. Chemical Analyses

Soil samples were air-dried and sieved at 5 mm before further analysis. Leaf samples were dried at 60 °C for 72 h and grounded with an electric grinder until obtaining a fine powder.

Soil samples were extracted with H_2O Milli-Q (high purity) at 1:10 ratio (weight/volume), to measure soluble anions and pH. After shaking for 1 h at 150 rpm in closed plastic tubes, the supernatant was separated by centrifugation at 4000 rpm for 4 min and filtered with 0.45 μm Cellulose Acetate Abluo syringe filters (GVS Filter Technology). The pH was measured with a pH electrode connected to an automatic titrator unit 877 Titrino-Plus (Methrom, Italy). Soil H_2O extracts were analyzed by Ion Chromatography (IC) with an ICS-1000 Dionex equipped with AS9-HC 4 × 250 mm anion column, AG9-HC 4 × 50 mm guard column, ADRS600 suppressor and AS-40 autosampler for the determination F^-, Cl^-, NO_2^-, Br^-, NO_3^-, PO_4^{3-}, and SO_4^{2-}. Calibration was performed with certified Thermo Fisher Scientific standards. Concerning anions, only the most significant results are shown in this paper, whereas all additional data are reported in Supplementary Material Table S1.

The Soil Organic Matter (SOM) was estimated by calculating the weight loss after heating 0.5 g of oven-dried soil at 550 °C according to [54].

The Total Nitrogen and Carbon (respectively, TN and TC) and the respective isotopic signature ($\delta^{15}N$ and $\delta^{13}C$) of soil and leaf samples were acquired with a Vario Micro Cube Elemental Analyser (EA) (Elementar, Langenselbold, Germany) connected to an Isoprime 100 Isotope Ratio Mass spectrometer (IRMS) (Isoprime, Cheadle, UK) operating in a continuous-flow mode. The EA-IRMS was calibrated with synthetic Sulfanilamide (provided by Isoprime Ltd.) and Carrara Marble (cross-calibrated at the Institute of Geoscience and Georesources of the National Council of Researches of Pisa) standards.

2.5. Vegetative Measurements

At all sites, one-year-old olive plants were provided by IBE nursery thus ensuring their growth uniformity, genetic correspondence and health status. The choice to study seven different cultivars is motivated by their different growth response (vigor). Vegetative growth parameters (plant height, number and length of branches including one-year shoots) were measured on 15 plants for each treatment and for each cultivar one year after the transplant. The sum of branch lengths for each plant was calculated. It is important to conduct these measurements during the first years of planting when the plant is left to grow without applying pruning techniques.

2.6. Statistical Analysis

All data were elaborated with R Studio 4.1.1 version. To address significant differences between the treatments due to the zeolite application, parametric and non-parametric tests were applied. Normality and homoscedasticity were tested through Shapiro–Wilk and Barlett tests ($p = 0.05$) for each variable. Data following normal distribution and with homogeneous variance were tested with a 1-way ANOVA and multiple comparison tests (Tukey HSD) at $p = 0.05$ in order to evaluate statistical differences for a whole three years. If normality or homoscedasticity were not reached (even after log or ln transformation), a non-parametric test (Kruskal–Wallis) was applied instead of ANOVA. Furthermore, Principal Component Analysis (PCA) was applied to discriminate groups of samples depending on the treatment variable (ZEO or CNT). "Ggplot2", Agricolae", "Ggally", "ggbiplot" and "ggfortify" [55–59] R-packages have been used for data analyses and figures in this paper.

3. Results and Discussion

Soils from BN and BG experimental sites are mainly characterized by silty-clay-loam textures, with a slightly higher silty fraction in BG (Figure 3). SL soils are characterized by an important sand fraction and were classified as sandy-loam and sandy-clay loam (Figure 3).

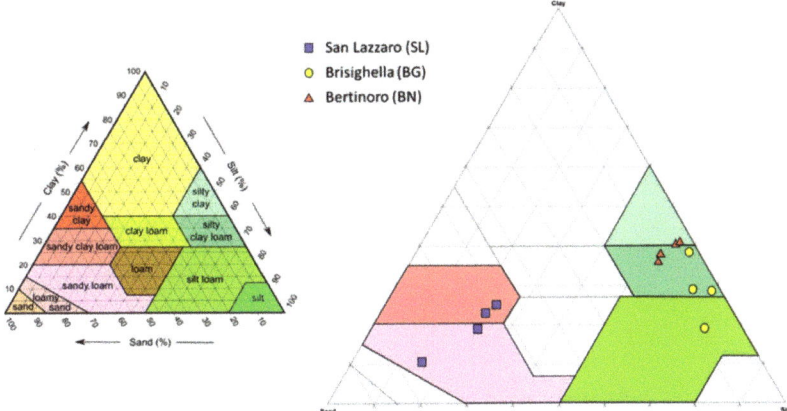

Figure 3. Particle size analysis and textural classification (USDA) of the soil samples from SL, BG and BN experimental fields.

3.1. Dynamics at Each Experimental Site

Given the large dataset, in the following we will discuss only annual trends and 3-year average significant observations. The complete dataset is available as Supplementary Material (Table S1).

In Table 2, the 3-year average of soil pH and SOM at each experimental site are reported. These basic parameters are indicators of soil quality and plant growth: SOM is the primary source of essential nutrients (N, P and S) and influences bulk density, water retention and soil temperature as well as biological activity, and buffers pH [60].

In the BG site, a slight decline in soil pH was observed in Post-Fert samples (Table 3), probably due to chemical fertilizer addition. In BG and BN, SOM is generally higher than in SL site, due to the different soil texture. The presence of silt and clay in fact maintains more C from primary production and increases SOM under certain environments [60]. SOM could be influenced by fertilization and irrigation of soil, and they are correlated with SO_4^{2-}, PO_4^{3-} (Figure 4), $\delta^{15}N$ (Figure 5) and Total Carbon (TC) (Figure 6). However, in SL and BN, SOM was higher in Post-Fert than in Pre-Fert and Harvest, while BG showed an opposite trend (Table 3). Moreover, Total Carbon (TC) in BG and BN sites confirmed the higher trend of SOM explained above, while SL showed an opposite trend (Figure 6). As far as pH and SOM are concerned, no significant effects related to the zeolitite addition to soil were observed over the 3 years of experimentation. The nutrient input reduction of 50% every year in ZEO treatment suggests a more favorable balance between inputs and outputs of SOM in the zeolitite-added soils.

Table 2. pH and Soil Organic matter (SOM) at each site (San Lazzaro, SL; Brisighella, BG; Bertinoro, BN). Data are divided by time of sampling (Pre-Fertilization, Post-Fertilization and Harvest) and treatment (ZEO and CNT). Average values represent a 3-year average (3 replicate/treatment per sampling, 3 sampling per year, 54 samples in total per site). Means in the same column followed by different letters are significantly different ($p < 0.05$) as a result of ANOVA and Tukey (HSD) tests. The complete dataset is shown in Supplementary Material Table S1.

		SL				BG				BN			
		pH		SOM (%)		pH		SOM (%)		pH		SOM (%)	
Pre-Fert	CNT	8.01°	±0.09	3.59°	±1.12	8.83°	±0.29	4.76ab	±1.32	8.84°	±1.21	6.44b	±0.03
	ZEO	7.93°	±0.37	4.04°	±0.48	8.77°	±0.22	5.51ab	±2.04	8.86°	±1.24	6.30b	±0.33
Post-Fert	CNT	7.71°	±0.43	5.05b	±0.80	8.66b	±0.19	5.26b	±0.30	8.65°	±0.92	6.97°	±0.38
	ZEO	8.27°	±0.55	3.18b	±2.72	8.67b	±0.08	5.22b	±0.94	8.45°	±0.96	7.64°	±0.63
Harvest	CNT	7.56°	±0.24	3.21c	±0.62	8.77ab	±0.05	5.91°	±0.51	8.47°	±1.11	5.35b	±2.37
	ZEO	7.56°	±0.48	3.02c	±0.51	8.84ab	±0.13	6.16°	±0.25	8.43°	±1.09	4.79b	±3.32

Table 3. Vegetative measurements of olive trees grown on soil treated with natural zeolite rich tuffs (ZEO) versus plants grown on unamended soil (CNT). Data are expressed as a mean of 15 replicates per thesis.

		Tree Height (cm)	Number of Branches	Average Branches Length (cm)	∑ Branches Length (cm)
Cv. Nostrana di Brisighella	ZEO	141.48	63.39	30.63	1918.05
	CNT	134.09	51.78	30.09	1597.65

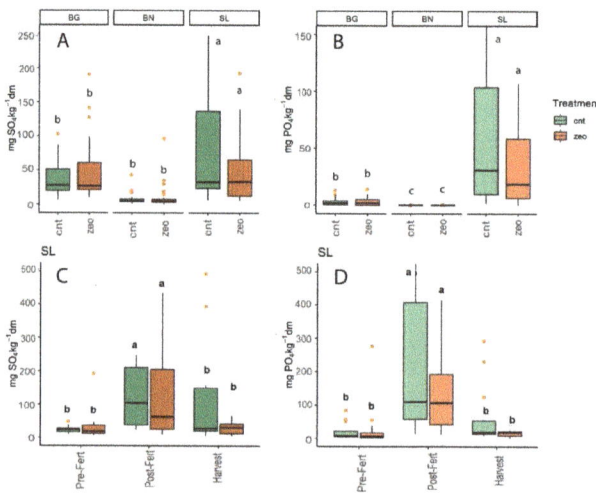

Figure 4. Box-plot of SO_4^{2-} (**A**), PO_4^{3-} (**B**) content of soil samples. The graphs are divided by experimental site (BG, BN, SL) and treatment (CNT and ZEO). Box-plot of SO_4^{2-} (**C**), PO_4^{3-} (**D**) content of soil samples from SL site. The graphs are divided by sampling (Pre-Fert, Post-Fert and Harvest) and Treatment (ZEO and CNT). (**A**,**B**): The graphs are constructed considering a 3-year average based on 27 samples per treatment at each site. (**C**,**D**): The graphs are constructed considering the site specific 3-year average (9 observations at each sampling time per each treatment, 54 total observations). Different letters represent significant differences ($p < 0.05$) as a result of ANOVA and Tukey (HSD) tests.

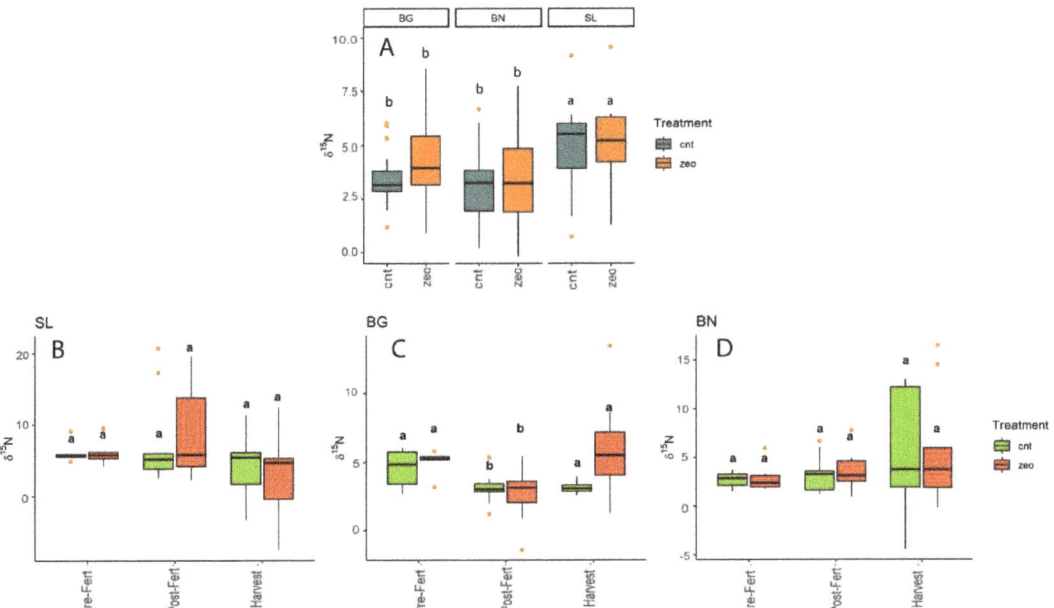

Figure 5. Box-plot of $\delta^{15}N$ (**A**) of soil samples of the three sites. The graph is divided by the experimental site (BG, BN, SL) and treatment. Boxplots of $\delta^{15}N$ of San Lazzaro (SL) (**B**), Brisighella (BG) (**C**) and Bertinoro (BN) (**D**). The graphs are divided by agronomic season (Pre-Fert, Post-Fert and Harvest) and treatment (CNT and ZEO). (**A**): the graph is constructed considering a 3-year average based on 27 samples per treatment at each site. (**B–D**): the graphs are constructed considering the site specific 3-year average (9 observations at each sampling time per each treatment, 54 total observations). Different letters represent significant differences ($p < 0.05$) as a result of ANOVA and Tukey (HSD) tests.

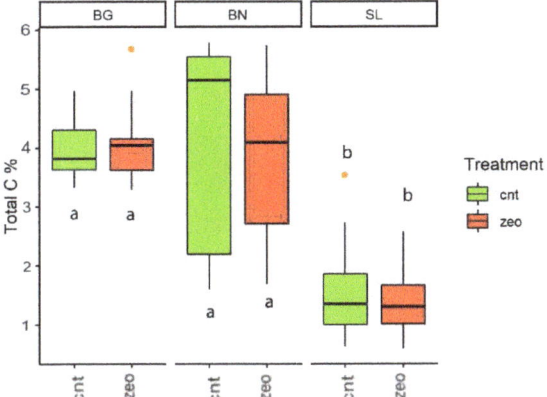

Figure 6. Box-plot of Total Carbon (TC) of soil samples in San Lazzaro (SL), Brisighella (BG) and Bertinoro (BN). The graphs are divided by treatment (CNT and ZEO). Data are the results of 3 years of experiment: For each year, 3 samplings with 3 replicates per treatment were sampled (54 samples per each site, divided in 27 samples per treatment). Different letters represent significant differences ($p < 0.05$) as a result of ANOVA and Tukey (HSD) tests.

Due to the different kind of fertilizer used, the SL site showed SO_4^{2-} and PO_4^{3-} values remarkably higher than BG and BN fields. The chemical fertilizer applied in SL in fact contained phosphate and sulfate, unlike the fertilizers used in BG and BN. Being both SO_4^{2-} and PO_4^{3-} negatively charged, they are unsuitable for cation exchange by natural zeolites, which led to non-significant differences in the retention of these ions in the soil between CNT and ZEO. However, given the lower amount of fertilizers applied to ZEO, a lower values of SO_4^{2-} and PO_4^{3-} were expected in this treatment, at least after fertilizer application. SL highlighted its highest values Post-Fertilization (Figure 4C,D), while BG and BN values showed no differences during the agronomic year.

The different kind of chemical fertilizers adopted in the experimental sites also influenced the N isotopic composition in the soil, as clearly shown in Figure 5A. On average, the $\delta^{15}N$ of SL soil is higher than in the other sites due to the use of organic fertilizers, which generally have higher ^{15}N content than the synthetic ones [61], but no significant variation over the agronomic year occurred (Figure 5B). At the BG site, after the addition of chemical fertilizers the $\delta^{15}N$ of soil tended to decrease (Figure 5C) while at the BN site, no differences were detected (Figure 5D). In natural ecosystems, soil $\delta^{15}N$ ranges from $-6‰$ to $16‰$ [62] and this high variability can be related to climate gradients and different atmospheric conditions. An inverse and a direct correlation between mean annual precipitation (MAP) and mean annual temperature (MAT) can be in fact observed with $\delta^{15}N$ [63]. BG presents the highest MAP, while both BG and BN sites show the lowest MAT during 2019–2021. An increase in the $\delta^{15}N$ values at BG is also observed in concomitance with the harvest (Figure 5C), probably due to the temperature peaks recorded during the summers of 2020 and 2021. The lower $\delta^{15}N$ values of BG and BN with respect to SL could thus result both from different N sources and climatic conditions.

To evaluate the influence of irrigation, the Carbon Discrimination Factor ($\Delta^{13}C$) was calculated from $\delta^{13}C$ data. Figure 7 shows the $\Delta^{13}C$ for leaves of each site, which means the $\delta^{13}C$ normalized for changes in atmospheric CO_2 concentration through Equation (1), where a and p refer to air and plant [64].

$$\Delta = \frac{\partial a - \partial p}{1 + \partial p} \qquad (1)$$

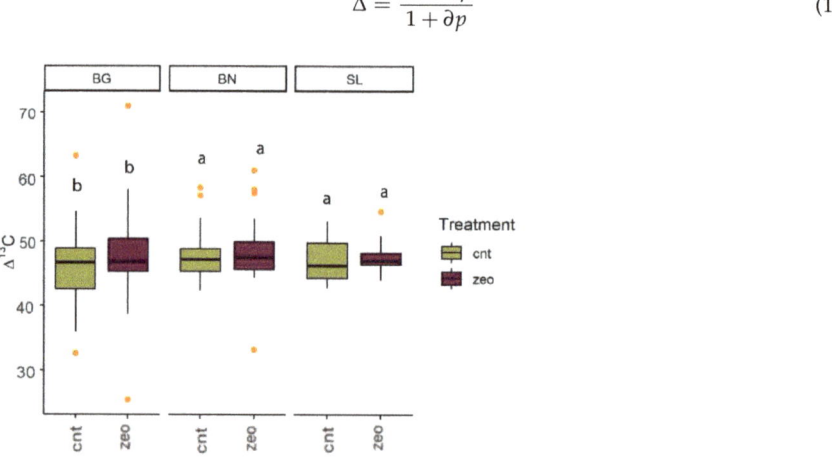

Figure 7. Box-plot of $\Delta^{13}C$ of leaves divided by area and treatment. The figure is constructed considering a 3-year average based on 27 samples per treatment at each site. Different letters represent significant differences ($p < 0.05$) as a result of ANOVA and Tukey (HSD) tests.

According to Riehl et al. (2014) [65] a 1‰ $\Delta^{13}C$ variation can be used to distinguish stressed from well-watered plants without accounting for soil fertility effects. Water stress conditions in fact causes a decrease in photosynthesis, transpiration and leaf conductance which in turn modify the carbon isotopic composition [64,66,67]. As it is known, zeolites can adsorb water molecules in their structure, which means an increase in the overall soil water holding capacity and the consequent possibility to reduce irrigation [68]. Nevertheless, no significant variations between ZEO and CNT treatments were highlighted by the Δ^{13}C data. This fact is partially in contrast with the results obtained by [26] where a change in Δ^{13}C in maize and wheat grown in soil amended with the same natural zeolite-rich tuff was observed. Although in that case, the authors ascribed the Δ^{13}C variations to the manuring effect. In our case, a significant difference in Δ^{13}C was observed in BG only, likely due to the additional water provided to the plants' trough irrigation. This site in fact is the only one which underwent artificial irrigation, added to the highest MAP over the three years of experimentation.

N is one of the most important nutrients for plants. Thus, analyses of its different inorganic speciation were performed to address the effects of natural zeolites on soil N cycling in the three experimental fields. Nitrite (NO_2^--N) usually does not accumulate in soils because it is an intermediate product of nitrification (that transforms NH_4^+ into NO_3^--N), or it is denitrified to NO and N_2O and N_2 gases. On the other hand, nitrate (NO_3^--N) is one of the main forms of N used by plants and can also be exploited by microbes to satisfy their N needs (immobilization processes) [69]. Nitrate, however, can follow various transformation pathways which may also lead to N losses in the atmosphere (as nitrous oxides due to incomplete denitrification) and/or can be leached into the water system as a result of anionic repulsion by soil particles.

The results of TN analyses of soils and leaves and NO_3^--N and NO_2^--N of soils are shown in Figures 8–10 for SL, BG and BN sites, respectively (3-year average).

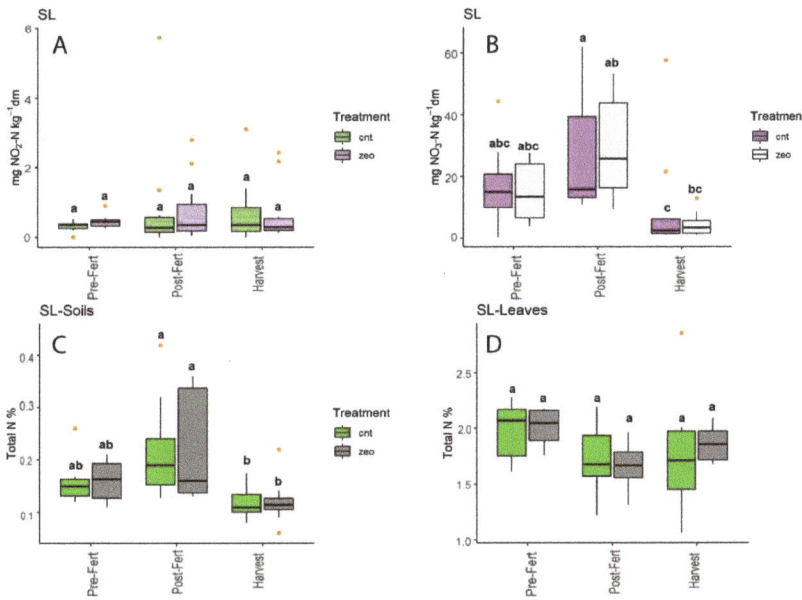

Figure 8. Box-plot of NO_2^--N (**A**), NO_3^--N (**B**) content of soil samples, Total Nitrogen (TN) of soils (**C**) and leaves (**D**) in San Lazzaro field (SL). Graphs consider the site specific 3-year average (9 observations at each sampling time per each treatment, 54 total observations). Different letters represent significant differences ($p < 0.05$) as a result of ANOVA and Tukey (HSD) tests.

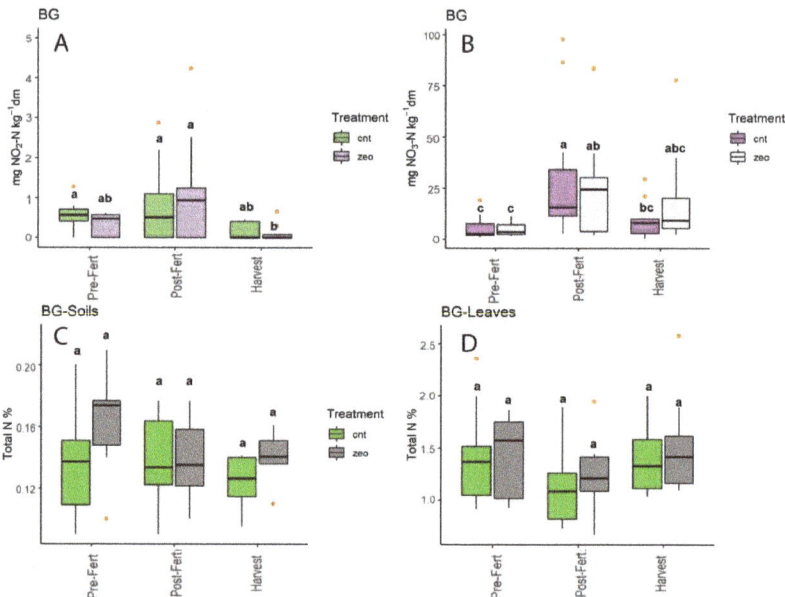

Figure 9. Box-plot of NO_2^--N (**A**) and NO_3^--N (**B**) content of soil samples, Total Nitrogen (TN) of soils (**C**) and leaves (**D**) in Brisighella field (BG). The graphs are constructed considering the site specific 3-year average (9 observations at each sampling time per each treatment, 54 total observations). Different letters represent significant differences ($p < 0.05$) as a result of ANOVA and Tukey (HSD) tests.

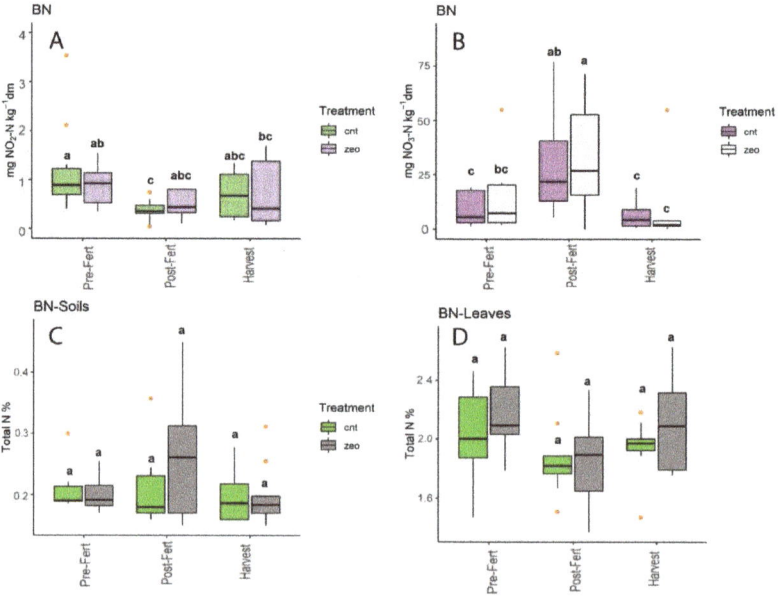

Figure 10. Box-plot of NO_2^--N (**A**) and NO_3^--N (**B**) content of soil samples, Total Nitrogen (TN) of soils (**C**) and leaves (**D**) in Bertinoro field (BN). The graphs are constructed considering the site specific 3-year average (9 observations at each sampling time per each treatment, 54 total observations). Different letters represent significant differences ($p < 0.05$) as a result of ANOVA and Tukey (HSD) tests.

In SL no differences were observed between the different treatments, although the N input in CNT treatment was twice that in ZEO treatment. NO_2^--N of soil (Figure 8A) did not show any difference between treatments (ZEO and CNT) or sampling time (Pre-Fert, Post-Fert and Harvest) among the 3 years of the project, showing values always below 10 mg kg^{-1}. Even NO_3^--N in soil samples (Figure 8B) showed no significant variations among the treatments. A remarkable difference between CNT treatment Post-Fertilization and Harvest can be observed, probably due to NO_3^--N removing processes (gaseous losses, leaching, microbial immobilization or Dissimilatory Nitrate Reduction to Ammonium). This evidence is partially sustained by a tendency to a lower N storage in olive leaves at the Harvest in the CNT (although not significant). Soil TN (Figure 8C), reflects the same trend for nitrate, showing no significant differences between treatments. The seasonal fluctuations of these N species (with higher values after fertilization) are related to the input of N brought by fertilizers. The TN of leaves (Figure 8D) likely supports this hypothesis because the leaves have shown no differences in N content due to treatments or time. However, they showed an opposite trend to that of soils, due to the different availability of N during the agronomic year in different environmental compartments. Immediately after fertilization, TN is concentrated in the soil, and it is lower in leaves while at the harvest the trend was opposite.

In the BG site, TN did not show any significant difference due to the treatment and sampling time for both soils (Figure 9C) and leaves (Figure 9D), coherently to the SL site. The NO_2^--N (Figure 9A) and NO_3^--N (Figure 9B) of BG soils showed a trend similar to SL and no differences were accounted for between ZEO (50% of fertilizer) and CNT (100% of fertilizer). However, sampling time significantly affected the amounts of N in the soil. NO_2^--N (Figure 9A) in Post-Fertilization ZEO samples showed significant differences with respect to ZEO at Harvest, suggesting lower nitrite production (or improved consumption) in this treatment. The NO_3^--N (Figure 9B) content in CNT treatment at Harvest was significantly lower than Post-Fertilization, but an increase in N uptake of plants is not able to explain the NO_3^--N reduction in soil. This decrease is probably due to a N loss in the surrounding environment which did not happen for ZEO treatments, as suggested by the tendency of ZEO leaves to have higher TN amounts for all sampling stages, although not statistically significant.

The BN samples showed a trend similar to the SL and BG areas during the three years of monitoring. NO_2^--N (Figure 10A) and NO_3^--N (Figure 10B) of soils showed no significant differences between ZEO and CNT. NO_2^--N showed significant differences between Pre-Fertilization and Post-Fertilization samplings, with higher values at Pre-Fert. NO_3^--N followed the trend linked to the fertilization, with higher values at Post-Fertilization right after the N input. The ZEO treatment in Pre-fertilization is similar to the CNT in Post-Fertilization (where twice the amount of fertilizer was applied with respect to the ZEO treatment), indicating that zeolite probably helped the soil to retain more N available to the plant during time. TN of soils (Figure 10C) revealed no variations due to the treatments or sampling time, and no other differences were highlighted neither for TN of leaves (Figure 10D) nor for the SL and BG sites. As for BG, also in BN a tendency for a higher N content of leaves was recorded although not significant from a statistical point of view.

In general, the results of N dynamics over the 3 years of monitoring in the 3 experimental sites indicate that notwithstanding 50% fewer N inputs, the soil N content was similar between CNT and ZEO. Given that no differences in N uptake by plants were observed, this evidence leads to the hypothesis that zeolite minerals helped to reduce N losses and promoted N storage in the soil, augmenting the fertilization efficiency.

3.2. Vegetative Measurements

The analysis of variance of the data collected in the BG field did not reveal any difference between the two treatments (Table 3), while some differences between ZEO and CNT were highlighted in both SL and BN fields. In the SL site, tree height, number and length of shoots were higher for Cv *Montebudello* and *Farneto* for ZEO treatment than for CNT (Table 4). The number of shoots was greater in the ZEO thesis for Cv *Colombina*, while other measurements exhibited no significant difference compared to CNT. The Cv. *Capolga* in the BN field showed no differences in the level of growth of the aerial part, while in the other two cultivars (*Colombina* and *Correggiolo*), a significantly greater development in the plants treated with natural zeolite-rich tuffs was observed, despite the reduced dose of fertilizer applied (Table 5).

Table 4. Vegetative measurements of olive trees grown on soil treated with natural zeolite rich tuffs (ZEO) versus plants grown on unamended soil (CNT). Data are expressed as a mean of 15 replicates per thesis. The bold font indicates statistically significant differences between the groups ($p < 0.05$).

		Tree Height (cm)	Number of Branches	Average Branches Length (cm)	$\sum$ Branches Length (cm)
Cv. Montebudello	ZEO	121.14	49.86	24.71	**1193.14**
	CNT	**92.00**	**20.00**	22.39	**441.86**
Cv. *Farneto*	ZEO	**114.31**	**69.69**	22.75	**1592.50**
	CNT	**84.54**	**44.38**	20.37	**972.15**
Cv. Montecapra	ZEO	104.50	**62.63**	22.71	1368.38
	CNT	99.29	**47.00**	21.55	1060.50

Table 5. Vegetative measurements of olive trees grown on soil treated with natural zeolite rich tuffs (ZEO) versus plants grown on unamended soil (CNT). Data are expressed as a mean of 15 replicates per thesis. The bold font indicates statistically significant differences between the groups ($p < 0.05$).

		Tree Height (cm)	Number of Branches	Average Branches Length (cm)	$\sum$ Branches Length (cm)
Cv. *Capolga*	ZEO	86	20.76	14.65	392.18
	CNT	83.1	18.95	13.92	347.18
Cv. *Colombina*	ZEO	74.88	**10.53**	15.63	233.82
	CNT	63.77	**6.46**	11.33	141.58
Cv. *Correggiolo*	ZEO	**102.38**	9.25	**21.04**	**291.38**
	CNT	**80.43**	7.79	**15.76**	**204.39**

The only field where no differences were observed between the two treatments is BG, the irrigated field. It is possible that the action of the zeolite, in addition to reducing N leaching and increasing the Nitrogen Use Efficiency (NUE), takes place at the water level (although no differences were observed by $\Delta^{13}C$), so in orchards without any water deficits, it is harder to account for differences in plant development.

These results are in agreement with those of Prisa (2020) [70], that found an increase in agronomic characteristics in plants of *Ranunculus asiaticus* treated with zeolites, and with Choo et al. (2020) [71] that found an increased number of fruits and greater fruit yield in papaya plants treated with zeolites.

3.3. Global Considerations

To evaluate the benefits of using zeolitite in olive growing, the general comparison of treatments year by year is presented in this chapter. pH, SOM and TC of soils are shown in Figure 11, while in Figure 12 each of the investigated N species is shown.

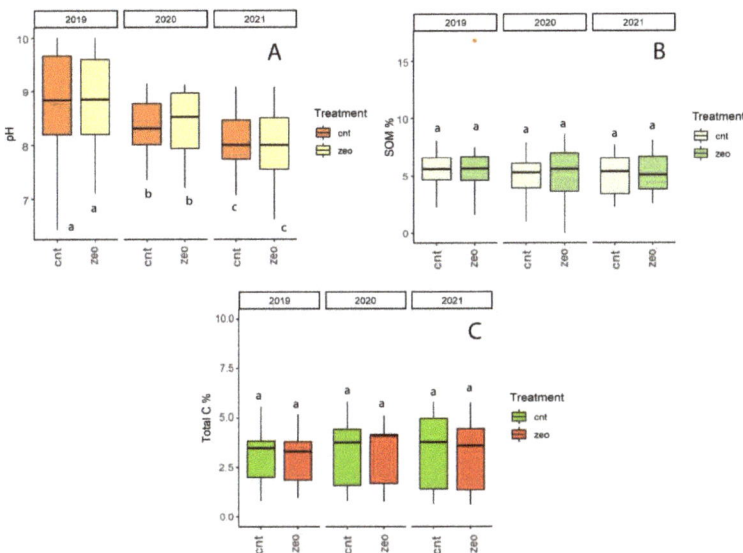

Figure 11. pH, Soil Organic Matter (SOM) and Total Carbon (TC) are shown for the three years of the project. Data are divided by year (2019, 2020 and 2021) and treatment (with zeolite and control for (**A**) pH, (**B**) Soil Organic Matter (SOM) and (**C**) Total Carbon (TC). The graphs are constructed considering the year specific average for all three sites (27 observations at each year per each treatment, 54 total observations per year). Different letters represent significant differences ($p < 0.05$) as a result of ANOVA and Tukey (HSD) tests.

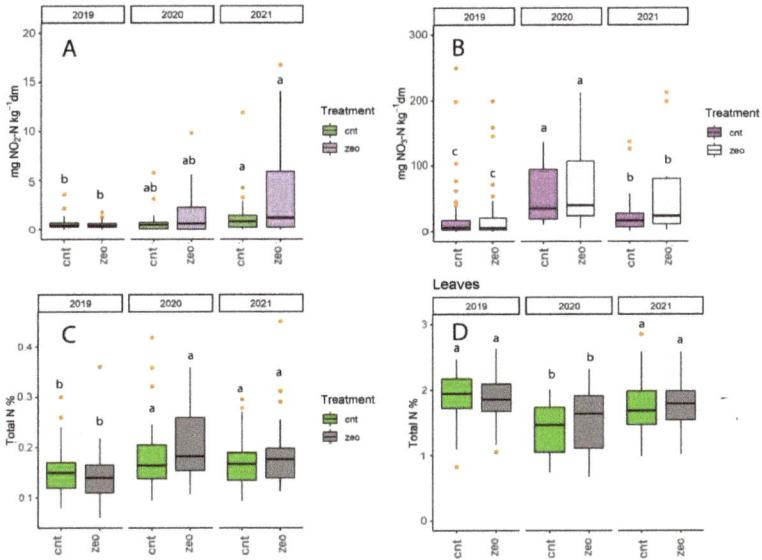

Figure 12. Box-plot of NO_2^--N (**A**) and NO_3^--N (**B**) content of soil samples, Total Nitrogen (TN) of soils (**C**) and leaves (**D**) divided by years and treatment. The graphs are constructed considering the year specific average for all three sites (27 observations at each year per each treatment, 54 total observations per year). Different letters represent significant differences ($p < 0.05$) as a result of ANOVA and Tukey (HSD) tests.

In all experimental fields, pH did not undergo any significant difference between CNT and ZEO treatments. A global trend towards an acidification of the soil from 2019 to 2021 can be however observed (Figure 11A), which can be a consequence of the leaching of exchangeable bases such as Mg^{2+}, K^+ and Ca^{2+} because of intense precipitation or irrigation practices. No significant differences were accounted between SOM and TC and they did not change in relation to different treatments and during time (Figure 11B,C). Although SOM and TC could be influenced by fertilization and irrigation practices, we did not observe any significant variation. At the same time, SOM, as well as TC, did not decrease over the 3 years of experimentation, proving that the use of zeolitite did not influence these parameters in soil but helps preserving SOM even with a reduced amount of nutrient inputs while maintaining or even improving the plant development.

The only difference that occurred in nitrogen species was linked to the time and to the type and amount of fertilizer applied to each field: (1) NO_2^--N showed a significant difference among years, with concentrations that increased from 2019 to 2021; (2) NO_3^--N in soil was significantly different during 3 years, with the lower values recorded in 2019 and the higher values recorded in 2020 and 2021; (3) TN in soil showed a very similar pattern to that of NO_3^--N with an increase after the first experimental year (2019) and (4) The TN of leaves was lower in 2020 (opposite trend to the NO_3^--N). For each N species, no differences were accounted for between CNT and ZEO treatments (notwithstanding the 50% reduction of fertilizers), as already demonstrated in detail for each experimental site.

TN of leaves strongly indicate that plants did not uptake more N in CNT than in ZEO treatment, although ZEO leaves showed a slight tendency in higher N uptake in 2020 and 2021 (not statistically significant), which can be caused by an augmented availability of N among the years.

Principal Component Analysis (PCA), which is often used to discriminate the groups of samples, reducing the dimensionality of the dataset without a large loss of information, was applied only to the data related to the ZEO and CNT treatments during the three years of the project.

PC1 and PC2 axes explained 48.59% of the total variance, divided into 30.61% of the First Principal Component (PC1) and 17.98% of the Second Principal Component (PC2). All the data showed a positive correlation in PC1, except Soil Organic Matter (SOM), Total Carbon (TC) and pH. Instead, in PC2 only Carbon Discrimination Factor ($\Delta^{13}C$) had a positive correlation, while all other parameters highlighted a negative correlation with PC2. This low value of total variance does not allow for distinguishing between the different treatments, thus further supporting the hypothesis that CNT and ZEO treatments were not different, notwithstanding the fertilizer input reduction of 50% in the ZEO treatment.

The similar N uptake recorded by the leaves in the three different experimental sites, as well as the tendency for a better development of plants grown on zeolite-amended soil, notwithstanding the 50% N input reduction, strongly suggest that in CNT treatment larger N losses occurred, leading to negative environmental and economic effects. On the other hand, the presence of zeolitite in the soil maintained the nutrient for a longer time contributing to a healthier condition for plants and yield production.

It is well known that zeolitite as a soil amendment reduces N leaching and increases Nitrogen Use Efficiency (NUE) and crop yield [72]. Since the addition of zeolitite probably influenced several pathways of N losses, it also allowed a more sustainable use of N fertilizers. Furthermore, the N in the topsoil is strongly related to agricultural practices and is influenced by the amount and form of the fertilizers used. This N can be easily lost by leaching, NH_3 volatilization and other N gas losses. Chemical fertilizers, such as urea, can lose even more than 30% of the applied N as NH_3 in the few hours after the spreading, if the conditions for volatilization are met [8]. Ferretti et al. (2017) found evidence of a higher FUE in zeolitite-amended soil after performing an isotopic tracing in the soil-plant system. In another study, it was demonstrated that in similar conditions, NH_3 emissions can be reduced up to 60% using the same type of zeolitite used in this work [26]. Consequently, the application of zeolitite to soil can be the key to reducing N losses in the environment, allowing a significant reduction in fertilizer N inputs (50%), maintaining or even increasing the vegetative development.

The mechanism through which zeolitite is able to maintain the nutrients in the soil for longer periods of time is, however, still a matter of debate. Ferretti et al. (2021) employed the ^{15}N pool dilution technique to measure gross N transformation rates in zeolitite-amended soil and found no evidence of increased ammonification in soil treated with natural zeolites in the short-term. Thus, the efficiency of zeolitites (at natural state) cannot be explained by an increased production of new mineral N from organic matter decomposition. However, from the same study emerges a slight "delay" effect on gross nitrification. Apparently, in zeolitite-amended soil, the ammonium is more slowly converted into nitrate. Thus, the mechanism that might be responsible for the improved NUE in the treated soil is the perturbation of various abiotic parameters after the addition of zeolite minerals (CEC, water retention) that is reflected in different biotic processes in the short-term, probably altering the quantity of N available for plant uptake. In another short-term incubation study at laboratory scale, it was observed that the exchange of N between minerals and the surrounding environment is very fast. Thus, N is accessible to microbial biomass in the short-term but only mild effects on the microbial community (fungal/bacterial ratio) and on N transformation rates were observed [73]. Thus, it is likely that the zeolites reduce the N mobility in the short-term and delay the transformation of ammonium into nitrate, resulting in "more time" for plants to uptake N and, by consequence, in a lower demand for N fertilizers and N losses.

4. Conclusions

Thanks to a 3-year experiment conducted in three sites within the Emilia-Romagna region, the efficiency of zeolite minerals in reducing the fertilizer N input up to 50% in olive growing was demonstrated.

N dynamics and all the observed variables were influenced by the fertilizer management (type, amount and timing of application), time, soil texture and irrigation.

However, no differences were observed owing to the different treatments (ZEO and CNT), neither in the detail of each experimental site nor from the general point of view, although in ZEO the fertilization had been reduced by 50%.

The vegetative measurements highlighted a greater development of the olive aerial parts in ZEO treatments compared to CNT. The vegetative measurements conducted in the first year indicate that the plants treated with zeolitite, despite the 50% reduction in fertilizers, have developed similarly to the CNT. In the two rainfed orchards (SL and BN), the ZEO-treated olive plants were characterized by a greater canopy development. Further studies are under way to evaluate the effects, in the long term, of the *una tantum* zeolitite addition as well as the influence on the fruit development and the chemical and sensorial quality of the oils.

In conclusion, the use of this specific Italian chabazite zeolitite in olive growing can allow a significant reduction in fertilizer N input, reducing the N losses and improving the plant's physiological status, with meaningful benefits under agronomic, environmental, economic and health aspects. It is indeed very important to specify that the effects of these minerals in the soil are long-lasting due to their long-term structural stability at an ambient temperature and pressure. Moreover, the reduction in the application of N fertilizers can be performed repeatedly over the years, with significant economic and environmental benefits which last forever.

Supplementary Materials: The following supporting information can be downloaded at https://www.mdpi.com/article/10.3390/land11091471/s1, Table S1: Supplementary Material S1.

Author Contributions: Conceptualization, V.M. and G.F.; methodology, V.M., G.F. and A.R.; formal analysis, V.M., G.G., L.M. and A.R.; field sampling, G.F., G.G., L.M. and A.R.; investigation, V.M.; resources, B.F. and M.C.; data curation, V.M.; writing-original draft preparation, V.M.; writing-review and editing, G.F. and M.C.; visualization, V.M.; supervision, G.F. and M.C.; project administration, M.C.; funding acquisition, M.C. All authors have read and agreed to the published version of the manuscript.

Funding: This research was funded by the ZeOliva project financed by the Italian Ministry of Agricultural, Food, and Forestry Policies (MIPAAF, Project CUP: F76C19000070001).

Institutional Review Board Statement: Not applicable.

Informed Consent Statement: Not applicable.

Data Availability Statement: All the data are available within the manuscript and supplementary materials.

Acknowledgments: We gratefully thank the "Azienda Agricola Bonazza", "Azienda Agricola Giorgia" and "Azienda Agricola Tenute Unite" for the land concession for the experiment. The authors are also grateful to Umberto Tessari for the textural analyses, Matteo Alberghini and Luca Adami for their help during laboratory procedures.

Conflicts of Interest: The authors declare no conflict of interest.

References

1. Mustafa, A.; Hu, X.; Shah, S.A.A.; Abrar, M.M.; Maitlo, A.A.; Kubar, K.A.; Saeed, Q.; Kamran, M.; Naveed, M.; Boren, W.; et al. Long-term fertilization alters chemical composition and stability of aggregate-associated organic carbon in a Chinese red soil: Evidence from aggregate fractionation, C mineralization, and 13C NMR analyses. *J. Soils Sediments* **2021**, *21*, 2483–2496. [CrossRef]
2. Drechsel, P.; Heffer, P.; Magen, H.; Mikkelsen, R.; Wichelns, D. *Managing Water and Fertilizer for Sustainable Agricultural Intensification*; IFA, IWMI, IPNI and IPI: Paris, France, 2015; ISBN 979-10-92366-02-0.
3. Basso, B.; Dumont, B.; Cammarano, D.; Pezzuolo, A.; Marinello, F.; Sartori, L. Environmental and economic benefits of variable rate nitrogen fertilization in a nitrate vulnerable zone. *Sci. Total Environ.* **2016**, *545*, 227–235. [CrossRef] [PubMed]
4. Peng, S.; Buresh, R.J.; Huang, J.; Yang, J.; Zou, Y.; Zhong, X.; Wang, G.; Zhang, F. Strategies for overcoming low agronomic nitrogen use efficiency in irrigated rice systems in China. *Field Crop. Res.* **2006**, *96*, 37–47. [CrossRef]
5. Dawson, J.C.; Huggins, D.R.; Jones, S.S. Characterizing nitrogen use efficiency in natural and agricultural ecosystems to improve the performance of cereal crops in low-input and organic agricultural systems. *Field Crop. Res.* **2008**, *107*, 89–101. [CrossRef]
6. Chien, S.H.; Teixeira, L.A.; Cantarella, H.; Rehm, G.W.; Grant, C.A.; Gearhart, M.M. Agronomic Effectiveness of Granular Nitrogen/Phosphorus Fertilizers Containing Elemental Sulfur with and without Ammonium Sulfate: A Review. *Agron. J.* **2016**, *108*, 1203–1213. [CrossRef]
7. Ferretti, G.; Keiblinger, K.M.; Zimmermann, M.; Di Giuseppe, D.; Faccini, B.; Colombani, N.; Mentler, A.; Zechmeister-Boltenstern, S.; Coltorti, M.; Mastrocicco, M. High resolution short-term investigation of soil CO_2, N_2O, NO_x and NH_3 emissions after different chabazite zeolite amendments. *Appl. Soil Ecol.* **2017**, *119*, 138–144. [CrossRef]
8. Soares, J.R.; Cantarella, H.; de Campos Menegale, M.L. Ammonia volatilization losses from surface-applied urea with urease and nitrification inhibitors. *Soil Biol. Biochem.* **2012**, *52*, 82–89. [CrossRef]
9. Sharma, L.K.; Bali, S.K. A review of methods to improve nitrogen use efficiency in agriculture. *Sustainability* **2017**, *10*, 58. [CrossRef]
10. Cataldo, E.; Salvi, L.; Paoli, F.; Fucile, M.; Masciandaro, G.; Manzi, D.; Masini, C.M.; Mattii, G.B. Application of Zeolites in Agriculture and Other Potential Uses: A Review. *Agronomy* **2021**, *11*, 1547. [CrossRef]
11. Shi, L.; Liu, L.; Yang, B.; Sheng, D.; Xu, T. Evaluation of Industrial Urea Energy Consumption (EC) Based on Life Cycle Assessment (LCA). *Sustainability* **2020**, *12*, 3793. [CrossRef]
12. Bremmer, J.; Gonzalez-Martinez, A.; Jongeneel, R.; Huiting, H.; Stokkers, R.; Ruijs, M. *Impact Assessment of EC 2030 Green Deal Targets for Sustainable Crop Production*; 2021; ISBN 9789464470413.
13. Reháková, M.; Čuvanová, S.; Dzivák, M.; Rimár, J.; Gaval'Ová, Z. Agricultural and agrochemical uses of natural zeolite of the clinoptilolite type. *Curr. Opin. Solid State Mater. Sci.* **2004**, *8*, 397–404. [CrossRef]
14. Ferretti, G.; Galamini, G.; Medoro, V.; Coltorti, M.; Giuseppe, D.D.; Faccini, B. Impact of sequential treatments with natural and na-exchanged chabazite zeolite-rich tuff on pig-slurry chemical composition. *Water* **2020**, *12*, 310. [CrossRef]
15. Rotondi, A.; Morrone, L.; Facini, O.; Faccini, B.; Ferretti, G.; Coltorti, M. Distinct particle films impacts on olive leaf optical properties and plant physiology. *Foods* **2021**, *10*, 1291. [CrossRef] [PubMed]
16. Eroglu, N.; Emekci, M.; Athanassiou, C.G. Applications of natural zeolites on agriculture and food production. *J. Sci. Food Agric.* **2017**, *97*, 3487–3499. [CrossRef] [PubMed]
17. Kesraoui-Ouki, S.; Cheeseman, C.R.; Perry, R. Natural zeolite utilisation in pollution control: A review of applications to metals' effluents. *J. Chem. Technol. Biotechnol.* **1994**, *59*, 121–126. [CrossRef]
18. Colombani, N.; Di Giuseppe, D.; Faccini, B.; Ferretti, G.; Mastrocicco, M.; Coltorti, M. Estimated Water Savings in an Agricultural Field Amended With Natural Zeolites. *Environ. Process.* **2016**, *3*, 617–628. [CrossRef]
19. Perez-Caballero, R.; Gil, J.; Benitez, C.; Gonzalez, J.L. The effect of adding zeolite to soils in order to improve the N-K nutrition of olive trees. Preliminary results. *Am. J. Agric. Biol. Sci.* **2008**, *3*, 321–324. [CrossRef]

20. Passaglia, E. *Zeoliti Naturali, Zeolititi e Loro Applicazioni*; Arvan: Gardena, CA, USA, 2008; ISBN 9788887801194.
21. Colombani, N.; Mastrocicco, M.; Giambastiani, B.M.S. Predicting Salinization Trends in a Lowland Coastal Aquifer: Comacchio (Italy). *Water Resour. Manag.* **2015**, *29*, 603–618. [CrossRef]
22. Gholamhoseini, M.; Ghalavand, A.; Khodaei-Joghan, A.; Dolatabadian, A.; Zakikhani, H.; Farmanbar, E. Zeolite-amended cattle manure effects on sunflower yield, seed quality, water use efficiency and nutrient leaching. *Soil Tillage Res.* **2013**, *126*, 193–202. [CrossRef]
23. Nakhli, S.A.A.; Delkash, M.; Bakhshayesh, B.E.; Kazemian, H. Application of Zeolites for Sustainable Agriculture: A Review on Water and Nutrient Retention. *Water Air Soil Pollut.* **2017**, *228*, 464. [CrossRef]
24. Mirzaei Aminiyan, M.; Safari Sinegani, A.A.; Sheklabadi, M. Aggregation stability and organic carbon fraction in a soil amended with some plant residues, nanozeolite, and natural zeolite. *Int. J. Recycl. Org. Waste Agric.* **2015**, *4*, 11–22. [CrossRef]
25. Eslami, M.; Khorassani, R.; Coltorti, M.; Malferrari, D.; Faccini, B.; Ferretti, G.; Di Giuseppe, D.; Fotovat, A.; Halajnia, A. Leaching behaviour of a sandy soil amended with natural and NH_4^+ and K^+ saturated clinoptilolite and chabazite. *Arch. Agron. Soil Sci.* **2017**, *64*, 1142–1151. [CrossRef]
26. Ferretti, G.; Di Giuseppe, D.; Natali, C.; Faccini, B.; Bianchini, G.; Coltorti, M. C-N elemental and isotopic investigation in agricultural soils: Insights on the effects of zeolitite amendments. *Chem. Der Erde* **2017**, *77*, 45–52. [CrossRef]
27. Tsintskaladze, G.; Eprikashvili, L.; Mumladze, N.; Gabunia, V.; Sharashenidze, T.; Zautashvili, M.; Kordzakhia, T.; Shatakishvili, T. Nitrogenous zeolite nanomaterial and the possibility of its application in agriculture. *Ann. Agrar. Sci.* **2017**, *15*, 365–369. [CrossRef]
28. Manjaiah, K.M.; Mukhopadhyay, R.; Paul, R.; Datta, S.C.; Kumararaja, P.; Sarkar, B. Clay minerals and zeolites for environmentally sustainable agriculture. *Modif. Clay Zeolite Nanocompos. Mater. Environ. Pharm. Appl.* **2019**, 309–329. [CrossRef]
29. Milosevic, T.; Milosevic, N. The effect of zeolite, organic and inorganic fertilizers on soil chemical properties, growth and biomass yield of apple trees. *Plant Soil Environ.* **2009**, *55*, 528–535. [CrossRef]
30. Ferretti, G.; Faccini, B.; Antisari, L.V.; Di Giuseppe, D.; Coltorti, M. 15N natural abundance, nitrogen and carbon pools in soil-sorghum system amended with natural and NH_4^+-enriched zeolitites. *Appl. Sci.* **2019**, *9*, 4524. [CrossRef]
31. Faccini, B.; Di Giuseppe, D.; Ferretti, G.; Coltorti, M.; Colombani, N.; Mastrocicco, M. Natural and NH_4^+-enriched zeolitite amendment effects on nitrate leaching from a reclaimed agricultural soil (Ferrara Province, Italy). *Nutr. Cycl. Agroecosyst.* **2018**, *110*, 327–341. [CrossRef]
32. Hazrati, S.; Tahmasebi-Sarvestani, Z.; Mokhtassi-Bidgoli, A.; Modarres-Sanavy, S.A.M.; Mohammadi, H.; Nicola, S. Effects of zeolite and water stress on growth, yield and chemical compositions of *Aloe vera* L. *Agric. Water Manag.* **2017**, *181*, 66–72. [CrossRef]
33. Ozbahce, A.; Tari, A.F.; Gönülal, E.; Simsekli, N.; Padem, H. The effect of zeolite applications on yield components and nutrient uptake of common bean under water stress. *Arch. Agron. Soil Sci.* **2014**, *61*, 615–626. [CrossRef]
34. El-Sherpiny, M.; Baddour, A.G.; El-Kafrawy, M.M. Effect of Zeolite Soil addition under Different Irrigation Intervals on Maize Yield (zea mays L.) and some Soil Properties. *J. Soil Sci. Agric. Eng.* **2020**, *11*, 793–799. [CrossRef]
35. Rodrigues, M.; Torres, L.D.N.D.; Damo, L.; Raimundo, S.; Sartor, L.; Cassol, L.C.; Arrobas, M. Nitrogen Use Efficiency and Crop Yield in Four Successive Crops Following Application of Biochar and Zeolites. *J. Soil Sci. Plant Nutr.* **2021**, *21*, 1053–1065. [CrossRef]
36. Minardi, S.; Haniati, I.L.; Nastiti, A.H.L. Adding manure and zeolite to improve soil chemical properties and increase soybean yield. *SainsTanah-J. Soil Sci. Agroclimatol.* **2020**, *17*, 1. [CrossRef]
37. Sun, Y.; Xia, G.; He, Z.; Wu, Q.; Zheng, J.; Li, Y.; Wang, Y.; Chen, T.; Chi, D. Zeolite amendment coupled with alternate wetting and drying to reduce nitrogen loss and enhance rice production. *Field Crop. Res.* **2019**, *235*, 95–103. [CrossRef]
38. Ferreira, I.Q.; Arrobas, M.; Moutinho-Pereira, J.M.; Correia, C.M.; Rodrigues, M.Â. The effect of nitrogen applications on the growth of young olive trees and nitrogen use efficiency. *Turk. J. Agric. For.* **2020**, *44*, 278–289. [CrossRef]
39. Fernández-Escobar, R.; Benlloch, M.; Herrera, E.; García-Novelo, J.M. Effect of traditional and slow-release N fertilizers on growth of olive nursery plants and N losses by leaching. *Sci. Hortic.* **2004**, *101*, 39–49. [CrossRef]
40. Fernández-Escobar, R. Use and abuse of nitrogen in olive fertilization. *Acta Hortic.* **2011**, *888*, 249–257. [CrossRef]
41. Othman, Y.A.; Leskovar, D. Nitrogen management influenced root length intensity of young olive trees. *Sci. Hortic.* **2019**, *246*, 726–733. [CrossRef]
42. Fernández-Escobar, R.; García-Novelo, J.M.; Molina-Soria, C.; Parra, M.A. An approach to nitrogen balance in olive orchards. *Sci. Hortic.* **2012**, *135*, 219–226. [CrossRef]
43. Lopes, J.I.; Arrobas, M.; Brito, C.; Gonçalves, A.; Silva, E.; Martins, S.; Raimundo, S.; Rodrigues, M.Â.; Correia, C.M. Mycorrhizal Fungi were More Effective than Zeolites in Increasing the Growth of Non-Irrigated Young Olive Trees. *Sustainability* **2020**, *12*, 10630. [CrossRef]
44. Ferretti, G.; Keiblinger, K.M.; Faccini, B.; Di Giuseppe, D.; Mentler, A.; Zechmeister-Boltenstern, S.; Coltorti, M. Effects of Different Chabazite Zeolite Amendments to Sorption of Nitrification Inhibitor 3,4-Dimethylpyrazole Phosphate (DMPP) in Soil. *J. Soil Sci. Plant Nutr.* **2020**, *20*, 973–978. [CrossRef]
45. Galamini, G.; Ferretti, G.; Medoro, V.; Tescaro, N.; Faccini, B.; Coltorti, M. Isotherms, Kinetics, and Thermodynamics of NH_4^+ Adsorption in Raw Liquid Manure by Using Natural Chabazite Zeolite-Rich Tuff. *Water* **2020**, *12*, 2944. [CrossRef]

46. Stamatakis, M.G.; Stamataki, I.S.; Giannatou, S.; Vasilatos, C.; Drakou, F.; Mitsis, I.; Xinou, K. Characterization and evaluation of chabazite- and mordenite-rich tuffs, and their mixtures as soil amendments and slow release fertilizers. *Arch. Agron. Soil Sci.* **2016**, *63*, 735–747. [CrossRef]
47. Malferrari, D.; Laurora, A.; Brigatti, M.F.; Coltorti, M.; Di Giuseppe, D.; Faccini, B.; Passaglia, E.; Vezzalini, M.G. Open-field experimentation of an innovative and integrated zeolitite cycle: Project definition and material characterization. *Rend. Lincei* **2013**, *24*, 141–150. [CrossRef]
48. GeoViewer—Geoportale. Available online: https://geoportale.regione.emilia-romagna.it/mappe/geo-viewer?layer_id=0f536ddad2c442d293b6afa0119edde5 (accessed on 22 August 2022).
49. World Reference Base. | FAO SOILS PORTAL | Food and Agriculture Organization of the United Nations. Available online: https://www.fao.org/soils-portal/data-hub/soil-classification/world-reference-base/en/ (accessed on 16 August 2022).
50. Tarocco, A.; Aprea, A. *Carta dei Suoli Della Regione Emilia-Romagna*; Regione Emilia-Romagna: Bologna, Italy, 2021.
51. Pavan, V.; Marletto, V. *Rapporto IdroMeteoClima Emilia-Romagna. Dati 2019*; Osservatorio Clima di ARPAE: Rimini, Italy, 2020.
52. Pavan, V. *Rapporto IdroMeteoClima Emilia-Romagna. Dati 2020*; Osservatorio Clima di ARPAE: Rimini, Italy, 2021.
53. Pavan, V. *Rapporto IdroMeteoClima Emilia-Romagna. Dati 2021*; Osservatorio Clima di ARPAE: Rimini, Italy, 2022.
54. Di Giuseppe, D.; Faccini, B.; Mastrocicco, M.; Colombani, N.; Coltorti, M. Reclamation influence and background geochemistry of neutral saline soils in the Po River Delta Plain (Northern Italy). *Environ. Earth Sci.* **2014**, *72*, 2457–2473. [CrossRef]
55. Wickham, H. *Ggplot2: Elegant Graphics for Data Analysis*; Springer International Publishing: Berlin/Heidelberg, Germany, 2016.
56. De Mendiburu, F. Agricolae: Statistical Procedures for Agricultural Research. R Package Version 1.2-0. 2016. Available online: http://CRAN.R-project.org/package=agricolae (accessed on 11 April 2022).
57. Schloerke, B.; Cook, D.; Larmarange, J.; Briatte, F.; Marbach, M.; Thoen, E.; Elberg, A.; Toomet, O.; Crowley, J.; Hofmann, H.; et al. GGally: Extension to Ggplot2 2021. Available online: https://CRAN.R-project.org/package=Ggally (accessed on 11 April 2022).
58. Vincent, Q. Vu Ggbiplot: A Ggplot2 Based Biplot 2011. Available online: http://github.com/vqv/ggbiplot (accessed on 11 April 2022).
59. Horikoshi, M.; Ggfortify, Y.T. *Data Visualization Tools for Statistical Analysis Results*; 2016; Available online: https://CRAN.R-project.org/package=ggfortify (accessed on 11 April 2022).
60. Eldor, A.P. *Soil Microbiology, Ecology, and Biochemistry*; Academic Press: Burlington, NJ, USA, 2007; ISBN 9780125468077.
61. Bateman, A.S.; Kelly, S.D. Fertilizer nitrogen isotope signatures. *Isot. Environ. Health Stud.* **2007**, *43*, 237–247. [CrossRef]
62. Shan, Y.; Huang, M.; Suo, L.; Zhao, X.; Wu, L. Composition and variation of soil δ15N stable isotope in natural ecosystems. *CATENA* **2019**, *183*, 104236. [CrossRef]
63. Amundson, R.; Austin, A.T.; Schuur, E.A.G.; Yoo, K.; Matzek, V.; Kendall, C.; Uebersax, A.; Brenner, D.; Baisden, W.T. Global patterns of the isotopic composition of soil and plant nitrogen. *Glob. Biogeochem. Cycles* **2003**, *17*, 1031. [CrossRef]
64. Farquhar, G.D.; Ehleringer, J.R.; Hubick, K.T. Carbon Isotope Discrimination and Photosynthesis. *Annu. Rev. Plant Physiol. Plant Mol. Biol.* **1989**, *40*, 503–537. [CrossRef]
65. Riehl, S.; Pustovoytov, K.E.; Weippert, H.; Klett, S.; Hole, F. Drought stress variability in ancient Near Eastern agricultural systems evidenced by δ13C in barley grain. *Proc. Natl. Acad. Sci. USA* **2014**, *111*, 12348–12353. [CrossRef]
66. Busch, F.A.; Holloway-Phillips, M.; Stuart-Williams, H.; Farquhar, G.D. Revisiting carbon isotope discrimination in C3 plants shows respiration rules when photosynthesis is low. *Nat. Plants* **2020**, *6*, 245–258. [CrossRef] [PubMed]
67. Kumar, S.; Singh, B. Effect of water stress on carbon isotope discrimination and Rubisco activity in bread and durum wheat genotypes. *Physiol. Mol. Biol. Plants* **2009**, *15*, 281–286. [CrossRef]
68. Jakkula, V.; Zeolites, W.S.P. Potential Soil Amendments for Improving Nutrient and Water Use Efficiency and Agriculture Productivity. Available online: https://www.researchgate.net/publication/325544295_Zeolites_Potential_soil_amendments_for_improving_nutrient_and_water_use_efficiency_and_agriculture_productivity (accessed on 11 April 2022).
69. Hatfield, J.L.; Follet, R.F. *Nitrogen in the Environment*; The Scientific World Journal: London, UK, 2018; ISBN 9780333227794.
70. Prisa, D. Optimised fertilisation with zeolitites containing Plant Growth Promoting Rhizobacteria (PGPR) in Ranunculus asiaticus. *GSC Biol. Pharm. Sci.* **2020**, *10*, 096–102. [CrossRef]
71. Choo, L.N.L.K.; Ahmed, O.H.; Talib, S.A.A.; Ghani, M.Z.A.; Sekot, S. Clinoptilolite zeolite on tropical peat soils nutrient, growth, fruit quality, and yield of Carica papaya L. CV. sekaki. *Agronomy* **2020**, *10*, 1320. [CrossRef]
72. Sepaskhah, A.R.; Barzegar, M. Yield, water and nitrogen-use response of rice to zeolite and nitrogen fertilization in a semi-arid environment. *Agric. Water Manag.* **2010**, *98*, 38–44. [CrossRef]
73. Ferretti, G.; Galamini, G.; Deltedesco, E.; Gorfer, M.; Fritz, J.; Faccini, B.; Mentler, A.; Zechmeister-Boltenstern, S.; Coltorti, M.; Keiblinger, K.M. Gross Ammonification and Nitrification Rates in Soil Amended with Natural and NH4-Enriched Chabazite Zeolite and Nitrification Inhibitor DMPP. *Appl. Sci.* **2021**, *11*, 2605. [CrossRef]

Article

Assessing Soil-like Materials for Ecosystem Services Provided by Constructed Technosols

Kristina Ivashchenko [1,2,*], Emanuela Lepore [2,3], Viacheslav Vasenev [2,4], Nadezhda Ananyeva [1], Sofiya Demina [2], Fluza Khabibullina [2], Inna Vaseneva [2], Alexandra Selezneva [1], Andrey Dolgikh [2,5], Sofia Sushko [1,2,6], Sara Marinari [3] and Elvira Dovletyarova [2]

1. Institute of Physicochemical and Biological Problems in Soil Science, Russian Academy of Sciences, 142290 Pushchino, Russia; ananyeva@rambler.ru (N.A.); alexandra_seleznyova@mail.ru (A.S.); sushko-sv@rudn.ru (S.S.)
2. Department of Landscape Design and Sustainable Ecosystems, Agrarian-Technological Institute, Peoples' Friendship University of Russia, 117198 Moscow, Russia; Emanuela.lepore@studenti.unitus.it (E.L.); vasenyov@mail.ru (V.V.); ibatulina_sa@pfur.ru (S.D.); khabibullina.fluza@mail.ru (F.K.); vaseneva-iz@rudn.ru (I.V.); dolgikh@igras.ru (A.D.); dovletyarova-ea@rudn.ru (E.D.)
3. Department for Innovation in Biological, Agro-Food and Forest Systems DIBAF, University of Tuscia, 01100 Viterbo, Italy; marinari@unitus.it
4. Soil Geography and Landscape Group, Wageningen University, 6700 Wageningen, The Netherlands
5. Institute of Geography, Russian Academy of Sciences, 119017 Moscow, Russia
6. Agrophysical Research Institute, 195220 Saint-Petersburg, Russia
* Correspondence: ivashchenko-kv@rudn.ru

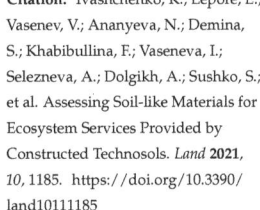

Citation: Ivashchenko, K.; Lepore, E.; Vasenev, V.; Ananyeva, N.; Demina, S.; Khabibullina, F.; Vaseneva, I.; Selezneva, A.; Dolgikh, A.; Sushko, S.; et al. Assessing Soil-like Materials for Ecosystem Services Provided by Constructed Technosols. Land 2021, 10, 1185. https://doi.org/10.3390/land10111185

Academic Editor: Cezary Kabala

Received: 26 September 2021
Accepted: 30 October 2021
Published: 4 November 2021

Publisher's Note: MDPI stays neutral with regard to jurisdictional claims in published maps and institutional affiliations.

Copyright: © 2021 by the authors. Licensee MDPI, Basel, Switzerland. This article is an open access article distributed under the terms and conditions of the Creative Commons Attribution (CC BY) license (https://creativecommons.org/licenses/by/4.0/).

Abstract: Urbanization results to a wide spread of Technosols. Various materials are used for Technosols' construction with a limited attention to their ecosystem services or disservices. The research focuses on the integral assessment of soil-like materials used for Technosols' construction in Moscow megalopolis from the ecosystem services' perspective. Four groups of materials (valley peats, sediments, cultural layers, and commercial manufactured soil mixtures) were assessed based on the indicators, which are integral, informative, and cost-effective. Microbial respiration, C-availability, specific respiration, community level physiological profile, and Shannon' diversity index in the materials were compared to the natural reference to assess and rank the ecosystem services and disservices. The assessment showed that sediments and low-peat mixtures (≤30% of peat in total volume) had a considerably higher capacity to provide C-sequestration, climate regulation and functional diversity services compared to peats and high-peat mixtures. Urban cultural layers provided ecosystem disservices due to pollution by potentially toxic elements and health risks from the pathogenic fungi. Mixtures comprising from the sediments with minor (≤30%) peat addition would have a high potential to increase C-sequestration and to enrich microbial functional diversity. Their implementation in urban landscaping will reduce management costs and increase sustainability of urban soils and ecosystem.

Keywords: urban soil; organo-mineral materials; ecosystem disservices; MicroResp technique; functional microbial diversity; fungi; Moscow megalopolis

1. Introduction

Urban ecosystems are to a great extent artificial by genesis and human-driven regarding their functions and services, and therefore highly variable and dynamic [1–3]. Considering the degree of anthropogenic impact, urban soils are identified as man-influenced, man-changed, or man-made [4]. From a variety of urban soils, soil constructions (constructed Technosols) are likely the most attractive and challenging for environmental assessment and modeling. Annually, thousands of tons of organic and mineral materials are imported into a big city and utilized for Technosols' construction [5–7]. The technologies and materials used for Technosols' construction are selected considering the management purposes, e.g., for reclamation, landscaping, establishment, and maintenance of urban green infrastructures [8–11]. A literature survey shows high diversity of materials used in urban soil engineering, including technic (e.g., rubble, sludge, wastes, and industrial by-products) and natural (e.g., peat, relocated topsoil, biosolids, and dredged (bottom) sediments) materials [9,12–15]. Various commercially manufactured materials are available on the market for landscaping and greening purposes. A variety of materials results in a unique diversity of chemical, physical, and biological properties of the constructed Technosols.

The quality of materials is usually legally regulated by state or municipal standards, which differ between countries and cities. For example, the mixture of agricultural topsoil, silt, clay, and sand is applied for greening public areas in Parma, Italy [16]. A mixture of compost, peat, and sand is recommended for planting trees at the roadsides in Rome [17]. In France, topsoil removed from agricultural and forest lands remains a key material for urban greening [15], whereas reusing the industrial by-products, wastes, and fine sediments is suggested as an environmentally friendly alternative [12,18]. In the United States, particularly in Chicago, biosolids and dredged sediments are implemented for renovation and greening works [9,14]. In Moscow, Russia, there are at least 50 companies, supplying about a hundred of different materials for greening, mainly comprised of peat, sand, compost, and excavated topsoil in different proportions [5,8].

Although the quality standards of materials for the Technosols' construction differ between the cities, most of them are focused on several chemical properties (e.g., pH or content of potentially toxic elements, PTE), whereas their capacity to provide the ecosystem services remains overlooked [19,20]. Soil microorganisms are responsible for such important ecosystem's service as nutrients' cycles [21], pollutants' biodegradation [22–24] and climate regulation [25]. The provisioning of these services can be projected based on the microbial functional indicators, e.g., specific microbial respiration, community-level physiological profile, abundance of metabolic genes or enzymatic activity [26–29]. Microbial diversity contributes to ecosystem resistance to external stress; however, the presence of pathogenic species evokes substantial health risks and shall be considered as an eco-system disservice [30,31].

The research aimed to analyze soil-like materials used for Technosols' construction in Moscow megalopolis (Russia) and assess their quality based on chemical and microbial properties. The study outcomes shall allow re-thinking the existing soil quality standards and regulations from the ecosystem services' perspective to support urban sustainable development.

2. Materials and Methods

2.1. Materials Used for Technosols' Construction in Moscow

Moscow megalopolis extents over 2500 km^2 and with the population above 12 million people is the largest city in Europe. Moscow is located in the Central part of East-European plain (56° N; 37° E) and has a temperate continental climate. Taiga and mixed forests on the Retisols dominating the natural areas of the region, in Moscow city, are to a great extent substituted by ornamental plants and green lawns on man-changed or man-made soils with a considerable portion of constructed Technosols [32,33]. More than 1 million m^3 of soil-like materials (e.g., organic and mineral components and commercially manufactured mixtures) are annually imported into the city for the needs of civil engineering and green infrastructures' development [5,8,34]. A major part of the commercially manufactured mixtures available on the market composes from the similar components: valley peat, topsoil from meadow or arable lands, urban topsoil, and subsoil excavated before building construction, excavated river valley topsoil, compost, sand, dredged sediments, and sludge from water treatment stations. For this study, the most representative groups of the materials were purchased and collected: valley peats, sediments, urban cultural layers, and commercially manufactured mixtures (Table 1).

Table 1. The origin, suppliers, and implementation of the studied materials for Technosols' construction (PTs, valley peat (n = 4); SDs, sediments (n = 2); CLs, cultural layers (n = 4); MIX_{LPT}, mixture (n = 3) with low peat content; MIX_{HPT}, mixture with high peat content (n = 3)).

Material	Origin	Suppliers	Implementation
PTs	peatlands	peat mining companies	high
SDs	water body	water management companies	low
CLs	urban subsoil		low
MIX_{LPT}		producers of soil mixtures for gardening and landscaping	high
MIX_{HPT}	man-made		high

The valley peats group included four samples from the major supplying companies. The sediments group was represented by sludge from surface water treatment stations (i.e., solid and non-soluble particles mechanically filtered prior the water supply) and bottom sediments (i.e., dredged sediments excavated from a lake bottom). Cultural layers included subsoil urban sediments accumulated during a long-term residential activity and frequently excavated during building and infrastructure constructing [35,36]. The commercially manufactured mixtures comprised from several components, including valley peat. Based on the portion of the valley peat in the mixtures' composition, they were subdivided to low-peat ($\leq$30% of peat in total volume) and high-peat ($\geq$75% of peat in total volume) (Table 2). All the materials are available on the market and were collected from the official suppliers—from two to four materials per group and three mixed samples (50 L bags) for each material. The topsoil (0–10 cm) of Retisols sampled in the four mixed forested parks of Moscow was considered as a natural soil reference.

Table 2. The composition of the investigated commercial mixtures.

Mixture	Number	Composition	Volume Portion, %
Low peat content	I	peat/excavated urban topsoil/sand/excavated river valley topsoil	30/30/30/10
	II	excavated urban topsoil/peat/compost/sand	25/25/25/25
	III	excavated urban topsoil/valley peat/sand	50/30/20
High peat content	IV	peat/sand	75/25
	V	peat/compost/sand	80/10/10
	VI	peat/sand	95/5

2.2. Integral Assessment of the Materials' Quality

To assess the quality of the materials and project the ecosystem services they can provide, sub-samples (300 g) were taken for each material and the natural soil reference. Sub-samples were sieved through a 2-mm mesh and subdivided into two parts. The first part was air-dried for chemical analysis. The second part was adjusted to 60% water-holding capacity and preincubated (150 g, 22 °C, 7 d) in the thermostat in a plastic bag with air exchange. The preincubation stage eliminated initial variation in materials' temperature and moisture, and excluded possible CO_2 efflux from the preparation procedures [37–39]. After preincubation, the sub-samples were analyzed for microbial properties.

2.3. Chemical Analysis

Total carbon (C) and nitrogen (N) contents were determined by spectrometry (CHNS-932, LECO Corp, USA) after oxygen combustion (1100 °C). The pH of peats, cultural layers, high-peat mixtures (the high organic material:water = 1:10) and soil, sediments, low-peat mixtures (material:water = 1:2.5) was measured by pH-meter (Basic Meter PB-11, Germany) [40]. Total contents of nickel (Ni), zinc (Zn), lead (Pb), cadmium (Cd) were measured by X-ray fluorescence spectroscopy (Spectroscan Max-GVM, Russia).

2.4. Microbiological Analysis

In the subsamples, the microbial respiration (MR) was evaluated by CO_2 production rate after its incubation for 24 h at the standardized condition at 22 °C [38,41]. The measurement of CO_2 was carried out by a gas chromatograph with a thermal conductivity detector (KrystaLLyuks 4000 M, Yoshkar-Ola, Russia). Microbial biomass carbon (MBC) was measured by substrate-induced respiration (SIR) method, which is based on the registration of the highest initial microbial CO_2 production after glucose addition [37,38]. The subsamples (1.0 g each) were placed in a vial (15 mL volume) and a glucose solution was added dropwise (10 mg glucose g^{-1}, volume was 0.1 mL). The vial was tightly closed and incubated at 22 °C during 3.5 h. The measured SIR was converted to MBC units ($\mu g\ C\ g^{-1}$) by the following equation: SIR ($\mu L\ CO_2\ g^{-1}\ h^{-1}$) × 40.04 + 0.37 [37]. The MBC:C and MR:MBC ratios were calculated to estimate microbial C-availability and specific respiration (qCO_2), respectively [25,42].

Community level physiological profile (CLPP) was measured by MicroResp^{TM} technique [27,28,43]. Briefly, samples were put to the 96-deep well (945 µL volume each) and solutions of four C-substrates' groups were added: amino acids (glycine, L-arginine, L-leucine, α-aminobutyric, L-aspartic acids), carbohydrate (D-galactose, D-fructose, D-glucose), carboxylic acids (L-ascorbic, citric, oxalic acids), and phenolic acid (vanillic and syringic acids). The response of microbial community was detected by CO_2 production by colorimetric method after 6 h of incubation with detection gel at 25 °C. The absorbance by the detection gel was analyzed at 595 nm wave length (microplate spectrophotometer FilterMax F5, USA) before and after incubation and expressed as $\mu g\ C\ g^{-1}\ h^{-1}$ [28]. Microbial functional diversity was assessed through Shannon index: H' = $-\sum pi \times \ln pi$ [44], where pi is the ratio of CO_2 response on the addition of single C-substrate to the sum of responses for all studied substrates.

The fungi species were cultivated on Getchinson's and Czapek's solid media, that allowed to cover the widely distributed fungi in materials consuming the cellulose and carbohydrates, respectively [45]. The Getchinson's solid medium consisted 2.5 g $NaNO_3$, 1.0 g K_2HPO_4, 0.3 $MgSO_4$, 0.1 g $CaCl_2$, 0.1 NaCl, 0.01 g $FeCl_3$, 7.5 g agar L^{-1} water. The Czapek's medium included 30 g sucrose, 2.0 g $NaNO_3$, 1.0 g K_2HPO_4, 0.5 g $MgSO_4$, 0.5 KCl, 0.01 g $FeSO_4$, 15 g agar L^{-1} water. Streptomycin sulfate (100 mg L^{-1}) was added to the media for bacterial growth inhibition. Briefly, sterile water (90 mL) was added to each 10 g soil and each subsample materials' group, and shaken for 10 min [46]. Serial dilutions (from 10^{-2} to 10^{-9}) were prepared by sequentially transferring 1 mL supernatant into glass tubes with 9 mL of sterile water. Subsamples (0.1 mL) at selected three dilutions were pipetted on the surface of three Petri dishes with each solid medium. The Getchinson's medium was covered by filter paper then. The dishes were incubated at 25 °C during 10 d [47]. Fungi genus and species identification was based on their morphological characteristics using the manual [48]. The occurrence of the fungi was calculated as the ratio of the number of Petri dishes with an identified species to their total number for each subsample. Occurrence was measured as follows: >83% is frequent, 33–83% is medium, and <33% is rare. Pathogenic potential was identified according to the atlas of clinical fungi [49].

2.5. Interpretation of Soil-like Materials' Properties from the Ecosystem Services' Perspective

The ecosystem services' assessment was based on the studied microbial properties, which are often used as soil quality indicators [26,50]. The organic matter decomposition rate based on MR was used to assess the nutrients' cycle service. The higher value could indicate a better performance of the service, however, could also show the acceleration of CO_2 production rate. Hence, the qCO_2 value was considered to indicate the balance between CO_2 production and C involved into microbial cells. The ratio of MBC to C determines C-availability to microbes and together with qCO_2 was used as indicators of the C-sequestration and climate regulation service. The microbial response to specific organic acids (e.g., phenolic) was considered as the capacity to biodegradation of organic pollutants, which include a benzene ring. Shannon functional diversity index was considered to assess the functional biodiversity supporting service. All the selected indicators were standardized to the natural soil reference values, which potential to provide the ecosystem services was considered the highest. The disservices of the materials were assessed based on PTE content compared to the health threshold level and pathogens occurrence in comparison to the natural soil reference. The ecosystem services' performance was ranked from 0 to 1 (where 1 is the best performance). The ecosystem disservices' performance was ranked by the same scale, where 1 means the minimal disservices provided.

2.6. Statistical Analysis

All measurements were performed in three replicates and calculated for the dry weight of subsamples. One-way analysis of variance and subsequent multiple comparisons by Tukey's test were performed for comparing the chemical and microbial properties among the materials' groups. The comparison between the materials' groups and the natural soil reference was done based on Dunnett's test. The relationships between chemical and microbial properties were analyzed using Spearman's correlation. Redundancy analysis (RDA) was used to examine the relations of fungal community composition to pH, nutrients (C, N), and PTE (Ni, Zn, Pb, Cd) contents among the studied materials. Data on fungi species occurrence were processed with Hellinger transformation [51]. Prior to RDA, a forward selection was performed to identify the best set of non-collinear explanatory variables with the highest adjusted multiple determination coefficient.

Significance level was accepted as 0.05. Statistical data analysis and visualization were processed in RStudio [52]. Data visualization was done by ggplot2 package [53]. Correlation matrix was visualized with the 'Performance Analytics' package. The RDA was performed using the 'vegan' package.

3. Results

3.1. Chemical Properties

The pH of all the materials was close to neutral with non-significant difference between the groups or with the natural soil reference (Table 3).

Table 3. Average pH and bulk of potentially toxic elements (PTE) (mg kg^{-1}) in the soil-like materials (PTs, valley peat); SDs, sediments; CLs, cultural layers; MIX$_{LPT}$, mixture with low peat content; MIX$_{HPT}$, mixture with high peat content).

PRP	PTs (n = 4)	SDs (n = 2)	CLs (n = 4)	MIX$_{HPT}$ (n = 3)	MIX$_{LPT}$ (n = 3)	THL
pH	6.4 ± 0.5 a	6.9 ± 0.5 a	7.2 ± 0.0 a	6.5 ± 0.6 a	6.7 ± 0.1 a	6.0–7.5
Ni	23 ± 6 a	23 ± 5 a	12 ± 1 b	29 ± 10 a	27 ± 2 a	80
Zn	215 ± 158 b	140 ± 12 b	**563 ± 212 a**	54 ± 14 c	53 ± 3 c	220
Pb	9 ± 3 a	12 ± 8 a	22 ± 3 a	15 ± 7 a	15 ± 2 a	130
Cd	0.5 ± 0.1 a	0.3 ± 0.0 b	0.6 ± 0.0 a	0.5 ± 0.1 a	0.2 ± 0.0 b	2.0

PRP, properties. Values are reported as mean ± standard error, different letters indicate a significant ($p < 0.05$) difference between the groups. Bold value represents the exceeding of threshold level (THL) for PTE (HS-514-11 regulation).

In contrast, C and N contents ranged more than one order of magnitude with the highest values in cultural layers and valley peats (Figure 1). Peat soil is widely recognized as the remarkable natural C stock, whereas high C and N content in cultural layers have an anthropogenic origin. They result from a long-term deposition during the residential activity and include organic wastes, wooden cheeps and other artifacts [35,54]. Only the low-peat mixtures contained a similar amount of C and N as a natural soil, whereas in all the other materials C and N contents exceeded the natural reference values 5 to 15 times. The C:N ratio for all the materials ranged between 10 and 20, indicating a balanced of C and N input. Compared to the other materials, cultural layers had higher contents of Cd and Zn, whereas Pb and Ni contents didn't differ significantly among the groups and were lower than the maximal permissible level recommended by Moscow' municipal regulations.

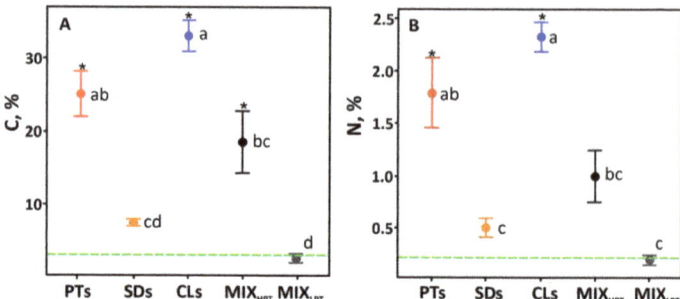

Figure 1. Mean (circles) and standard errors (bars) of total carbon (**A**) and nitrogen (**B**) in peats (PTs), sediments (SDs), cultural layers (CLs), high-peat (MIX$_{HPT}$) and low-peat (MIX$_{LPT}$) mixtures. Dotted green line represents the mean for the natural soil reference. Letters indicate the significantly different groups (Tukey's test). Means with * indicate a significant difference from the natural soil (Dunnett's test).

3.2. Microbial Properties

An extremely high MR obtained for the cultural layers was two orders of magnitude above other materials and the natural soil reference (Figure 2A). The highest specific respiration was reported for the materials rich in easily mineralizable organic matter (peats, cultural layers and high-peat mixtures), whereas sediments and low-peat mixtures were not significantly different from the natural soil (Figure 2B). Assuming a balanced specific respiration in natural soils as (i.e., qCO_2), sediments and low-peat mixtures were balanced as well, whereas the other substrates were not. The C-availability in the natural soils was significantly higher than in any soil-like material, and the lowest values were obtained for the peats and high-peat mixtures (Figure 2C). High qCO_2 and low C-availability in peats and high-peat mixtures indicate their low capacity for C sequestration. Only the small part of C stored in these materials could be consumed by microbes for anabolism and accumulated in microbial cells, whereas the major part was released as CO_2.

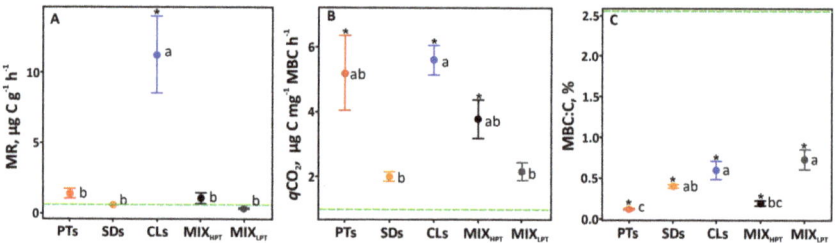

Figure 2. Microbial respiration (MR, **A**), microbial metabolic quotient (qCO_2, **B**) and ratio of microbial biomass carbon to total carbon (MBC:C, **C**) in soil-like materials. Dotted green line represents the mean for the natural soil reference. Letters indicate the significantly different groups (Tukey's test). Means with * indicate a significant difference from the natural soil (Dunnett's test).

The CLPP results showed that microbial structure in the peats was mostly shifted to groups consuming the ascorbic acid, whereas in the cultural layers and sediments, the highest response was obtained on the citric and ascorbic acids (Figure 3A). The response of microbial community on the arginine addition was found only for the peats and high-peat mixtures. For all materials except cultural layers, the capacity of microbial community to decompose complex organic compounds with benzene ring such as phenolic acids (vanillic and syringic) was lower compared to the natural soil. The highest microbial diversity was also reported for the natural soils, for which the Shannon index was considerably higher than in any of the soil-like materials. Among the materials, the index increased in a row: high-peat mixtures, sediments, peats, cultural layers, low-peat mixtures (Figure 3B).

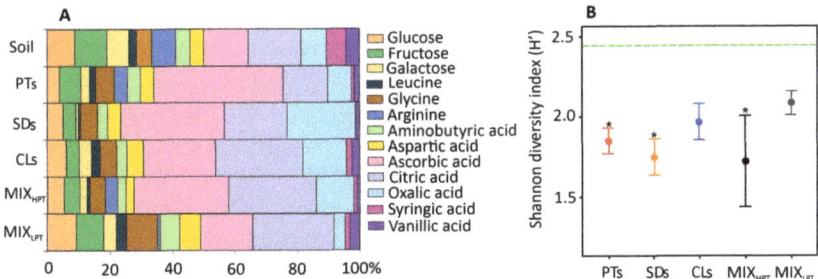

Figure 3. Community level physiological profile (**A**) expressed by contribution of respiration on addition of individual substrates to total substrate-induced respiration in the natural soil and soil-like materials. The Shannon diversity index (**B**) presents as mean (circles) are standard error (bars) for materials, dotted line represents the mean for the soil. Means with * indicate significantly differences from the natural soil (Dunnett's test).

A more detailed analysis of the fungal diversity in the materials allowed identifying 31 species from 16 genera (Table 4). Between 8 and 11 fungi species were identified in soil, peats, and mixtures, which was 1.6–4.0 times more than in the sediments and cultural layers. The identified species differed between the natural soil and soil-like materials, as well as among the materials' groups. The highest frequency of the opportunistic fungi genera (e.g., *Aspergillus*, *Chaetomium* and *Geomyces*) and plant pathogenic fungi genera (e.g., *Verticillium* genus) were found in cultural layers. These species could cause mycoses in individuals having a weakened immunity. They are also harmful for the plant leaves and stems. Considering the potential risks for human and plant health, the implementation of cultural layers for urban greening purposes is questionable. Some opportunistic fungi genera were also found in low-peat mixtures and in the natural soil reference; however, the frequency of occurrence was less compared to the cultural layers.

3.3. Relationships between Microbial and Chemical Properties

The difference in microbial properties between the investigated materials was partly driven by C and N contents and the polluting level by Cd and Zn (Figure 4). A positive significant strong correlation was shown between MR, C, N, Zn, and Cd contents, whereas a significant negative effect of contaminants on microbial properties was not shown.

The negative effects of the PTE on the microbial community (reflected in high MR) were reported for urban soils before [54,55] and indicated stressful conditions for microbiome; however, the opposite effect of C and N input was expected. Apparently, C and N contents in some materials (e.g., in valley peats and high-peat mixtures) were so high that they could not be taken due to the exceeded capacity of their assimilation by microbial community, and therefore resulted in a qCO_2 increase. The most optimal microbial functional capacity (low qCO_2) was found at a range of 1.6 to 8.0% for C and 0.1 to 0.6% for N (Figure 5A,B).

Based on the RDA ordination of fungi species in the studied materials (Figure 6), the forward selection indicated that the best fitted model included Pb and Ni as factors, which explained 51.3% of the variance in fungal composition. Among these factors, Pb content was significant, explained 30.2% of variance, and was considered as gradient for RDA1 (pseudo F = 1.7, p = 0.006; 999 permutations). The ordination showed that fungi of cultural layers were more exclusive and less diverse compared to the other studied materials. The occurrence of *Verticillium* and *Aspergillus niger* pathogens increased along the Pb contamination gradient and associated with cultural layers.

Table 4. The occurrence of the fungi in the natural soil and soil-like materials and the potential health risks from the pathogenic fungi. Bold font represents the opportunistic fungi according to risk groups 1, which indicates the dangerous of fungi's impact on immunocompromised people (de Hoog et al., 2019).

Fungi	Abbreviation	Health Risks	Soil	PTs	CLs	SDs	MIX$_{HPT}$	MIX$_{LPT}$
Acremonium strictum Gams	Astr	pulmonary, pleuritis, fungemia	-	-	-	-	-	++
Aspergillus niger Tiegh.	Anig	otomycosis, aspergillosis	-	-	++	-	-	+
Aspergillus sp.	Asper		-	+	-	-	-	-
Acremonium charticola Lindau	Achar		-	-	-	-	-	++
Chaetomium globosum Kunze	Cglob	onychomycosis, cutaneous lesions	-	+	-	+	-	-
Chaetomium indicum Corda	Cind		-	+	-	-	-	-
Chaetomium spiralliforum Bainier	Cspi		-	-	-	+	-	-
Chaetomium spirale Zopf	Cspir		-	-	-	+	-	-
Chaetomium sp.	Csp		++	+	-	+	-	-
Geomyces pannorum Link	Gpan	onychomycosis	++	-	-	-	-	-
Gliocladium catenulatum Gilman & Abbott	Gcat		-	-	-	+	-	-
Gliocladium roseum Bainier	Gros		-	-	+++	-	-	-
Monocillium sp.	Mon		-	+	-	-	-	-
Monocillium pygmaea Chalab.	Mpyg		++	-	-	-	-	-
Mortierella polycephala Coem.	Mpol		+	-	-	-	-	-
Mortierella sp.	Mor		+	-	-	-	-	++
Mucor sp.	Muc		-	+	-	-	-	-
Paecilomyces farinosus Holm	Pfar		-	-	+++	-	-	-
Penicillium islandicum Sopp	Pisl		-	-	-	-	+	-
Penicillium steckii Zaleski	Pst		-	-	-	-	+	-
Penicillium sclerotiorum Beyma	Pscl		-	-	-	-	-	++
Penicillium rubrum Stoll	Prub		-	-	-	-	++	-
Penicillium terlikowskii Zaleski	Pter		-	-	-	-	++	-
Penicillium sp.	Pen		-	+	-	-	-	++
Stachybotrys parvispora Hughes	Spar		-	-	-	-	+	-
Stachybotrys lobulatus Berk.	Slob		-	+	-	+	-	-
Trichoderma sp.	Trich		++	-	-	-	-	-
Verticillium sp.	Vert	plant diseases	-	-	+	-	+	-
Moniliaceae sp.1	Mon1		+++	+++	-	+	+	-
Moniliaceae sp.2	Mon2		+	++	-	+	+++	+++
Micelia sterilia dark-colored	Msdc		-	++	-	-	-	+

Occurrence was measured as follows: +++, frequent (>83%); ++, medium (33–83%); +, rare (<33%); –, no.

3.4. From Properties towards Ecosystem Services

Chemical and microbial properties of the analyzed soil-like materials were integrated and interpreted to assess the ecosystem services or disservices, which they can provide. A high capacity to provide functional biodiversity and nutrient cycles' services was shown for all the materials. For the cultural layers, however, the biodiversity service was hampered by the health risk disservice induced by the occurrence of the pathogenic fungi. At the same time, only a few materials (sediments and low-peat mixtures) had the potential to provide C-sequestration and climate regulation services. For all the other materials with very high contents of easily mineralizable organic matter, the risks of CO_2 emissions were much higher than in the natural soils, considered as a reference for the ecosystem services' assessment. In result, the capacity of cultural layers, peats, and high-peat mixtures to provide the service was assessed 20% lower than for sediments and low-peat mixtures and 80% lower than for the natural soil. An opposite pattern was shown the pollutants' biodegradation services, which was performed by peats and high-peat mixtures 20 to 30% better than by the sediments with low C and N contents. For cultural layers, an optimal performance of the pollutants' biodegradation services coincided with the disservice evoked by PTE pollution; therefore, cultural layers can be considered quite an ambiguous material for Technosols' construction. Services' and disservices' assessment aggregated on Figure 7 clearly illustrate the multi-functionality of the materials. Likely, the preliminary idea of the target service to obtain (or disservice to avoid) shall be developed prior to selecting the particular material for Technosols' construction.

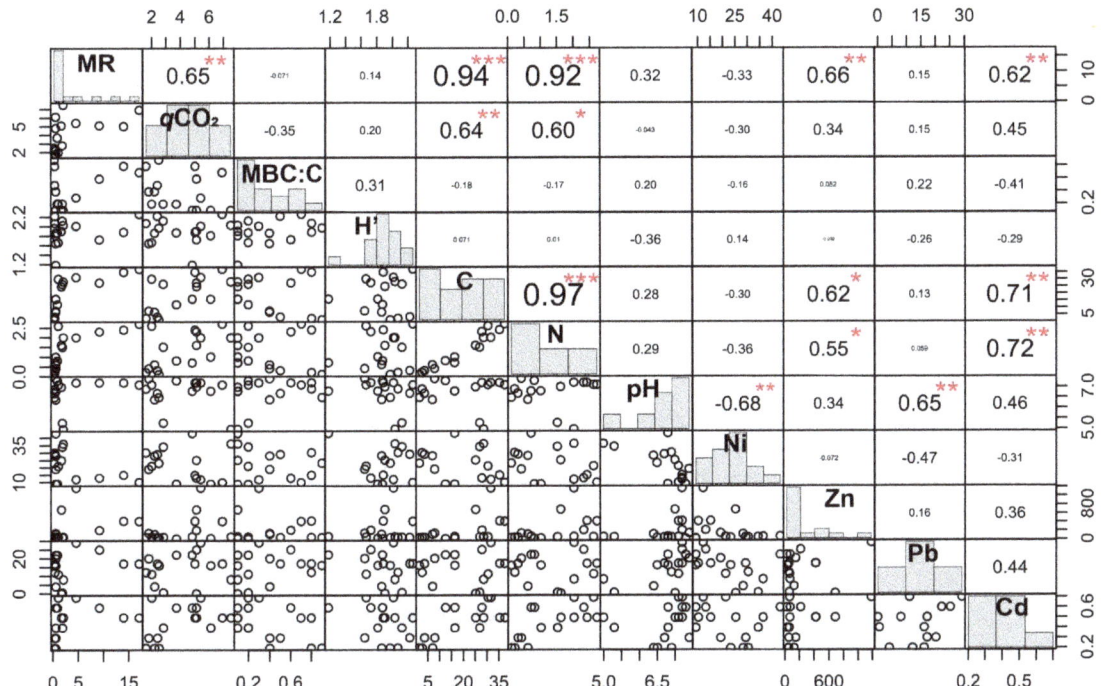

Figure 4. Relationships between microbial (MR, qCO₂, MBC:C, H') and chemical (C, N, pH, Ni, Zn, Pb, Cd) properties of the soil-like materials (n = 16). Significant correlation coefficients are indicated with * α ≤ 0.05; ** 0.01; *** 0.001.

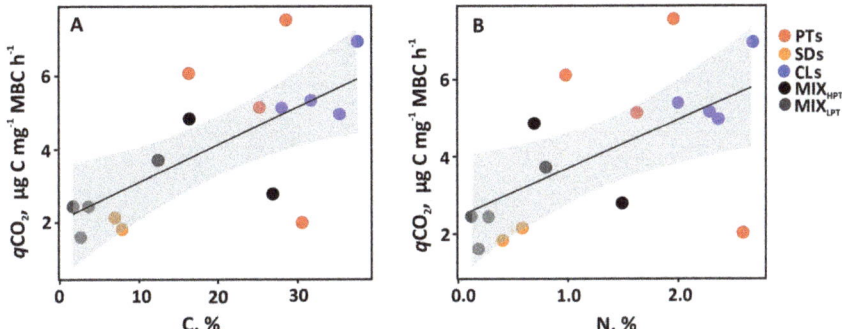

Figure 5. Scatter plot for microbial (qCO_2) and carbon (**A**) and nitrogen (**B**) of the soil-like materials. The gray 'bands' represent the standard error of the regression line.

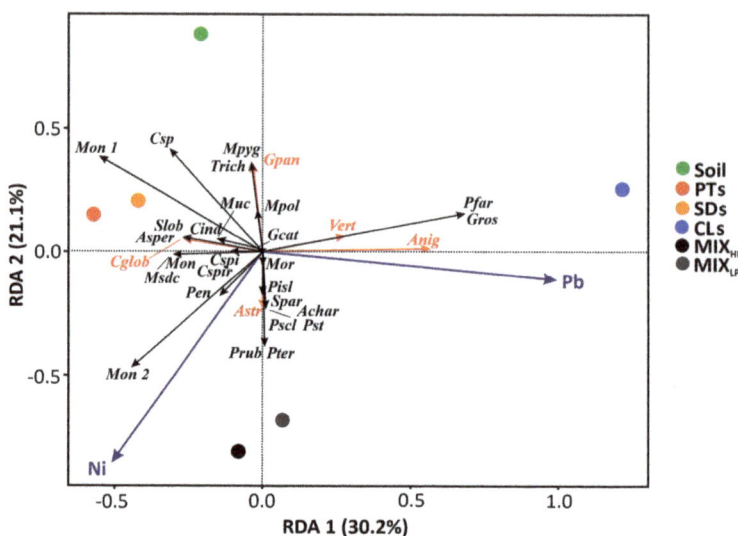

Figure 6. Ordination biplot of redundancy analysis for the fungi composition explained by Pb and Ni contents for the natural soil and soil-like materials. Non-pathogenic and pathogenic fungi are plotted as black and red arrows, respectively (fungi species abbreviations see in the Table 4).

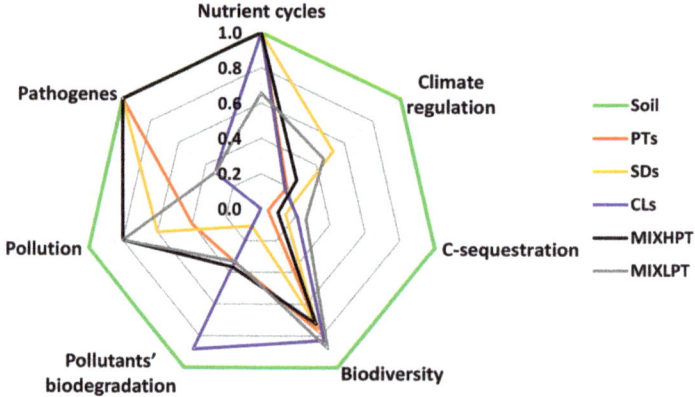

Figure 7. Estimation of ecosystem services (1 is the highest) and disservices (1 is the lowest) provided by different groups of materials: peats (PTs), sediments (SDs), cultural layers (CLs), high-peat (MIX$_{HPT}$), and low-peat (MIX$_{LPT}$) mixtures in relation to soil.

4. Discussion

4.1. Advantages and Disadvantages of the Soil-like Materials from the Ecosystem Services' Perspective

A comprehensive analysis of chemical and microbial properties of the materials projected into the ecosystem services'/disservices' assessment allowed ranking their quality and applicability for Technosols' construction. A high rank of the sediments, which balanced most of the analyzed services, is one of the principal and unexpected research outcomes. So far, dredged sediments are frequently used in agriculture as amendments [56–58], but in urban greening and landscaping, preference is traditionally given to C-rich 'dark' materials, which are supposed to be more fertile [5]. In fact, dredged sediments combined with biosolids can considerably improve soil fertility and support plant growth as it was shown for Chicago [9]. Technosols constructed from water treatment station sediments and composts had high nutrient contents and showed a positive dynamics in formation of stable aggregates [18,59]. Mixing sediments with clay loam, sand, and peat in volumetric proportions 25/30/40/5 or 15/40/40/5 at the pilot project in Moscow allowed constructing Technosols, whose quality satisfied governmental ecological and health standards [59–61]. These examples confirm our conclusion that non-polluted sediments could make a good alternative for the excavated natural and arable soils in composed mixtures for Technosols' construction, especially in the regions where the sedimentation in the water reservoirs is an important problem and dredging activities are needed.

In comparison to sediments, a considerably lower capacity to provide climate regulation and C-sequestration services was shown for the valley peat and high-peat mixtures, which so far dominate the greening markets of Moscow [5,34] and many other cities in Europe [62]. Although the nutrient content in peat materials is high, their vast implementation for Technosols' construction can result in a dramatic increase in CO_2 release to the atmosphere due to intensive mineralization of easily mineralizable organic matter [63,64], which can be even more facilitated by urban heat island effect [65]. We do not appeal for a complete ban for peat implementation in Technosols' construction; however, the proportions shall be thoroughly verified. Based on the research outcomes, a minor addition ($\leq$30%) of peat in the mixture composition didn't have a negative impact on the climate regulation and C-sequestration services and contributed to microbial functional diversity, which is in agreement with the previous studies [66].

Urban cultural layers were probably the most "exotic" group of materials we tested, due to specific genesis, properties, and limited implementation for greening and landscaping needs. Cultural layers include various deposits that reflected the anthropogenic activity in the past: wood chips, wastes, excavated bedrock, bricks, and gardening traces [67]. We are not aware of a widely spread practical application of the cultural layers in soil engineering; however, the nutrients' richness could make them attractive for this purpose. Urban cultural layers showed high microbial activity and C-availability indicated by a high potential to accumulate C in microbial cells. As a result, the high capacity in pollutants' biodegradation, C-sequestration, and functional biodiversity was also observed. However, the intensive mineralization of organic matter increases the risks of CO_2 emission and depletes the climate regulation service. Presence of pathogens (*Aspergillus niger* and *Verticillium genera*) and pollution by PTE (copper, zinc, lead), likely inherited from the historical land-use [35] are the principal disservices of cultural layer, which limit their application for Technosols' construction and urban greening.

4.2. Microbial Properties of the Materials in Relation to Nutrients and PTE Contents

Chemical and microbial properties in the studied materials were interrelated, and therefore, the variation in microbial indices and the values of corresponding ecosystem services were partly explained by nutrients' and PTE contents. Commonly, soil C and N contents stimulate microbial biomass growth [68,69]; however, in our study, a positive correlation between C and N contents and qCO_2 was shown. Apparently, the energy costs for microorganisms to maintain their biomass under intensive input of C and N are too high. There is a threshold level of saturation, above which an additional input of organic matter doesn't stimulate microbial activity [70]. Apparently, in peats, cultural layers and high-peat mixtures this threshold was exceeded. This outcome doubts existing municipal regulations, which allow or even recommend the high content of organic matter in the materials used for soil construction. For instance, the permissible content of organic matter in materials used for landscaping in Moscow range from 10 to 25% [60], which is completely unsustainable and can result in intensive CO_2 emission.

Soil microbial properties are quite sensitive to pollution by PTE. However, in our study, such a negative effect was not evident, that is likely due to relatively low concentrations of the pollutants (for most of the materials, their contents were below health thresholds, Table 3). Moreover, based on the correlation analysis, MR was positively related to Cd and Zn contents. This unexpected outcome is likely explained by the specific properties of the cultural layers, where considerable contents of heave metals coincide with a very high C and N contents, and correspondingly, which a high microbial activity.

4.3. Perspectives of Microbial Indicators for the Materials Quality Control

Existing material quality standards often ignore the fundamental view on soil quality and ecosystem services [71,72]. For instance, Moscow government regulates the permissible values of organic matter, pH, nutrients, as well as pollutants' content, pathogens, and weed seeds in the soil-like materials used for Technosols' construction and urban greening [60]. City of Evans municipality (Colorado, CO, USA) regulates pH, nitrogen, phosphorus, organic matter contents, bulk density, texture, moisture, and soluble salt concentration in the amendments used in landscaping [73]. The British standard for topsoil cut and translocated in building construction considers texture, nitrogen, phosphorus, potassium, organic matter contents, and a wide range of pollutants for quality control [74]. None of these and other reviewed regulations consider microbial properties as an important criteria of urban soil quality. Today, even a shortlist of microbial indicators within the standardized protocols includes microbial (basal) respiration [41], microbial biomass [75], enzymes activity [76], nitrogen mineralization, and nitrification in soils [77]. Partly, implementation of these microbial indicators in urban soils' assessment is constrained by high temporal dynamics, especially during the first years after Technosols' construction [78,79]. From the other perspective, monitoring dynamics of these indicators can reflect the evolution and pedogenesis processes in the constructed soils. For instance, a positive dynamics of microbial biomass carbon in Technosols constructed from the mining wastes to remediate an industrial barren indicated their effectiveness for the ecosystem restoration [11]. Assessing microbial properties of soil-like materials could be a promising tool to project functions and ecosystem services of the constructed Technosols' and therefore shall not be ignored in urban landscaping, planning, and management.

5. Conclusions

An artificial origin of the constructed Technosols gives a unique opportunity to project their functions and ecosystem services based on selecting soil-like materials with particular chemical and microbial properties. Assessment of the materials used for Technosols' construction in Moscow showed the highest performance for the sediments, which so far are almost completely ignored in urban greening. Regarding nutrient contents and balanced microbial functioning, they can be recommended as a promising replacement of native soils in organo-mineral mixtures used in soil constructions. Much lower ranks were given to peats due to very high risks of CO_2 emissions. Their implementation in Technosols' constructions shall be limited to minor ($\leq$30%) amendments to mixtures composed from the sediments or native soil. Cultural layers were exposed to high biological (pathogens) and chemical (PTE) pollution, which was considered an ecosystem disservice. Therefore, they shall not be recommended for urban greening and landscaping. Although the research outcomes and recommendations are based on the analysis obtained for Moscow megapolis, they are applicable for many other world cities since most of the investigated materials (e.g., peat, sediments, and organo-mineral mixtures) are universal and widely spread. Our study showed the efficiency of microbial properties for testing the quality of materials and their potential to contribute to the ecosystem services provided by constructed Technosols already at the planning stage. Assessment of microbial functional capacity can be an important factor for developing recommendations on materials and technologies to enhance ecosystem services of urban soil constructions and support urban sustainable development.

Author Contributions: Conceptualization, K.I., E.L., S.M. and V.V.; methodology, K.I., V.V., N.A., F.K.; software, K.I.; validation, A.D., S.S., I.V., A.S. and S.M.; formal analysis, K.I., I.V.; investigation, K.I., E.L., S.D.; resources, V.V., E.D.; data curation, K.I., E.L. and V.V.; writing—original draft preparation, K.I. and. V.V.; writing—review and editing, V.V., K.I., N.A., S.D.; visualization, K.I.; supervision, V.V., K.I., S.M.; project administration, E.D.; funding acquisition, V.V., K.I. All authors have read and agreed to the published version of the manuscript.

Funding: Functions of soil mixtures and their components were studied as a part of the Russian Science Foundation No 17-77-20046. Field investigations, sampling, and microbial analysis were supported by RFBR project No 19-34-90133. Data synthesis and paper preparation were supported by the RUDN University Strategic Leadership Program.

Institutional Review Board Statement: Not applicable.

Informed Consent Statement: Not applicable.

Data Availability Statement: The data presented in the study are available upon request from the corresponding authors.

Conflicts of Interest: The authors declare no conflict of interest.

References

1. Grimm, N.B.; Faeth, S.H.; Golubiewski, N.E.; Redman, C.L.; Wu, J.; Bai, X.; Briggs, J.M. Global change and the ecology of cities. *Science* **2008**, *319*, 756–760. [CrossRef] [PubMed]
2. Pickett, S.T.A.; Cadenasso, M.L.; Grove, J.M.; Boone, C.G.; Groffman, P.M.; Irwin, E.; Kaushal, S.S.; Marshall, V.; McGrath, B.P.; Nilon, C.H.; et al. Urban ecological systems: Scientific foundations and a decade of progress. *J. Environ. Manag.* **2011**, *92*, 331–362. [CrossRef] [PubMed]
3. Gómez-Baggethun, E.; Barton, D.N. Classifying and valuing ecosystem services for urban planning. *Ecol. Econ.* **2013**, *86*, 235–245. [CrossRef]
4. Lehmann, A.; Stahr, K. Nature and significance of anthropogenic urban soils. *J. Soils Sediments* **2007**, *7*, 247–260. [CrossRef]
5. Brianskaia, I.P.; Vasenev, V.I.; Brykova, R.A.; Markelova, V.N.; Ushakova, N.V.; Gosse, D.D.; Gavrilenko, E.V.; Blagodatskaya, E.V. Analysis of volume and properties of imported soils for prediction of carbon stocks in soil constructions in the Moscow metropolis. *Eurasian Soil Sci.* **2020**, *53*, 1809–1817. [CrossRef]
6. Deeb, M.; Groffman, P.M.; Blouin, M.; Egendorf, S.P.; Vergnes, A.; Vasenev, V.; Cao, D.L.; Walsh, D.; Morin, T.; Séré, G. Using constructed soils for green infrastructure—Challenges and limitations. *Soil* **2020**, *6*, 413–434. [CrossRef]
7. Fabbri, D.; Pizzol, R.; Calza, P.; Malandrino, M.; Gaggero, E.; Padoan, E.; Ajmone-Marsan, F. Constructed technosols: A strategy toward a circular economy. *Appl. Sci. (Switzerland)* **2021**, *11*, 3432. [CrossRef]
8. Smagin, A.V.; Sadovnikova, N.B. Creation of soil-like constructions. *Eurasian Soil Sci.* **2015**, *48*, 981–990. [CrossRef]
9. Brose, D.A.; Hundal, L.S.; Oladeji, O.O.; Kumar, K.; Granato, T.C.; Cox, A.; Abedin, Z. Greening a steel mill slag brownfield with biosolids and sediments: A case study. *J. Environ. Qual.* **2016**, *45*, 53–61. [CrossRef] [PubMed]
10. Vasenev, V.I.; Smagin, A.V.; Ananyeva, N.D.; Ivashchenko, K.V.; Gavrilenko, E.G.; Prokofeva, T.V.; Patlseva, A.; Stoorvogel, J.J.; Gosse, D.D.; Valentini, R. Urban soil's functions: Monitoring, assessment, and management. In *Adaptive Soil Management: From Theory to Practices*; Rakshit, A., Abhilash, P., Singh, H., Ghosh, S., Eds.; Springer: Singapore, 2017; pp. 359–409.
11. Slukovskaya, M.V.; Vasenev, V.I.; Ivashchenko, K.V.; Morev, D.V.; Drogobuzhskaya, S.V.; Ivanova, L.A.; Kremenetskaya, I.P. Technosols on mining wastes in the subarctic: Efficiency of remediation under Cu-Ni atmospheric pollution. *Int. Soil Water Conserv. Res.* **2019**, *7*, 297–307. [CrossRef]
12. Séré, G.; Schwartz, C.; Ouvrard, S.; Sauvage, C.; Renat, J.C.; Morel, J.L. Soil construction: A step for ecological reclamation of derelict lands. *J. Soils Sediments* **2008**, *8*, 130–136. [CrossRef]
13. Burghardt, W.; Morel, J.L.; Zhang, G.L. Development of the soil research about urban, industrial, traffic, mining and military areas (SUITMA). *Soil Sci. Plant Nutr.* **2015**, *61*, 3–21. [CrossRef]
14. Kumar, K.; Hundal, L.S. Soil in the city: Sustainably improving urban soils. *J. Environ. Qual.* **2016**, *45*, 2–8. [CrossRef] [PubMed]
15. Vidal-Beaudet, L.; Schwartz, C.; Séré, G. Using Wastes for Fertile Urban Soil Construction—The French Research Project SITERRE. In *Soils within Cities*; Levin, M.J., Kim, K.-H.J., Morel, J.L., Burghardt, W., Charzynski, P., Shaw, R.K., IUSS Working Group SUITMA, Eds.; Schweizerbart Science Publishers: Stuttgart, Germany, 2017; pp. 159–168.
16. Municipality of Parma. *Municipal Regulation of Public and Private Green Areas*; Municipality of Parma: Parma, Italy, 2016. (In Italian)
17. Malenkovska Todorova, M.; Donceva, R.; Bunevska, J. Road Regulation and Functional Classification of the Urban Streets of Rome Capital. *Transp. Probl.* **2009**, *4*, 97–104.
18. Fourvel, G.; Vidal-Beaudet, L.; le Bocq, A.; Brochier, V.; Théry, F.; Landry, D.; Kumarasamy, T.; Cannavo, P. Early structural stability of fine dam sediment in soil construction. *J. Soils Sediments* **2018**, *18*, 2647–2663. [CrossRef]
19. Morel, J.L.; Chenu, C.; Lorenz, K. Ecosystem services provided by soils of urban, industrial, traffic, mining, and military areas (SUITMAs). *J. Soils Sediments* **2015**, *15*, 1659–1666. [CrossRef]

20. Vasenev, V.I.; van Oudenhoven, A.P.E.; Romzaykina, O.N.; Hajiaghaeva, R.A. The ecological functions and ecosystem services of urban and technogenic soils: From theory to practice (a review). *Eurasian Soil Sci.* **2018**, *51*, 1119–1132. [CrossRef]
21. Saccá, M.L.; Caracciolo, A.B.; di Lenola, M.; Grenni, P. Ecosystem services provided by soil microorganisms. In *Soil Biological Communities and Ecosystem Resilience*; Springer: Cham, Switzerland, 2017. [CrossRef]
22. Frey, S.D.; Drijber, R.; Smith, H.; Melillo, J. Microbial biomass, functional capacity, and community structure after 12 years of soil warming. *Soil Biol. Biochem.* **2008**, *40*, 2904–2907. [CrossRef]
23. Diakhaté, S.; Gueye, M.; Chevallier, T.; Diallo, N.H.; Assigbetse, K.; Abadie, J.; Diouf, M.; Masse, D.; Sembène, M.; Ndour, Y.B.; et al. Soil microbial functional capacity and diversity in a millet-shrub intercropping system of semi-arid senegal. *J. Arid. Environ.* **2016**, *129*, 71–79. [CrossRef]
24. Kumaresan, D.; Cross, A.T.; Moreira-Grez, B.; Kariman, K.; Nevill, P.; Stevens, J.; Allcock, R.J.N.; O'Donnell, A.G.; DIxon, K.W.; Whiteley, A.S. Microbial functional capacity is preserved within engineered soil formulations used in mine site restoration. *Sci. Rep.* **2017**, *7*, 1–9. [CrossRef] [PubMed]
25. Joergensen, R.G.; Emmerling, C. Methods for evaluating human impact on soil microorganisms based on their activity, biomass, and diversity in agricultural soils. *J. Plant Nutr. Soil Sci.* **2006**, *169*, 295–309. [CrossRef]
26. Ritz, K.; Black, H.I.J.; Campbell, C.D.; Harris, J.A.; Wood, C. Selecting biological indicators for monitoring soils: A framework for balancing scientific and technical opinion to assist policy development. *Ecol. Indic.* **2009**, *9*, 1212–1221. [CrossRef]
27. Marinari, S.; Bonifacio, E.; Moscatelli, M.C.; Falsone, G.; Antisari, L.V.; Vianello, G. Soil development and microbial functional diversity: Proposal for a methodological approach. *Geoderma* **2013**, *192*, 437–445. [CrossRef]
28. Moscatelli, M.C.; Secondi, L.; Marabottini, R.; Papp, R.; Stazi, S.R.; Mania, E.; Marinari, S. Assessment of soil microbial functional diversity: Land use and soil properties affect CLPP-MicroResp and enzymes responses. *Pedobiologia* **2018**, *66*, 36–42. [CrossRef]
29. Escalas, A.; Hale, L.; Voordeckers, J.W.; Yang, Y.; Firestone, M.K.; Alvarez-Cohen, L.; Zhou, J. Microbial functional diversity: From concepts to applications. *Ecol. Evol.* **2019**, *9*, 12000–12016. [CrossRef] [PubMed]
30. Marfenina, O.E.; Makarova, N.V.; Ivanova, A.E. Opportunistic fungi in soils and surface air of a megalopolis (for the Tushino Region, Moscow). *Microbiology* **2011**, *80*, 870–876. [CrossRef]
31. Korneykova, M.V.; Myazin, V.A.; Fokina, N.V.; Chaporgina, A.A. Bioremediation of soil of the kola peninsula (Murmansk region) contaminated with diesel fuel. *Geogr. Environ. Sustain.* **2021**, *14*, 171–176. [CrossRef]
32. Stroganova, M.N.; Myagkova, A.D.; Prokof'eva, T.V. The role of soils in urban ecosystems. *Eurasian Soil Sci.* **1997**, *30*, 82–86.
33. Prokofyeva, T.V.; Martynenko, I.A.; Ivannikov, F.A. Classification of Moscow soils and parent materials and its possible inclusion in the classification system of Russian soils. *Eurasian Soil Sci.* **2011**, *44*, 561–571. [CrossRef]
34. Prokhorov, I.S.; Karev, S.Y. Particularities of artificial soil-ground production for landscape and shade gardening. *Agrochem. Her.* **2012**, *3*, 21–25. (In Russian)
35. Alexandrovskiy, A.L.; Dolgikh, A.V.; Alexandrovskaya, E.I. Pedogenetic features of habitation deposits in ancient towns of european russia and their alteration under different natural conditions. *Bol. De La Soc. Geol. Mex.* **2012**, *64*, 71–77. [CrossRef]
36. Vasenev, V.I.; Stoorvogel, J.J.; Dolgikh, A.V.; Ananyeva, N.D.; Ivashchenko, K.V. Changes in soil organic carbon stocks by urbanization. In *Advance in Soil Science "Urban Soils"*; Lal, R., Stewart, B.A., Eds.; CRC Press Taylor and Francis Group: New York, NY, USA, 2017; pp. 61–93.
37. Anderson, J.P.E.; Domsch, K.H. A physiological method for the quantitative measurement of microbial biomass in soils. *Soil Biol. Biochem.* **1978**, *10*, 215–221. [CrossRef]
38. Ananyeva, N.D.; Susyan, E.A.; Chernova, O.V.; Wirth, S. Microbial respiration activities of soils from different climatic regions of European Russia. *Eur. J. Soil Biol.* **2008**, *44*, 147–157. [CrossRef]
39. Creamer, R.E.; Schulte, R.P.O.; Stone, D.; Gal, A.; Krogh, P.H.; lo Papa, G.; Murray, P.J.; Pérès, G.; Foerster, B.; Rutgers, M.; et al. Measuring basal soil respiration across europe: Do incubation temperature and incubation period matter? *Ecol. Indic.* **2014**, *36*, 409–418. [CrossRef]
40. Kroetsch, D.; Wang, C. *Soil Sampling and Methods of Analysis*, 2nd ed.; Carter, M.R., Gregorich, E.G., Eds.; CRC Press: Boca Raton, FL, USA, 2007. [CrossRef]
41. ISO. *Soil quality–laboratory methods for determination of microbial soil respiration*; International Organization for Standardization: Geneva, Switzerland, 2002.
42. Dilly, O. Microbial energetics in soils. In *Microorganisms in Soils: Roles in Genesis and Functions*; Springer: Berlin, Heidelberg, 2005. [CrossRef]
43. Campbell, C.D.; Chapman, S.J.; Cameron, C.M.; Davidson, M.S.; Potts, J.M. A rapid microtiter plate method to measure carbon dioxide evolved from carbon substrate amendments so as to determine the physiological profiles of soil microbial communities by using whole soil. *Appl. Environ. Microbiol.* **2003**, *69*, 3593–3599. [CrossRef] [PubMed]
44. Shannon, C.; Weaver, W. *The mathematical theory of communication*; The University of Illinois Press: Champaign, IL, USA, 1964; 131p.
45. Davet, P.; Rouxel, F. *Detection and Isolation of Soil Fungi*; Science Publishers: Plymouth, UK, 2000.
46. Acea, M.J.; Carballas, T. Changes in physiological groups of microorganisms in soil following wildfire. *FEMS Microbiol. Ecol.* **1996**, *20*, 33–39. [CrossRef]
47. Marfenina, O.E.; Danilogorskaya, A.A. Effect of elevated temperatures on composition and diversity of microfungal communities in natural and urban boreal soils, with emphasis on potentially pathogenic species. *Pedobiologia* **2017**, *60*, 11–19. [CrossRef]
48. Seifert, K.A. Compendium of soil fungi—by Domsch, K.H.; Gams, W.; Anderson, T.H. *Eur. J. Soil Sci.* **2008**, *59*, 1007. [CrossRef]

49. De Hoog, G.S.; Guarro, J.; Gené, J.; Ahmed, S.A.; Al-Hatmi, A.M.S.; Figueras, M.J.; Vitale, R.G. *Atlas of Clinical Fungi*, 4th ed.; Centraalbureau voor Schimmelcultures: Utrecht, The Netherlands, 2019.
50. Thiele-Bruhn, S.; Schloter, M.; Wilke, B.M.; Beaudette, L.A.; Martin-Laurent, F.; Cheviron, N.; Mougin, C.; Römbke, J. Identification of new microbial functional standards for soil quality assessment. *SOIL* **2020**, *6*, 17–34. [CrossRef]
51. Legendre, P.; Gallagher, E.D. Ecologically meaningful transformations for ordination of species data. *Oecologia* **2001**, *129*, 271–280. [CrossRef]
52. TeamCore, R.C. *Language and Environment for Statistical Computing: R Foundation for Statistical Computing*; The R Development Core Team: Vienna, Austria, 2018.
53. Wickham, H. *Ggplot2 Elegant Graphics for Data Analysis (Use R!)*; Springer: New York, NY, USA, 2016.
54. Vasenev, V.I.; Stoorvogel, J.J.; Vasenev, I.I. Urban soil organic carbon and its spatial heterogeneity in comparison with natural and agricultural areas in the Moscow region. *Catena* **2013**, *107*, 96–102. [CrossRef]
55. Ivashchenko, K.; Ananyeva, N.; Vasenev, V.; Sushko, S.; Seleznyova, A.; Kudeyarov, V. Microbial C–availability and organic matter decomposition in urban soils of megapolis depend on functional zoning. *Soil Environ.* **2019**, *38*, 31–41. [CrossRef]
56. Darmody, R.G.; Marlin, J.C.; Talbott, J.; Green, R.A.; Brewer, E.F.; Stohr, C. Dredged Illinois river sediments: Plant growth and metal uptake. *J. Environ. Qual.* **2004**, *33*, 458–464. [CrossRef] [PubMed]
57. Vasbieva, M.T.; Kosolapova, A.I. Changes in fertility parameters and contents of heavy metals of soddy-podzolic soils upon the long-term application of sewage sludge. *Eurasian Soil Sci.* **2015**, *48*, 518–523. [CrossRef]
58. Plekhanova, I.O. Self-purification of agrosoddy-podzolic sandy loamy soils fertilized with sewage sludge. *Eurasian Soil Sci.* **2017**, *50*, 491–497. [CrossRef]
59. Shchegolkova, N.M.; Smagin, A.V.; Rybka, K.Y. Methodological aspects of soil engineering: Agrophysical properties (in Russian). *Voda Khimiya Ekol.* **2013**, *7*, 9–17.
60. HS-514-11 Moscow Government Document Regulates Quality of the Materials Applied for Greening in Moscow. 2019.
61. Khrenov, K.E.; Kozlov, M.N.; Shchegolkova, N.M.; Vanyushina, A.Y.; Grachev, V.A. Properties investigation of the created soils made from sediments of water treatment station (in Russian). *Water Supply Sanit. Tech.* **2011**, *10*, 20–25.
62. Kitir, N.; Yildirim, E.; Şahin, Ü.; Turan, M.; Ekinci, M.; Ors, S.; Kul, R.; Ünlü, H.; Ünlü, H. Peat use in horticulture. In *Peat*; IntechOpen: London, UK, 2018. [CrossRef]
63. Shchepeleva, A.S.; Vasenev, V.I.; Mazirov, I.M.; Vasenev, I.I.; Prokhorov, I.S.; Gosse, D.D. Changes of soil organic carbon stocks and CO_2 emissions at the early stages of urban turf grasses' development. *Urban Ecosyst.* **2017**, *20*, 309–321. [CrossRef]
64. Smagin, A.V.; Sadovnikova, N.B.; Vasenev, V.I.; Smagina, M.V. Biodegradation of some organic materials in soils and soil constructions: Experiments, modeling and prevention. *Materials* **2018**, *11*, 1889. [CrossRef] [PubMed]
65. Vasenev, V.; Varentsov, M.; Konstantinov, P.; Romzaykina, O.; Kanareykina, I.; Dvornikov, Y.; Manukyan, V. Projecting urban heat island effect on the spatial-temporal variation of microbial respiration in urban soils of moscow megalopolis. *Sci. Total. Environ.* **2021**, *786*, 147457. [CrossRef]
66. Kelly, J.J.; Favila, E.; Hundal, L.S.; Marlin, J.C. Assessment of soil microbial communities in surface applied mixtures of Illinois river sediments and biosolids. *Appl. Soil Ecol.* **2007**, *36*, 176–183. [CrossRef]
67. Alexandrovskaya, E.I.; Alexandrovskiy, A.L. History of the cultural layer in moscow and accumulation of anthropogenic substances in it. *Catena* **2000**, *41*, 249–259. [CrossRef]
68. Ivashchenko, K.V.; Ananyeva, N.D.; Vasenev, V.I.; Kudeyarov, V.N.; Valentini, R. Biomass and respiration activity of soil microorganisms in anthropogenically transformed ecosystems (Moscow region). *Eurasian Soil Sci.* **2014**, *47*, 892–903. [CrossRef]
69. Ovsepyan, L.; Kurganova, I.; de Gerenyu, V.L.; Kuzyakov, Y. Recovery of organic matter and microbial biomass after abandonment of degraded agricultural soils: The influence of climate. *Land Degrad. Dev.* **2019**, *30*, 1861–1874. [CrossRef]
70. Shahbaz, M.; Kuzyakov, Y.; Sanaullah, M.; Heitkamp, F.; Zelenev, V.; Kumar, A.; Blagodatskaya, E. Microbial decomposition of soil organic matter is mediated by quality and quantity of crop residues: Mechanisms and thresholds. *Biol. Fertil. Soils* **2017**, *53*, 287–301. [CrossRef]
71. Karlen, D.L.; Mausbach, M.J.; Doran, J.W.; Cline, R.G.; Harris, R.F.; Schuman, G.E. Soil quality: A concept, definition, and framework for evaluation (a guest editorial). *Soil Sci. Soc. Am. J.* **1997**, *61*, 4–10. [CrossRef]
72. Doran, J.W.; Zeiss, M.R. Soil health and sustainability: Managing the biotic component of soil quality. *Appl. Soil Ecol.* **2000**, *15*, 3–11. [CrossRef]
73. Aqua Engineering, Inc. *City of Evans Standard: Lawn and Grass Specification Standard-02930*; Aqua Engineering, Inc.: City of Evans, CO, USA, 2000.
74. British Standards Institution. *BS 3882 Specification for Topsoil*; British Standards Institution: London, UK, 2015.
75. ISO. *ISO 14240-1 Soil Quality—Determination of Soil Microbial Biomass—Part 1: Substrate-Induced Respiration Method*; International Organization for Standardization: Geneva, Switzerland, 1997.
76. ISO. *ISO/TS 22939 (2019): Soil Quality—Measurement of Enzyme Activity Patterns in Soil Samples Using Fluorogenic Substrates in Micro-Well Plates*; International Organization for Standardization: Geneva, Switzerland, 2019.

77. ISO. *ISO 14238 Soil Quality—Biological Methods—Determination of Nitrogen Mineralization and Nitrification in Soils and the Influence of Chemicals on These Processes*; International Organization for Standardization: Geneva, Switzerland, 2012.
78. Hafeez, F.; Martin-Laurent, F.; Béguet, J.; Bru, D.; Cortet, J.; Schwartz, C.; Morel, J.L.; Philippot, L. Taxonomic and functional characterization of microbial communities in technosols constructed for remediation of a contaminated industrial wasteland. *J. Soils Sediments* **2012**, *12*, 1396–1406. [CrossRef]
79. Vidal-Beaudet, L.; Galopin, G.; Grosbellet, C. Effect of organic amendment for the construction of favourable urban soils for tree growth. *Eur. J. Hortic. Sci.* **2018**, *83*, 173–186. [CrossRef]

Article

Influence of Leguminous Cover Crops on Soil Chemical and Biological Properties in a No-Till Tropical Fruit Orchard

Ariel Freidenreich [1,2], Sanku Dattamudi [1], Yuncong Li [2,3] and Krishnaswamy Jayachandran [1,*]

[1] Department of Earth and Environment, Florida International University, Miami, FL 33199, USA; afreidenreich@ufl.edu (A.F.); sdattamu@fiu.edu (S.D.)
[2] Department of Soil, Water, and Ecosystem Sciences, University of Florida, Gainesville, FL 32611, USA; yunli@ufl.edu
[3] Tropical Research and Education Center, University of Florida, Homestead, FL 33031, USA
* Correspondence: jayachan@fiu.edu; Tel.: +1-(305)-348-6553

Abstract: South Florida's agricultural soils are traditionally low in organic matter (OM) and high in carbonate rock fragments. These calcareous soils are inherently nutrient-poor and require management for successful crop production. Sunn hemp (SH, *Crotalaria juncea*) and velvet bean (VB, *Mucuna pruriens*) are highly productive leguminous cover crops (CCs) that have shown potential to add large quantities of dry biomass to nutrient- and organic-matter-limited systems. This study focuses on intercropping these two CCs with young carambola (*Averrhoa carambola*) trees. The objective was to test the effectiveness of green manure crops in providing nutrients and supplementing traditional fertilizer regimes with a sustainable soil-building option. Typically, poultry manure (PM) is the standard fertilizer used in organic or sustainable production in the study area. As such, PM treatments and fallow were included for comparison. The treatments were fallow control (F), fallow with PM (FM), sunn hemp (SH), SH with PM (SHM), velvet bean (VB), and VB with PM (VBM). Sunn hemp and VB were grown for two summer growing seasons. At the end of each 90-day growing period, the CCs were terminated and left on the soil surface to decompose in a no-till fashion. The results suggest that SH treatments produced the greatest amount of dry biomass material ranging from 48 to 71% higher than VB over two growing seasons. As a result, SH CCs also accumulated significantly higher amounts of total carbon (TC) and total nitrogen (TN) within their dry biomass that was added to the soil. Sunn hemp, SHM, and FM treatments showed the greatest accumulation of soil OM, TC, and TN. Soil inorganic N ($NH_4^+ + NO_3^- + NO_2$) fluctuated throughout the experiment. Our results indicate that generally, VB-treated soils had their highest available N around 2 months post termination, while SH-treated soils exhibited significantly higher N values at CC termination time. Sunn hemp + PM (SHM) treatments had highest soil N availability around 4 months after CC termination. Soil enzyme activity results indicate that at CC termination, SHM exhibited the highest levels of β-1-4- glucosidase and β-N-acetylglucosaminidase among all treatments. Overall, SH, SHM, and FM treatments showed the greatest potential for supplementing soil nutrients and organic matter in a no-till fruit production setting.

Keywords: sunn hemp (*Crotalaria juncea*); velvet bean (*Mucuna pruriens*); carambola (*Averrhoa carambola*); soil health; soil enzyme activity

1. Introduction

Conservation agriculture, a sustainable approach to maximize crop production while preserving environmental quality, has become an increasingly popular subject of research. This is largely attributable to the lack of applicable information regarding sustainable agriculture components and their harmony in organic agroecosystems. Cover cropping is a widely recognized conservation agricultural strategy [1–3], specifically for low-input agricultural systems. The soil ecosystem services provided by cover crops are substantial, specifically in the form of organic carbon additions [4]. No-till (NT) farming, combined with cover cropping, is often considered as an ideal system for judicious resource utilization in maximizing return on investment (ROI) and maintaining soil biodiversity [5]. Fruit orchards are traditionally NT systems once trees are planted, as mechanical tillage would cause damage to surface feeder roots that remain in the top centimeters of the soil. Therefore, soil surrounding perennial tree crops is often left unstimulated, leaving trees to ultimately experience a reduction in productivity, as crop rotation and tillage is not possible in most cases [6]. Perennial NT systems can significantly benefit from cover crop species that produce large amounts of dry biomass to achieve high C inputs to soil [7].

In tropical fruit production settings, farmers face various challenges which stem from land management in warm and wet climates. This phenomenon is reflected through quick decomposition of organic matter [8–10] and higher pressure from pests and diseases [11]. Leguminous cover crops, specifically varieties suited for tropical climates, have great potential to ameliorate these issues by improving soil resilience and enhancing farmland diversity [12]. Sunn hemp (SH, *Crotalaria juncea*) and velvet bean (VB, *Mucuna pruriens*), two commonly used leguminous cover crops, have been shown to fix 40 to 80 kg N ha^{-1} in tropical climates [13,14]. No-till organic ecosystems can significantly benefit from cover crop species such as SH and VB which produce large amounts of dry biomass to achieve high C inputs to soil [7]. Previous studies found that SH and VB can significantly increase soil C and N fractions, improve soil aggregate stability, and influence the abundance of beneficial soil microbes in no-till tropical production [15–18]). Consequently, it is of great importance to study soil nutrient cycling in harmony with soil microbiota. Microbial functional diversity is a driver for a plethora of ecological and environmental interactions [19]. Soil enzymes are directly related to soil microbial activity and overall soil fertility [20], making them a dynamic indicator for the effectiveness of soil amendments in stimulating nutrient cycling and mineralization rates.

The popularity of organic agriculture has increased more than 550% worldwide within the last couple of decades [21]. Consumers are more interested in 'healthy food habits' and relate that concept to products coming from organic farms. Although organic consumables are becoming increasingly more available, sustainable production of organic commodities can be challenging. Additionally, due to rapid integration of organic farming into large commercial food-production systems, small and medium-size organic growers are struggling to make a minimum profit. Certified organic farms in the US saw a ~39% increase from 2012 to 2017, while Florida's organic farms only increased by ~4.5% within the same time frame [22]. However, the numbers of small and mid-size farmers in South Florida are decreasing, and currently, fewer than ten certified organic vegetable growers can be found in Miami–Dade [23], the major fruit- and vegetable-producing county in South Florida. In addition, soils in South Florida are predominantly porous sandy loam with very low organic matter content (less than 2%, [24]), which often causes production problems for the local growers.

Cover-cropping practices combined with low-cost organic nutrient sources could potentially help small and mid-size farmers achieve better economic return and promote environmental and economic sustainability. Composted poultry manure (PM) is commonly applied as a fertilizer in organic production systems [25,26]. Nutrients such as C and N in poultry manure are in organic forms [27,28] and are released slowly as the materials decompose. This process is similar to that employed in commercially available slow-release fertilizers that have become standard for reducing nutrient leaching and protecting water quality.

Carambola (*Averrhoa carambola*), more commonly known as starfruit, is a tropical fruit tree native to Southeast Asia. The carambola tree is small to medium in height and produces fruit mainly in the mid-canopy area. Carambola is accustomed to hot, humid weather, making it ideal for growth in (sub)tropical climates, and consequently, South Florida is the only location in the contiguous US where carambola is produced commercially [29]. Carambola production has been estimated to contribute ~$3.7 million to Florida's economy, most of which comes from South Florida [30]. Carambola has huge potential for South Florida growers as a lucrative cash crop. For many years, the avocado has been the staple tropical fruit crop for Miami–Dade County growers (over 6000 planted acres, [31]). However, within the last decade, the aggressive emergence of a devastating fungal pathogen, commonly known as laurel wilt (*Raffaelea lauricola*), has led to the mandatory eradication of many infected avocado groves [32–34]. As such, many local growers are looking towards alternative tropical crops to populate their groves. Carambola is a promising candidate, as its current individual tree value is highest for all maturity increments (1–3 years: $567, 4–6 years: $860, 7+ years: $984) as compared to avocado and other feasible alternatives [35].

In an effort to target current concerns within the Miami–Dade County agricultural scope and to explore solutions to improve management practices for sustainable production of tropical fruits worldwide, we developed a study to test the effectiveness of SH and VB as cover crops. The goal was to quantify the response of dynamic soil characteristics to cover crop incorporation in a young carambola grove by exploring responsive soil parameters. The specific objectives of this 2-season study were to (1) assess carbon and nitrogen inputs from cover crops and poultry manure incorporated into an organic carambola grove, (2) monitor physiochemical soil responses to these inputs in a no-till setting, and (3) assess soil enzymatic activity in response to these added amendments.

2. Materials and Methods

2.1. Site Location and Characteristics

This 2-season (May 2018 to December 2019) field experiment took place in a certified organic (as listed by USDA-AMS) fruit orchard (6.07 ha) located in the Redlands Agricultural Area (RAA) of South Florida, United States. The RAA is subtropical and located in plant hardiness zone 10b [36] (USDA, 2012). This subtropical climate is characterized by a typical wet summer season (May–October, 26.6 °C average temperature, and 18 cm average annual rainfall) and a dry winter season (November–April, 21.1 °C average temperature, 4.6 cm average annual rainfall) with warm weather year-round (23.6 °C average temperature) [37].

The USDA NRCS National Cooperative Soil Survey categorizes the soil in the RAA as Krome series soil, high in calcium carbonate ($CaCO_3$) rock fragments and generally recognized as gravelly loam [38]. The soil profile is shallow, with a plowed layer ranging from 0 to 18 cm. The limestone parent material has resulted in well-drained and slightly alkaline soil. As a result of this shallow soil profile, rock plowing is a common practice in agricultural fields to create enough depth (10–20 cm) for root growth and establishment. In addition to rock plowing, tropical fruit managers typically trench their land (46 to 61 cm deep and 41 to 46 cm wide) to ensure enough depth for tree root growth and anchoring for protection during tropical cyclones [39].

2.2. Experimental Design

One year before the start of the experiment, two rows of young carambola trees were planted, extending 122 m long with 7 m spacing between rows (Figure 1). The trees used for this study were ~three-year-old 'Hawaiian Super Sweet' trees grafted onto 'Golden Star' seedling rootstocks. Carambola saplings were planted with 3.8 m between each tree, resulting in 30 trees per row, and 54 trees were randomly selected for treatment. Experimental sites were arranged in a completely randomized design (CRD) with two cover crop treatments: sunn hemp (SH) and velvet bean (VB); two cover crop + manure treatments: sunn hemp + poultry manure (SHM) and velvet bean + poultry manure (VBM); and two fallow control treatments: fallow (F) and fallow + poultry manure (FM). The design included six treatments with nine replications for each treatment, involving a total of 54 trees. Twenty-seven trees were treated with an organic composted fertilizer amendment (5N-3P-2K USDA Organic Certified poultry manure), and the other 27 trees did not receive any fertilizer treatments. Details about the experimental timeframe and treatments can be found in our other published work [40].

Figure 1. (**A**) The location of the experiment was a multi-use tropical fruit grove (experimental area highlighted in red), (**B**) seeding cover crops around carambola trees in a circular fashion, (**C**) velvet bean established in experimental plots, (**D**) sunn hemp established in experimental plots.

2.3. Field Sampling and Laboratory Analyses

2.3.1. Field Methodology

Weeds were physically removed from each plot before the start of the experiment in preparation for cover crop seeding. A planting area of 8.8 m^2 was established, and CCs were seeded directly in a concentric pattern starting at the dripline (approx. 0.5 m radius from the trunk) and circling around the tree in a 1.25 m radius (Figure 1). Carambola trees treated with cover crops received either 33 kg ha^{-1} (89 g/plot) of SH (*Crotalaria juncea* L. cv. 'Tropic Sun') seed or 25 kg ha^{-1} (67 g/plot) VB (*Mucuna pruriens* var. *'pruriens'*) seed. Seeding rates were calculated following the Miami–Dade County Extension recommendations for CC seeding in vegetable crop scenarios and adjusted to a 33% grove coverage rate. The CC coverage used for this study was determined based on size of trees, spacing, and management equipment. Cover crop seeds were treated with OMRI-certified Guard'n Seed Inoculant (Verdesian Life Sciences, Cary, NC, USA), which contains a variety of rhizobium species (*Bradyrhizobium japonicum, Bradyrhizobium sp. (Vigna), Rhizobium leguminorsarum biovar viceae,* and *Rhizobium leguminosarum biovar phaseoli*), to facilitate root nodulation and N fixation.

Sunn hemp and VB treatments were planted simultaneously. All plots were irrigated for one hour per day via sprinkler system. Sunn hemp and VB were terminated 90 days after germination. Sunn hemp was terminated mechanically via hedge trimmer, and VB was hand clipped, leaving root systems intact. The cover crop biomass was laid around the base of each respective tree to decompose on the soil surface. Following termination, fertilizer treatments (1.4 kg poultry manure per tree) were applied every two months except for the CC growing season (4 times per year), resulting in 120 kg ha^{-1} N added per year from PM, as per recommendation by Crane, 2001 [41]. The carbon content of the poultry manure was 33.76 ± 0.29%, and the N content was 5.53 ± 0.13%. The cover crop treatments were planted and terminated two times over this study period. Individual treatments for each tree remained the same for both years.

Composite soil samples (0–15 cm depth, four per plot per sampling time) were collected over the course of the experiment from the planting area of each tree. Soil samples were taken before cover crop planting, before termination, and every 2 months following, except for the last 4 months in which sampling occurred once per month. Temperature, moisture percentage, and electrical conductivity (EC) (STEVENS Hydraprobe, Portland, OR, USA) were measured once per month and at corresponding soil-sampling times. At 90 days after cover crop seed germination, a 40 cm^2 area of plant matter (cover crop and weed) was collected from each plot, including control plots. Aboveground biomass was measured to determine organic matter and nutrient additions to the soil.

2.3.2. Soil Physicochemical Properties

Composite soil samples were oven-dried (30 °C for 72 h), sieved (2 mm), and ground to prepare for analysis of chemical properties. Soil organic matter (SOM) was determined through the standard loss-on-ignition method (550 °C for 4 h). Soil total carbon and nitrogen (CN) were measured via dry combustion using a Truspec Carbon/Nitrogen analyzer (LECO Corporation, St. Joseph, MI, USA).

Inorganic N was extracted using a 2M KCl extraction method. Extracts were then analyzed for nitrate (NO_3^-) and nitrite (NO_2^-) following USEPA Nitrate-Nitrite by Automated Colorimetry Method 353.2, Revision 2.0 (1993) [42]. These same extracts were used for ammonium (NH_4^+) determination following USEPA Method 350.1, Revision 2.0 (1993) [43]. All readings were quantified with a SEAL Analytical AQ2 Discrete Auto Analyzer (Mequon, WI, USA).

Soil moisture content was determined via the gravimetric method (dried at 105 °C for 24 h) and bulk density via cylinder method. Soil textural analysis was performed using the standard hydrometer method. All analysis was conducted at Florida International University within the Soil–Plant–Microbiology Laboratories (Miami, FL, USA).

2.3.3. Soil Enzyme Activity

Soil enzyme analysis was conducted for determination of β-1-4-glucosidase (C) and β-N-acetylglucosaminidase (N). A methodology adopted from Sinsbaugh et al. (1997), Hoppe (1993), and Chróst and Kambeck (1986) [44–46] was utilized to determine soil enzyme activity using differences in concentration of fluorescent substrate released during incubation time compared with no incubation. Soil slurries with a 2:1 water-to-soil ratio (4 g distilled deionized water to 2 g fresh soil) were made, and pH readings were taken. Substrates were prepared using morpholinoethanesulfonic acid (MES) in combination with 4-methylumbelliferone (MUF) β-D-glucosidase (MUF-C) and MUF-N-acetyl- β-D-glucosaminide (MUF-N). Soil floc was prepared at varying dilutions according to concentration. For C and N, 10^{-2} dilutions were analyzed using a Synergy HT Multi-Mode 96 Well Plate Reader (Biotek Inc., Winooski, VT, USA).

2.3.4. Statistical Analyses

Statistical analyses were conducted using IBM SPSS Statistics for Windows (IBM Corp., 1968. Version 25.0, Armonk, NY, USA) and SAS (SAS Institute Inc., 1976. Base SAS® 9.4. SAS Institute Inc., Cary, NC, USA) software. Data was analyzed via one-way ANOVA to distinguish differences between treatments at each sampling time. Repeated measures (two-way ANOVA) were run to determine significant interactions between individual parameters and sampling time for appropriate groupings. Duncan's post hoc test was used to distinguish differences and considered significant at $p < 0.05$.

3. Results and Discussion

3.1. Background Soil Characteristics, Climatic Conditions, and Cover Crop Contributions

Soil at the experimental site was a sandy loam (73% sand, 17% clay, and 10% slit) with an average pH of 7.6. The experiment was conducted at a certified-organic farm where the average SOM content was ~17%, fairly high compared to the mineral soils (average < 2%; [24]) in the same region. Additionally, at the start of the experiment, inorganic N levels were also higher than expected for the area at 34.24 g kg^{-1}. These values are unusual for soils within the Redland area and can be explained by land use practices by the farm manager. Prior to carambola being planted, mature sapodilla trees were growing in this area. These trees were treated with the same 5–3–2 poultry manure (N–P–K) as used in our experiment at a rate of ~224 N kg ha^{-1} per year, nearly double the amount added to juvenile carambola for our study. Mature trees also contribute leaf litter, which can add supplemental OM and N to soil as it decomposes [47]. As a result, these cultural practices had a great impact on soil health parameters and as such were reflected through enhanced OM and nutrient content within the soil at the start of the experiment and throughout season 1.

The average air and soil temperatures were 26.4 °C and 28.8 °C, respectively, for summer and 21.4 °C and 23.7 °C, respectively, for winter months (Figure 2). Precipitation trends were highest from July to August (18 to 23.7 cm average) in both years (Figure 2), as expected for South Florida, given climatic trends that result in wet summers and dry winters [37]. As such, a reduction in precipitation can be observed beginning in the fall months (September to November) and continuing throughout the year until summer. The first-year cover crop growing season (May 2018 to August 2018) received 64% higher rainfall than the second season (May 2019 to August 2019). Relative humidity (%) remained consistent (range 81 to 86%; average ~83% per month) throughout the experiment except for March 2019 to May 2019, when increased temperature and low rainfall resulted in lower humidity (average 78%).

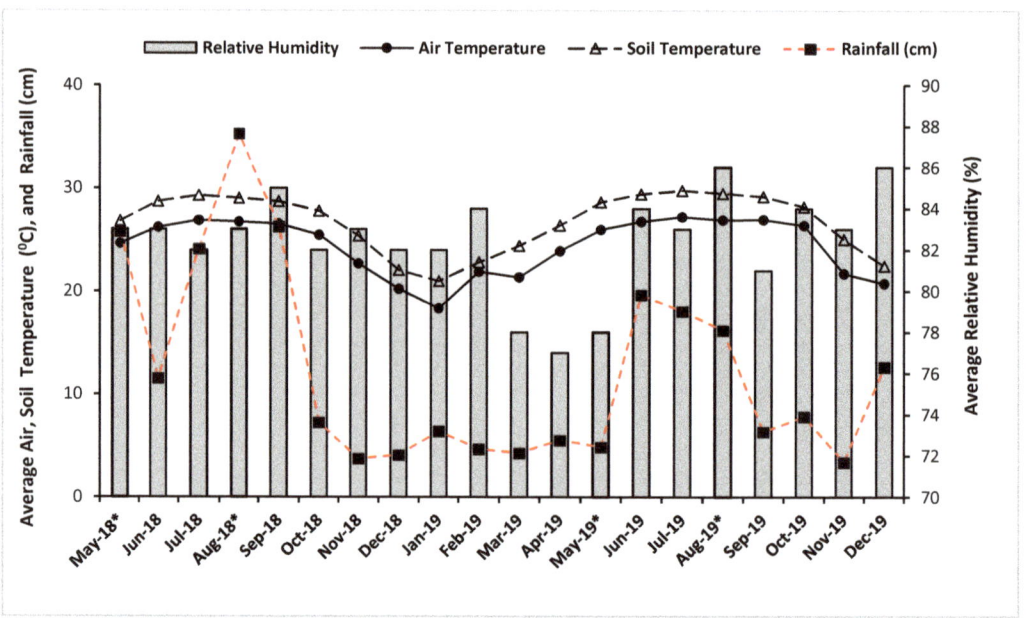

Figure 2. Climatic conditions of the sampling sites over the 1.5-year trial period. This graph represents average relative humidity (%), air temperature (°C), soil temperature (°C), and rainfall (cm). Months with (*) indicate cover crop planting in May and termination in August of each season.

Our results indicate that SH (with or without manure) produced significantly higher ($p < 0.05$) dry shoot biomass (range 10.7 to 5.4 Mg ha^{-1}) than VB (with or without manure; range 5.0 to 1.9 Mg ha^{-1}) for both growing seasons (Table 1). Consequently, SH biomass produced 33% more C than VB treatments throughout the experiment [40]. As a non-wood fiber crop, the SH stem can become strong and woody in its later growth stages, providing higher C additions over time [48]. As such, on average throughout our experiment, SH leaf material had a C:N ratio of 11.05, while stem material had a C:N of 32.44, which explains the higher carbon inputs to the system. Additional information on carbon inputs and accumulation related to cover crop biomass from this study can be found in our previously published work [40].

Table 1. Above-ground cover crop biomass, total N cover crop residue added to soil throughout season 1 and season 2, and total weed biomass. Three cover crop treatments (F = fallow, SH = sunn hemp, and VB = velvet bean) and three manure treatments (FM = fallow + poultry manure, SHM = sunn hemp + poultry manure, and VBM = velvet bean + poultry manure) were analyzed for these parameters. Values within a column followed by different letters denote statistical difference at $p < 0.05$ within the same season.

Treatment	Season 1			Season 2		
	Biomass (Mg ha^{-1})	N (kg ha^{-1})	Weed Biomass (kg ha^{-1})	Biomass (Mg ha^{-1})	N (kg ha^{-1})	Weed Biomass (kg ha^{-1})
F	-	-	9475 a	-	-	6496 a
FM	-	-	7802 a	-	-	6531 a
SH	8.8 a	177 ab	1778 b	9.4 a	213 a	1930 b
SHM	10.7 a	238 a	2322 b	5.4 b	135 b	3302 b
VB	4.5 b	88 c	3928 b	2.7 c	84 bc	2742 b
VBM	5.0 b	136 bc	2384 b	1.9 c	65 c	3156 b

When comparing total N contributed by cover crop dry matter, the SH treatments contributed up to ~92% (238 kg ha^{-1}) more biomass N than VB (88 kg ha^{-1}) treatments in season 1 ($p < 0.05$). Sunn hemp treatments produced up to 106% more biomass N than VB treatments ($p < 0.05$) in season 2. The difference in N accumulation by cover crops was a direct result of biomass production. There was a large variation in biomass production between SH and VB, likely attributable to their growth habits. Once SH is established, it grows and develops vertically, which is conducive to producing high biomass in confined spaces. Velvet bean produces large quantities of biomass through its vining growth habit and spreading surface roots [49]. As our experimental growing area around the carambola trees was limited to 8.8 m^2, VB produced less biomass than it would have if grown in a typical vegetable field setting. When comparing biomass production in a tomato production setting, Wang et al. (2009) [38] found that SH and VB produced similar quantities of biomass for both growing seasons in a 2-year study within the same region. However, when comparing SH and VB in a potted study within the RAA, Wang et al. (2006) [50] found that VB produced less biomass and TC when compared to SH for two consecutive growing seasons, findings that align with our study.

In addition to cover crop biomass rates, Table 1 reports weed biomass in the form of shoot dry matter production. The results reveal that weed biomass was up to 81% higher in the fallow plots where cover crops were not planted (F and FM). Weed biomass quantities reflect the ability of SH and VB to suppress weed growth and proliferation, a common environmental benefit of CC usage [51]. In general, CCs and weeds accumulated greater biomass in season 1 when compared to season 2, as reflected by shoot dry matter quantity (Table 1). Increased biomass in season 1 may have been the result of higher rainfall amounts during the seeding and germination period (May 2018) when compared to season 2 (April 2019) and throughout growing season 1 overall (Figure 2). Additionally, legumes tend to be more successful in non-amended soil. This phenomenon occurs because their nitrogen-fixing capacity is enhanced, correlating directly with plant productivity and, in turn, dry biomass production [52]. This is a possible explanation for lack of CC biomass production in season 2, particularly for SHM and VBM which were treated with PM throughout the previous season.

3.2. Effect of Cover Crops on Soil Chemical Properties

Precise CC management has been shown to improve soil quality through increasing soil organic matter (SOM), providing nutrients to cash crops, and enhancing overall physical, chemical, and biological properties without the addition of synthetic inputs [53]. In our experiment, when considering soil pH within treatments, FM plots had significantly lower pH than all other treatments, remaining slightly alkaline throughout the experiment (7.75–7.83, $p < 0.05$, Table 2). This is likely the result of PM being added directly to the soil surface without the hindrance of CC residue. Conversely, the F plots had the highest pH consistently throughout the experiment (8.03–8.08, $p < 0.05$, Table 2), which can be attributed to the lack of OM residue added to the soil. All CC treatments remained similar to one another within the moderately alkaline range for both seasons, fluctuating from 7.85–8.03 ($p > 0.05$). Organic matter addition from the CCs may have played a role in acidifying soil in the short-term but was not reflected by data grouped by season. Calcareous soils generally have a high pH-buffering capacity; however, long-term fertilization can reduce the calcium carbonate ($CaCO_3$) content of naturally calcareous soils, ultimately lowering the buffering capacity [54]. While there were not obvious changes in soil pH throughout the study period, it is possible that over time, the loss of $CaCO_3$ and the increase of SOM could result in the acidification of soil with cover crop and manure additions.

Table 2. Displays average soil pH, organic matter percentage (SOM%), total carbon (TC), total nitrogen (TN), and C:N ratios throughout the experiment (n = 40 for each treatment). Three cover crop treatments (F = fallow, SH = sunn hemp, and VB = velvet bean) and three manure treatments (FM= fallow + poultry manure, SHM = sunn hemp + poultry manure, and VBM = velvet bean + poultry manure) were analyzed for these parameters. Values within a column followed by different letters denote statistical difference at $p < 0.05$.

Treatment	pH		SOM%		TC (g kg^{-1})		TN (g kg^{-1})		C:N (mol:mol)	
	Season 1	Season 2	Season 1	Season 2	Season 1	Season 2	Season 1	Season 2	Season 1	Season 2
F	8.03 a	8.08 a	15.00 bc	13.99 b	158.63 c	159.81 c	7.08 b	7.10 bc	26.55 a	26.88 ab
FM	7.75 b	7.83 b	17.86 a	16.45 a	183.63 a	181.37 a	9.17 a	8.31 a	23.38 a	24.93 b
SH	7.98 ab	8.03 ab	17.32 ab	15.64 a	175.32 ab	169.67 abc	8.95 a	7.86 ab	24.27 a	25.18 ab
SHM	7.85 ab	7.99 ab	17.45 a	15.48 ab	188.05 a	170.79 abc	9.25 a	7.41 abc	24.55 a	27.37 ab
VB	7.95 ab	7.93 ab	16.31 abc	14.14 b	178.11 ab	173.89 ab	7.70 ab	6.55 c	26.52 a	28.10 a
VBM	7.87 ab	7.87 ab	14.70 c	15.48 ab	166.42 bc	168.94 bc	7.89 ab	6.98 bc	25.38 a	24.91 b

Soil treated with FM and SHM accumulated the highest SOM% (17.45–17.86%, $p < 0.05$, Table 2), while the F and VBM (14.70–15%) treatments exhibited the lowest SOM% within season 1. Total carbon results also reflected this trend for season 1 which aligns with high input of aboveground shoot biomass and PM additions to the soil for the SHM treatment and direct addition of PM to the soil surface as applied in the FM treatment (Tables 1 and 2). The significant contribution of OM to the soil by SH treatments suggests that SH was effective for rapid addition of SOM during season 1, when CC biomass production was highest. Relatively lower SOM% and TC is expected for the F treatment with no manure or CC addition, and that treatment yielded consistently less C throughout the experiment when compared to other treatments. The comparably lower SOM% and TC within the VBM treatment throughout the experiment is likely attributed to a lesser contribution of cover crop biomass (as compared to SH treatments) along with the possibility of high decomposition/volatilization rates resulting from no-till management [55].

Soil organic matter results varied between treatments during season 2, as FM (16.45%) and SH (15.64%) had the highest SOM%, with that of SHM being marginally lower (15.48%,

$p < 0.05$, Table 2). This was also reflected in high soil TC levels for FM when compared to CC treatments. As previously mentioned, it is possible that because FM plots received manure treatments without the hindrance of CC mulch, change in SOM was more obvious in the short-term. Additionally, plots without CCs planted had a significantly higher degree of weed establishment throughout the experiment (Table 1). Mowing of weeds around plots occurred approximately every three months, making it possible for decaying weeds to contribute to soil building and nutrition for both fallow treatments. Fallow land within the F and FM treatments was advantageous for weed growth throughout the study, which may have contributed carbon to the soil through above- and belowground activities. High SOM values in SH plots are based on large contributions of CC biomass during season 2 growth (9.4 Mg ha^{-1}, $p < 0.05$). Overall, in regard to CC treatments, SH- and VB-treated soil were similar in TC to each other, yet lower than the FM treatment for season 2, which may have been a result of reduced CC biomass production from season to season.

Soil TN results from season 1 indicated that soil treated with F (7.08 g kg^{-1}) had the lowest TN. The FM, SH, and SHM treatments had the highest TN ($p < 0.05$, 9.17, 8.95, 9.25 g kg^{-1}, respectively), which corresponds with N inputs from biomass (Table 1). Soil TN was highest within soils treated with FM (8.31 g kg^{-1}) and lowest in those treated with VB (6.55 g kg^{-1}), with all other treatments being similar during season 2. Total N results for VB are likely a reflection of less successful biomass contribution by VB treatments during the second growing season (Table 1). These soil TN findings are consistent with other studies that have shown VB treatments add lower quantities of TN to soil than do SH treatments within the RAA subtropical climate [50,56]. Generally, although not statistically significant, in most cases ($p > 0.05$), a trend can be seen throughout both seasons in which treatments that received PM reflected higher TN.

Throughout season 1, there was no significant difference between treatments for soil C:N ratios ($p > 0.05$). Season 2 showed significantly lower C:N ratios for soils treated with FM (24.93) and VBM (24.91), indicating an environment more conducive to OM breakdown and nutrient cycling by microorganisms when compared to other treatments [57] (Eiland et al., 2001). The FM treatment consistently had the highest contributions of TC and TN to the soil throughout both growing seasons, which coincides with these results.

3.3. Soil Inorganic Nitrogen

Soil nutrient cycling, specifically nitrogen mineralization, is critical to crop health, as plant metabolism and vital processes are fueled by uptake of N [58]. When utilizing legume CCs for the purpose of green manure, residue added to the soil after termination adds N as a food source for soil organisms and subsequent crops [59].

When comparing soil inorganic N in the form of ammonium (NH_4^+) + nitrate (NO_3^-) + nitrite (NO_2) throughout the experiment, significant variability was observed at numerous sampling points and between treatments (Table A1, Figure 3). Soil inorganic N was highest, regardless of treatment, at the October 2018 (38.26 g kg^{-1}) sampling time and lowest in September 2019 (4.03 g kg^{-1}, $p < 0.05$). These results show a trend in which, two months after cover crop termination, inorganic N increases; this can also be observed in season 2, where N increases from September 2019 to October 2019 (Table A1, $p < 0.05$). During both growing seasons, air and soil temperature remained elevated from August–September during termination and throughout the first month of organic matter decomposition. These climatic conditions are characteristic of South Florida, as significant rainfall and high temperatures are conducive to organic matter decomposition and N mineralization [60]. Furthermore, the few months after CC termination are crucial for decomposition and N mineralization, which is apparent, as CC treatments showed significantly higher N at these time points.

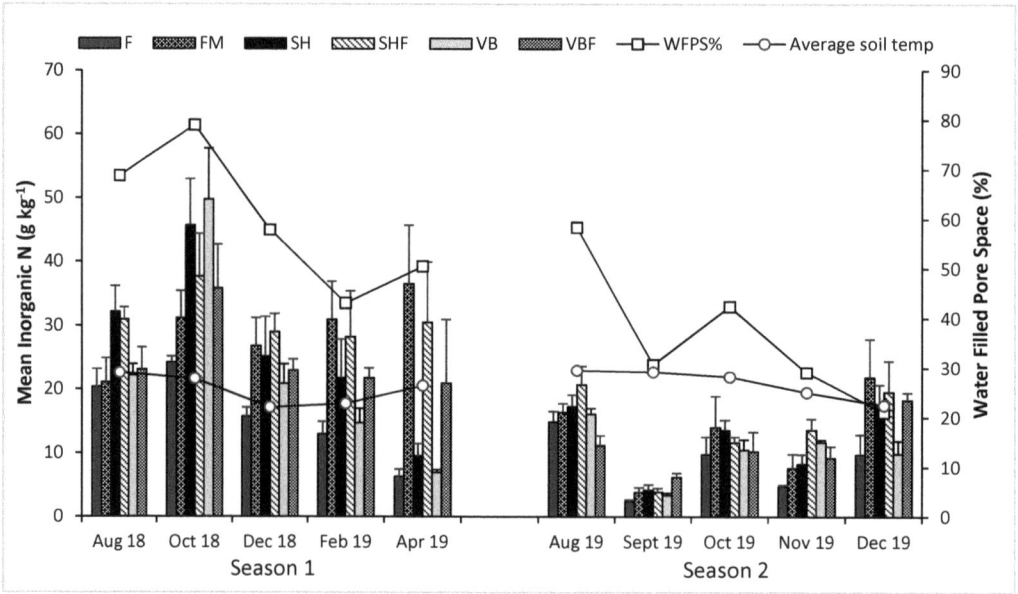

Figure 3. Displays average soil inorganic N concentrations by season (n = 4 for each treatment at each time point) with ambient soil temperature and WFPS. Three cover crop treatments (F = fallow, SH = sunn hemp, and VB = velvet bean) and three manure treatments (FM= fallow + poultry manure, SHM = sunn hemp + poultry manure, and VBM = velvet bean + poultry manure) were analyzed for these parameters. Error bars denote standard error.

At the September 2019 sampling period (1 month after the second CC termination), the average inorganic N was low, corresponding with lower water-filled pore space (WFPS)% as compared to other sampling periods (Figure 3). Water-filled pore space played a significant role in the N mineralization process throughout the experiment. In general, it was observed that as WFPS% increases, so does available N content and vice versa (Figure 3). Soils under no-till management are typically less aerobic and can have higher WFPS than those that undergo traditional tillage [61]. Tillage practices have a large influence on N_2O emissions, which are generally higher in soil under no-till management, as anaerobic conditions are more common for such soil [62]. A water-holding capacity around 60% is the threshold for maximum aerobic activity ideal for ammonification and nitrification [63]. Plant-available N can be compared with WFPS, as previous studies have shown a link between WFPS, soil moisture, and N_2O emission [64–66]. Results indicated that inorganic N content was elevated even when WFPS surpassed the 60% threshold at the October 2018 sampling time, although generally, high inorganic N was observed with WFPS at ~60%, ideal conditions for this parameter. This finding suggests that anerobic bacteria may have played a role in nutrient cycling throughout the study when WFPS% was high.

There are two groups of organisms responsible for N transformations in soils: ammonia-oxidizing bacteria (AOB) and ammonia-oxidizing archaea (AOA) [67,68]. The population size and response of AOA and AOB are highly related to soil type and management strategies. Traditionally, it has been found that AOB are more likely to contribute N additions in agricultural soils, as their populations are generally more elevated when N supply is higher, enhancing nitrification potential, while AOA are more commonly dominant in soils from more natural or diverse ecosystems [69]. Shen et al. (2008) [70] compared the abundance of AOB and AOA communities in an alkaline sandy loam (similar to the tested soil type) with various fertilizer treatments. They found significantly higher communities of AOB in soils treated with traditional N fertilizer when compared to organic manure treatments, which is possibly explained by competition with heterotrophic bacteria, commonly present in soils amended with carbon (green or organic manures) [71]. Although not measured in this study, it is possible that an increased presence of AOA contributed significantly to N cycling in the present experiment, as these organisms are highly adaptable to extreme environmental conditions like low oxygen levels and are more common in diversified soils [72]. Because additions of OM to soil are favorable for microbial diversity, it is likely that the combination of CC inputs, paired with occasional anaerobic soil conditions, created a diversified microbial environment that facilitated N fixation and mobilization.

While WFPS may have played a role in N mineralization, it is probable that BNF had a significant impact on soil inorganic N content at the time of CC termination. At the August 2018 termination time, all soils treated with CCs had significantly higher inorganic N ($p < 0.05$) than the F soils (Table A1). Specifically, both SH and SHM exhibited the highest levels (32.15 and 30.94 g kg^{-1}, respectively). This sampling time is specifically interesting because this was before any PM fertilizer was applied, indicating successful N fixation due to the legume treatments. Legume symbiosis with rhizobium bacteria works to reduce N_2 to NH_{4+} and NO_{3-} in ideal climatic soil conditions [73]. Biological nitrogen fixation is dependent on many factors and can vary by species and effectiveness of rhizobium type/inoculation success [74]. Nezomba et al. (2008) [75] found that Crotalaria spp. had high potential to fix N in sandy soil. Specifically, Crotalaria juncea (SH) was estimated to have a 90% N fixation rate, resulting in 58 kg ha^{-1} N provided to soil. Within our experiment, both SH and VB seeds were inoculated with the recommended cowpea-type rhizobium before planting to encourage nodulation. Genus *Crotalaria* (SH) has been shown to create symbiotic relationships with many strains of rhizobium bacteria, resulting in high potential for N fixation and biomass accumulation [76]. Conversely, much less is known about the genus *Mucuna* (VB) and its rhizobial-host-plant interactions. Cowpea-type rhizobium is compatible with genus *Mucuna*; however, successful nodulation has not been shown with a wide variety of rhizobium species [76]. These factors may have had an impact on the differences in N content in soil between the SH and VB treatments, specifically during and directly after the CC growing seasons (August 2018 and August 2019) in which SH-treated soil showed more success in providing available N.

In addition to termination, there were multiple sampling times in which significant differences in soil inorganic N were observed between treatments. At the October 2018 sampling time, two months after CC termination, VB-treated soil had higher inorganic N (49.74 g kg^{-1}, $p < 0.05$) than the F treatment, with all other treatments being similar to one another. The VB treatment also received higher biomass contribution in season 1 than VBM (~9% higher), which may have contributed to significantly higher levels of inorganic N. It is apparent that at this sampling time (two months after termination and first PM application), N from CC residue and PM application was in various stages of decomposition, and N was being utilized by the carambola tree, as it was readily available. Generally, we saw that VB/VBM residue had a significantly lower C:N ratio than SH [40], which enhanced N cycling from VBM treatments soon after CC incorporation. This pattern can be seen again at the October 2019 sampling time, where the VBM-treated soils had significantly higher levels of inorganic N than all other treatments (6.23, $p < 0.05$), which was a reflection of PM addition and the low C:N ratio of the residue contributing to higher levels of N mineralization.

At the December 2018 sampling time, soils treated with SHM (29.00 g kg^{-1}, $p < 0.05$) were higher in inorganic N than the F-treated soils (15.71 g kg^{-1}), with all other treatments being statistically similar to one another. This result also corresponds to season 2 results at the November 2019 sampling time in which SHM (13.63, $p < 0.05$) soils showed significantly higher N than all other treatments (Table A1). These results indicate that, 3 to 4 months after cover crop termination, plots that received SH and PM combined had the greatest capacity for N mineralization. This corresponds with the findings of Rao and Li (2003) [60], who observed that SH had the highest level of cumulative N mineralization in calcareous Krome soil at 16 to 20 weeks after CC residue was added at ambient South Florida temperature conditions.

At six months (February 2019) and eight months (April 2019) after termination, the impact of CC residue on soil inorganic N begins to taper off, and added PM becomes the main source of plant-available N. At six months after termination (February 2019), the FM-treated soils (30.94 g kg^{-1}, $p < 0.05$) had the highest levels of inorganic N compared to F-treated soil (12.96 g kg^{-1}), with all other treatments being similar to one another (28.23–14.71 g kg^{-1}). However, at the April 2018 sampling time, eight months after termination, a distinct difference can be seen between treatments that received PM and those that did not. Figure 3 shows this distinct phenomenon, in which FM, SHM, and VBM had higher levels of inorganic N (36.49, 30.50, and 20.90 g kg^{-1}, respectively, $p < 0.05$) when compared to treatments that did not receive any PM additions. These results suggest that at six to eight months after termination, CC residue ceases to contribute to N availability in the soils at our research site.

3.4. Soil Enzyme Activity

We chose to study two enzymes, β-1-4-glucosidase, which is associated with C cycling, and β-N-acetylglucosaminidase, which is associated with N cycling [77]. Soil β-1-4-glucosidase rates differed over time and between treatments (Figure 4). Throughout the sampling times, with all treatments considered, the August 2019 time (second CC termination) had the highest rates of β-1-4-glucosidase. There was no significant difference between treatments at individual sampling times except for the August 2018 (the end of cover crop growing season 1) and August 2019 (the end of cover crop growing season 2). At both August times, the SHM treatment had the highest β-1-4-glucosidase level (0.0386 and 0.1393 µmol m 2 s, respectively), with the levels of all other treatments significantly lower ($p < 0.05$), which may be indicative of sunn hemp excreting carbon into the rhizosphere, enhancing overall microbial diversity. Overall, soils treated with SHM reflected the highest numeric β-1-4-glucosidase, although no significant difference was discernable ($p > 0.05$, Figure 4).

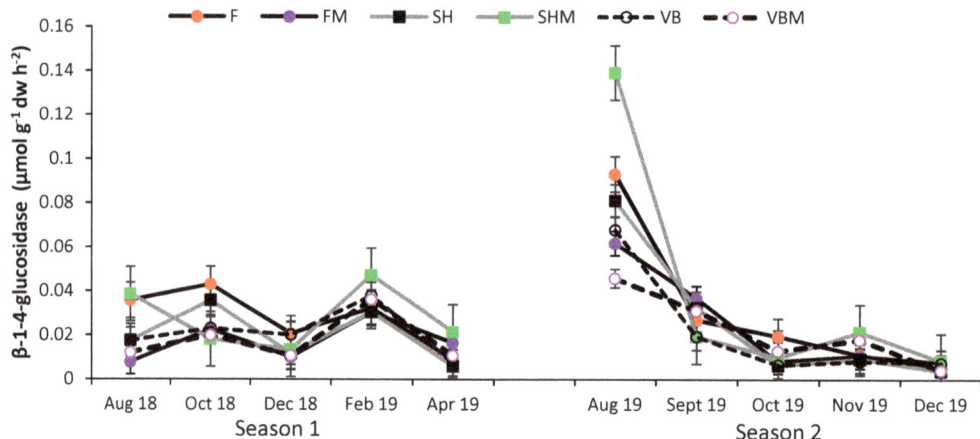

Figure 4. Displays average soil β-1-4-glucosidase rates (n = 4) over a two-season period after cover crop growth and termination. The lines at each sampling time represent three cover crop treatments (F = fallow, SH = sunn hemp, and VB = velvet bean) and three manure treatments (FM= fallow + poultry manure, SHM = sunn hemp + poultry manure, and VBM = velvet bean + poultry manure). Error bars denote standard error.

The enzyme β-N-acetylglucosaminidase reflects N cycling by microbial biomass and overall breakdown of OM [78]. Soil β-N-acetylglucosaminidase results indicated change over time, regardless of treatment. Like β-1-4-glucosidase, the highest occurrences were in August 2019 (0.0094 µmol m 2 s), at the second CC termination (Figure 5). There was no significant difference in β-N-acetylglucosaminidase between treatments at any of the individual sampling times except for August 2019. At the August 2019 time, soils treated with SHM (0.0169 µmol m 2 s) showed significantly higher β-N-acetylglucosaminidase activity than the rest of the treatments (Figure 5).

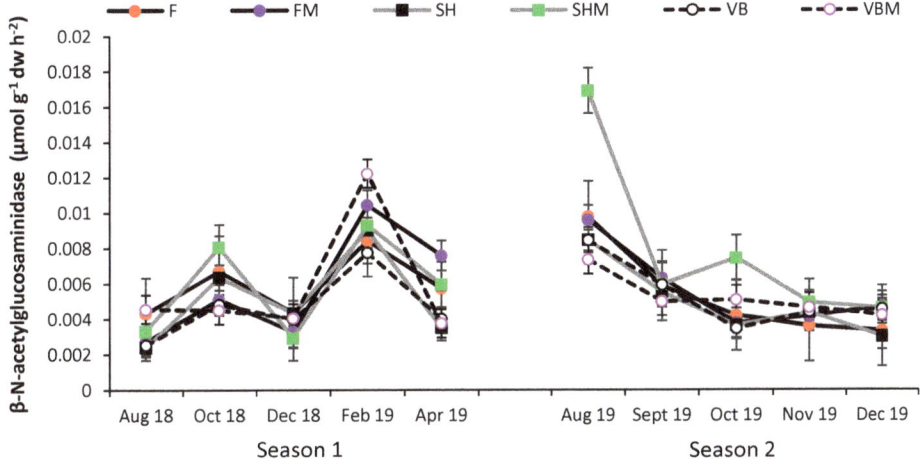

Figure 5. Displays average β-N-acetylglucosaminidase rates (n = 4) in soil over a two-season period after cover crop growth and termination. The lines at each sampling time represent three cover crop treatments (F = fallow, SH = sunn hemp, and VB = velvet bean) and three manure treatments (FM= fallow + poultry manure, SHM = sunn hemp + poultry manure, and VBM = velvet bean + poultry manure). Error bars denote standard error.

These enzyme activity results can be reflective of two processes. After the 90 days of the second growing season, it appears that the SHM treatment was effective in providing belowground stimulation to soil, and as such, the breakdown of glucose and transformation of N occurred, as shown by increased enzyme activity in this sampling period (August 2019). These results coincide with higher concentrations of inorganic N in the corresponding sampling period in which SHM had the highest levels (Figure 3, $p < 0.05$). A study conducted by Maltais-Landry (2014) [79] concluded that legume CCs had high β−glucosidase activity in the rhizosphere by the end of their growing season when compared to non-legumes, which was especially apparent when legume cover crops were combined with composted poultry manure. Our study shows that this is true in a no-till field setting for SH treated with PM (SHM). Shoot and root contributions supplied by cover crops are specifically important in no-till systems, as root exudates added during the growing season and organic material added after termination stimulate soil microbial communities [80].

4. Conclusions

Cover crops have rarely been explored as a soil management method for tropical fruits. This study shows that tropical leguminous cover crops have potential as beneficial soil amendments to add OM and nutrients and promote nutrient cycling by stimulating microbial activity. This experiment was conducted in an organic production farm where cover crops were intercropped with young carambola trees. Carambola trees require ~90–270 N kg ha^{-1} per year [41]. With the seeding rate utilized in this experiment, SH and VB treatments potentially provided sufficient amounts of total dry matter N to supply carambola trees with the N that they require. The results suggest that SH treatments produced the greatest amount of dry biomass material, ranging from 48 to 71% higher than VB treatments over two growing seasons. Consequently, SH also accumulated significantly higher amounts of TC and TN within its dry biomass that was added to the soil. Sunn hemp, SHM, and FM treatments showed the greatest accumulation of soil organic matter, TC, and TN. Soil inorganic N fluctuated throughout the experiment. Our results indicate that, generally, VB-treated soils had their highest available N around 2 months post termination, while SH-treated soils exhibited significantly higher N values at CC termination time. Sunn hemp + PM treatments had highest soil N availability around 4 months after CC termination. Soil enzyme activity results indicate that at CC termination, SHM exhibited the highest β-1-4-glucosidase and β-N-acetylglucosaminidase levels among all treatments.

With all results considered, SH and VB both have the potential to act as soil enhancers for fruit production in tropical and subtropical settings. Applying the findings to tropical fruit production, these cover crops can provide chemical and biological benefits to enhance soil for the successful growth of tropical fruit trees. These cover crops can be utilized in combination with PM or other organic fertilizers for ideal crop development and soil improvement. With the growing issues of soil erosion and organic matter depletion, it is imperative that farmers consider these matters and incorporate management strategies that ensure the long-term sustainable productivity of their land.

Author Contributions: Conceptualization: A.F., Y.L. and K.J.; methodology: A.F., Y.L. and K.J.; formal analysis: A.F.; investigation and data curation: A.F. and S.D.; resources: K.J.; writing—original draft preparation: A.F.; visualization: A.F. and S.D.; writing—review and editing: A.F., S.D., Y.L. and K.J.; funding acquisition: K.J. All authors have read and agreed to the published version of the manuscript.

Funding: This research was funded through FIU's Agroecology Program via the U.S. Department of Agriculture–National Institute of Food and Agriculture, National Needs Fellowship (2015-38420-23702), FIU Tropics and the Susan S. Levine Trust, and the FIU Graduate School Doctoral Evidence Acquisition and Dissertation Year Fellowships.

Institutional Review Board Statement: Not applicable.

Informed Consent Statement: Not applicable.

Data Availability Statement: Not applicable.

Acknowledgments: We would like to acknowledge Marc Ellenby and family for supplying the field site for this experiment.

Conflicts of Interest: The authors declare no conflict of interest.

Appendix A

Table A1. Displays average soil inorganic N concentrations at each sampling time ($n = 4$ for each treatment at each time point). Three cover crop treatments (F = fallow, SH = sunn hemp, and VB = velvet bean) and three manure treatments (FM= fallow + poultry manure, SHM = sunn hemp + poultry manure, VBM = velvet bean + poultry manure) were analyzed for these parameters. Different uppercase letters denote statistical difference between sampling times, different lowercase letters denote statistical differences between treatments at each individual sampling time ($p < 0.05$).

	Soil Inorganic N (g kg^{-1})						
Season 1	Overall	F	FM	SH	SHM	VB	VBM
August 2018	25.48 B	20.32 c	21.04 bc	32.15 a	30.94 ab	22.16 bc	23.08 abc
October 2018	38.26 A	24.22 b	31.15 ab	45.70 ab	37.62 ab	49.74 a	35.82 ab
December 2018	23.42 B	15.71 b	26.76 ab	25.18 ab	29.00 a	20.83 ab	22.93 ab
February 2019	20.90 B C	12.96 b	30.94 a	21.76 ab	28.23 ab	14.71 ab	21.77 ab
April 2019	17.58 C	6.32 b	36.49 a	9.55 b	30.50 a	6.97 b	20.90 ab

Table A1. Cont.

	Soil Inorganic N (g kg^{-1})							
Season 2								
August 2019	16.30 CD	14.89 ab	16.39 ab	17.28 ab	20.70 a	16.10 ab	11.19 b	
September 2019	4.03 F	2.45 b	3.90 b	4.23 b	3.94 b	3.41 b	6.23 a	
October 2019	11.54 DE	9.81 a	14.02 a	13.56 a	11.67 a	10.51 a	10.24 a	
November 2019	9.25 EF	4.82 c	7.65 bc	8.27 bc	13.63 a	11.69 ab	9.24 abc	
December 2019	16.36 CD	9.76 a	21.83 a	17.43 a	19.52 a	9.87 a	18.30 a	

Table A2. Displays bivariate correlations for recorded parameters throughout the study.

	SOM [a]	TC [b]	TN [c]	C:N	Inorganic N [d]	MUFN [e]	MUFC [f]	Moisture [g]
pH	−0.326 ** [i]	−0.165 *	−0.315 **	0.299 **	−0.511 **	−0.273 **	−0.171 **	−0.240 **
SOM		0.690 **	0.764 **	−0.541 **	0.506 **	0.081	0.130 *	0.294 **
TC			0.740 **	−0.353 **	0.404 **	−0.034	0.060	0.234 **
TN				−0.783 **	0.532 **	0.055	0.124	0.410 **
C:N					−0.346 **	−0.101	−0.141	−0.268 **
Inorganic N						0.243 **	0.212 **	0.613 **
MUFN							0.430 **	0.313 **
MUFC								0.214 **

[a] Soil organic matter. [b] Soil total carbon. [c] Soil total nitrogen. [d] Soil inorganic nitrogen. [e] β-N-acetylglucosaminidase. [f] β-1-4-glucosidase. [g] Soil moisture. [i] Representing Pearson's correlation coefficient (r) significant at $p \leq 0.05$ (*) or $p \leq 0.01$ (**).

References

1. Camarotto, C.; Dal Ferro, N.; Piccoli, I.; Polese, R.; Furlan, L.; Chiarini, F.; Morari, F. Conservation agriculture and cover crop practices to regulate water, carbon and nitrogen cycles in the low-lying venetian plain. *Catena* **2018**, *167*, 236–249. [CrossRef]
2. Kassam, A.; Friedrich, T.; Derpsch, R. Global spread of conservation agriculture. *Int. J. Environ. Stud.* **2019**, *76*, 29–51. [CrossRef]
3. Hobbs, P.R.; Sayre, K.; Gupta, R. The role of conservation agriculture in sustainable agriculture. *Philos. Trans. R. Soc. B Biol. Sci.* **2008**, *363*, 543–555. [CrossRef] [PubMed]
4. Ghimire, B.; Ghimire, R.; VanLeeuwen, D.; Mesbah, A. Cover crop residue amount and quality effects on soil organic carbon mineralization. *Sustainability* **2017**, *9*, 2316. [CrossRef]
5. Horowitz, J.K.; Ebel, R.M.; Ueda, K. "No-Till" Farming Is a Growing Practice; Report No. 1476-2016-120976; USDA: Washington, DC, USA, 2010.
6. Vukicevich, E.; Lowery, T.; Bowen, P.; Úrbez-Torres, J.R.; Hart, M. Cover crops to increase soil microbial diversity and mitigate decline in perennial agriculture. A review. *Agron. Sustain. Dev.* **2016**, *36*, 48. [CrossRef]
7. Bayer, C.; Dieckow, J.; Amado, T.J.C.; Eltz, F.L.F.; Vieira, F.C.B. Cover crop effects increasing carbon storage in a subtropical no-till sandy acrisol. *Commun. Soil Sci. Plant Anal.* **2009**, *40*, 1499–1511. [CrossRef]
8. Chen, Q.; Sun, Y.; Shen, C.; Peng, S.; Yi, W.; Li, Z.; Jiang, M. Organic matter turnover rates and CO_2 flux from organic matter decomposition of mountain soil profiles in the subtropical area, south China. *Catena* **2002**, *49*, 217–229. [CrossRef]
9. Tang, J.; Qi, Y.; Xu, M.; Misson, L.; Goldstein, A.H. Forest thinning and soil respiration in a ponderosa pine plantation in the sierra nevada. *Tree Physiol.* **2005**, *25*, 57–66. [CrossRef]
10. Curiel Yuste, J.; Baldocchi, D.D.; Gershenson, A.; Goldstein, A.; Misson, L.; Wong, S. Microbial soil respiration and its dependency on carbon inputs, soil temperature and moisture. *Glob. Chang. Biol.* **2007**, *13*, 2018–2035. [CrossRef]
11. Abang, A.F.; Kouamé, C.M.; Abang, M.; Hanna, R.; Fotso, A.K. Assessing vegetable farmer knowledge of diseases and insect pests of vegetable and management practices under tropical conditions. *Int. J. Veg. Sci.* **2014**, *20*, 240–253. [CrossRef]
12. Vincent, C.; Schaffer, B.; Rowland, D.L.; Migliaccio, K.W.; Crane, J.H.; Li, Y. Sunn hemp intercrop and mulch increases papaya growth and reduces wind speed and virus damage. *Sci. Hortic.* **2017**, *218*, 304–315. [CrossRef]
13. Ramos, M.G.; Villatoro, M.A.A.; Urquiaga, S.; Alves, B.J.; Boddey, R.M. Quantification of the contribution of biological nitrogen fixation to tropical green manure crops and the residual benefit to a subsequent maize crop using 15N-isotope techniques. *J. Biotechnol.* **2001**, *91*, 105–115. [CrossRef]
14. Perin, A.; Santos, R.H.S.; Urquiaga, S.S.; Cecon, P.R.; Guerra, J.G.M.; Freitas, G.B.d. Sunnhemp and millet as green manure for tropical maize production. *Sci. Agric.* **2006**, *63*, 453–459. [CrossRef]
15. Rigon, J.; Franzluebbers, A.J.; Calonego, J.C. Soil aggregation and potential carbon and nitrogen mineralization with cover crops under tropical no-till. *J. Soil Water Conserv.* **2020**, *75*, 601–609. [CrossRef]

16. Oliveira, F.C.C.; Ferreira, G.W.D.; Souza, J.L.S.; Vieira, M.E.O.; Pedrotti, A. Soil physical properties and soil organic carbon content in northeast brazil: Long-term tillage systems effects. *Sci. Agric.* **2019**, *77*. [CrossRef]
17. Comin, J.J.; Ferreira, L.B.; dos Santos, L.H.; de Paula Koucher, L.; Machado, L.N.; dos Santos, E., Jr.; Mafra, Á.L.; Kurtz, C.; Souza, M.; Brunetto, G.; et al. Carbon and nitrogen contents and aggregation index of soil cultivated with onion for seven years using crop successions and rotations. *Soil Tillage Res.* **2018**, *184*, 195–202. [CrossRef]
18. Calonego, J.C.; Raphael, J.P.; Rigon, J.P.; de Oliveira Neto, L.; Rosolem, C.A. Soil compaction management and soybean yields with cover crops under no-till and occasional chiseling. *Eur. J. Agron.* **2017**, *85*, 31–37. [CrossRef]
19. Nannipieri, P.; Kandeler, E.; Ruggiero, P. Enzyme activities and microbiological and biochemical processes in soil. In *Enzymes in the Environment*; Marcel Dekker: New York, NY, USA, 2002; pp. 1–33.
20. Hamido, S.A.; Kpomblekou-A, K. Cover crop and tillage effects on soil enzyme activities following tomato. *Soil Tillage Res.* **2009**, *105*, 269–274. [CrossRef]
21. Willer, H.; Lernoud, J. *The World of Organic Agriculture: Statistics and Emerging Trends 2019*; Research Institute of Organic Agriculture FiBL: Frick, Switzerland; IFOAM Organics International: Bonn, Germany, 2019.
22. USDA National Agricultural Statistics Service. *Census of Agriculture*; USDA National Agricultural Statistics Service: Washington, DC, USA, 2017.
23. USDA Agricultural Marketing Service. Organic Integrity Database. 2022. Available online: https://organic.ams.usda.gov/integrity/ (accessed on 16 May 2022).
24. Li, Y. Calcareous Soils in Miami Dade County. Fact Sheet 183. Soil and Water Science Department, Florida Coorperative Extension Services, University of Florida. 2001. Available online: Http://Edis.Ifas.Ufl.Edu/Pdffiles/TR/TR00400.Pdf (accessed on 1 October 2007).
25. Farhad, W.; Saleem, M.F.; Cheema, M.A.; Hammad, H.M. Effect of poultry manure levels on the productivity of spring maize (*Zea mays* L.). *J. Anim. Plant Sci.* **2009**, *19*, 122–125.
26. Hochmuth, G.; Hochmuth, R.; Mylavarapu, R. Using composted poultry manure (litter) in mulched vegetable production. *EDIS* **2009**, *2009*. [CrossRef]
27. Nyakatawa, E.Z.; Reddy, K.C.; Sistani, K.R. Tillage, cover cropping, and poultry litter effects on selected soil chemical properties. *Soil Tillage Res.* **2001**, *58*, 69–79. [CrossRef]
28. Reddy, C.K.; Nyakatawa, E.Z.; Reeves, D.W. Tillage and poultry litter application effects on cotton growth and yield. *Agron. J.* **2004**, *96*, 1641–1650. [CrossRef]
29. Núñez-Elisea, R.; Crane, J.H. Selective pruning and crop removal increase early-season fruit production of carambola (*Averrhoa carambola* L.). *Sci. Hortic.* **2000**, *86*, 115–126. [CrossRef]
30. Ballen, F.H.; Evans, E.; Crane, J.; Singh, A. Sample profitability and cost estimates of producing sweet flavored carambola (*Averrhoa carambola*) in south florida. *EDIS* **2020**, *2020*, 7. [CrossRef]
31. Crane, J.H.; Wasielewski, J. Tropical Fruit Acerage in Florida. UF/IFAS Extension Miami Dade County. 2018. Available online: https://sfyl.ifas.ufl.edu/media/sfylifasufledu/miami-dade/documents/tropical-fruit/Tropical-Fruit-Acreage.pdf (accessed on 16 May 2022).
32. Menocal, O.; Kendra, P.E.; Montgomery, W.S.; Crane, J.H.; Carrillo, D. Vertical distribution and daily flight periodicity of ambrosia beetles (coleoptera: Curculionidae) in florida avocado orchards affected by laurel wilt. *J. Econ. Entomol.* **2018**, *111*, 1190–1196. [CrossRef] [PubMed]
33. Menocal, O.; Kendra, P.E.; Padilla, A.; Chagas, P.C.; Chagas, E.A.; Crane, J.H.; Carrillo, D. Influence of Canopy Cover and Meteorological Factors on the Abundance of Bark and Ambrosia Beetles (Coleoptera: Curculionidae) in Avocado Orchards Affected by Laurel Wilt. *Agronomy* **2022**, *12*, 547. [CrossRef]
34. Carrillo, D.; Duncan, R.E.; Peña, J.E. Ambrosia beetles (coleoptera: Curculionidae: Scolytinae) that breed in avocado wood in florida. *Fla. Entomol.* **2012**, *95*, 573–579. [CrossRef]
35. Evans, E.; Wasielewski, J.; Crane, J.H. Monetary Value of Tropical and Sub-Tropical Fruit Trees in US Dollars, UF/IFAS Extension Miami-Dade County. 2017. Available online: https://sfyl.ifas.ufl.edu/media/sfylifasufledu/miami-dade/documents/disaster-preparation/post-hurricane-and-disaster/TropicalFruitTreeWorthFactsheet2017.pdf (accessed on 16 May 2022).
36. USDA Plant Hardiness Zone Map. Agricultural Research Service, U.S. Department of Agriculture. 2012. Available online: https://planthardiness.ars.usda.gov (accessed on 16 May 2022).
37. Kwon, H.; Lall, U.; Obeysekera, J. Simulation of daily rainfall scenarios with interannual and multidecadal climate cycles for south florida. *Stoch. Environ. Res. Risk Assess.* **2009**, *23*, 879–896. [CrossRef]
38. Wang, Q.; Klassen, W.; Li, Y.; Codallo, M. Cover crops and organic mulch to improve tomato yields and soil fertility. *Agron. J.* **2009**, *101*, 345–351. [CrossRef]
39. Crane, J.H.; Balerdi, C.F. Preparing for and recovering from hurricane and tropical storm damage to tropical fruit groves in Florida1. *EDIS* **2006**, *2006*.
40. Freidenreich, A.; Dattamudi, S.; Li, Y.C.; Jayachandran, K. Soil Respiration and Carbon Balance Under Cover Crop in a no-Till Tropical Fruit Orchard. *Front. Environ. Sci.* **2021**, *9*, 653. [CrossRef]
41. Crane, J.H. The Carambola (Star Fruit). University of Florida Cooperative Extension Service, Institute of Food and Agriculture Sciences, EDIS 2001. Available online: https://www.doc-developpement-durable.org/file/Culture/Arbres-Fruitiers/FICHES_ARBRES/Carambolier/Carambola%20Star%20Fruit.pdf (accessed on 12 October 2020).

42. USEPA—US Environmental Protection Agency. *Method 350.1. Determination of Ammonia Nitrogen by Semi-Automated Colorimetry*; Environmental Monitoring Systems Laboratory, Office of Research and Development: Cincinnati, OH, USA, 1993.
43. USEPA—US Environmental Protection Agency. *Method 353.2. Determination of Nitrate Nitrite Nitrogen by Automated Colorimetry*; Environmental Monitoring Systems Laboratory, Office of Research and Development: Cincinnati, OH, USA, 1993.
44. Sinsabaugh, R.L.; Findlay, S.; Franchini, P.; Fischer, D. Enzymatic analysis of riverine bacterioplankton production. *Limnol. Oceanogr.* **1997**, *42*, 29–38. [CrossRef]
45. Hoppe, H. Use of fluorogenic model substrates for extracellular enzyme activity (EEA) measurement of bacteria. In *Handbook of Methods in Aquatic Microbial Ecology*; CRC Press: Boca Raton, FL, USA, 1993; pp. 423–431.
46. Chróst, R.J.; Krambeck, H. Fluorescence correction for measurements of enzyme activity in natural waters using methylumbelliferyl-substrates. *Arch. Hydrobiol.* **1986**, *106*, 79–90.
47. Xiong, Y.; Zeng, H.; Xia, H.; Guo, D. Interactions between leaf litter and soil organic matter on carbon and nitrogen mineralization in six forest litter-soil systems. *Plant Soil* **2014**, *379*, 217–229. [CrossRef]
48. Sheahan, C.M. Plant Guide for Sunn Hemp (Crotalaria Juncea). USDA-Natural Resources Conservation Service, Cape May Plant Materials Center, Cape May, NJ, 8210. 2012. Available online: https://www.nrcs.usda.gov/Internet/FSE_PLANTMATERIALS/publications/njpmcpg11706.pdf (accessed on 16 June 2020).
49. Baligar, V.C.; Fageria, N.K. Agronomy and physiology of tropical cover crops. *J. Plant Nutr.* **2007**, *30*, 1287–1339. [CrossRef]
50. Wang, Q.; Li, Y.; Klassen, W. Summer cover crops and soil amendments to improve growth and nutrient uptake of okra. *HortTechnology* **2006**, *16*, 328–338. [CrossRef]
51. Wayman, S.; Cogger, C.; Benedict, C.; Collins, D.; Burke, I.; Bary, A. Cover crop effects on light, nitrogen, and weeds in organic reduced tillage. *Agroecol. Sustain. Food Syst.* **2015**, *39*, 647–665. [CrossRef]
52. Romanyà, J.; Casals, P. Biological nitrogen fixation response to soil fertility is species-dependent in annual legumes. *J. Soil Sci. Plant Nutr.* **2019**, *20*, 546–556. [CrossRef]
53. Garcia, R.A.; Li, Y.; Rosolem, C.A. Soil organic matter and physical attributes affected by crop rotation under no-till. *Soil Sci. Soc. Am. J.* **2013**, *77*, 1724–1731. [CrossRef]
54. Zhang, Y.; Zhang, S.; Wang, R.; Cai, J.; Zhang, Y.; Li, H.; Huang, S.; Jiang, Y. Impacts of fertilization practices on pH and the pH buffering capacity of calcareous soil. *Soil Sci. Plant Nutr.* **2016**, *62*, 432–439. [CrossRef]
55. Janzen, H.H.; McGinn, S.M. Volatile loss of nitrogen during decomposition of legume green manure. *Soil Biol. Biochem.* **1991**, *23*, 291–297. [CrossRef]
56. Wang, Q.; Li, Y.; Alva, A. Cover crops in mono-and biculture for accumulation of biomass and soil organic carbon. *J. Sustain. Agric.* **2012**, *36*, 423–439. [CrossRef]
57. Eiland, F.; Klamer, M.; Lind, A.M.; Leth, M.; Bååth, E. Influence of initial C/N ratio on chemical and microbial composition during long term composting of straw. *Microb. Ecol.* **2001**, *41*, 272–280. [CrossRef]
58. Leghari, S.J.; Wahocho, N.A.; Laghari, G.M.; HafeezLaghari, A.; MustafaBhabhan, G.; HussainTalpur, K.; Bhutto, T.A.; Wahocho, S.A.; Lashari, A. Role of nitrogen for plant growth and development: A review. *Adv. Environ. Biol.* **2016**, *10*, 209–219.
59. Dias, T.; Dukes, A.; Antunes, P.M. Accounting for soil biotic effects on soil health and crop productivity in the design of crop rotations. *J. Sci. Food Agric.* **2015**, *95*, 447–454. [CrossRef]
60. Rao, R.B.; Li, Y.C. Nitrogen mineralization of cover crop residues in calcareous gravelly soil. *Commun. Soil Sci. Plant Anal.* **2003**, *34*, 299–313. [CrossRef]
61. Linn, D.M.; Doran, J.W. Effect of water-filled pore space on carbon dioxide and nitrous oxide production in tilled and nontilled soils. *Soil Sci. Soc. Am. J.* **1984**, *48*, 1267–1272. [CrossRef]
62. Ball, B.C.; Crichton, I.; Horgan, G.W. Dynamics of upward and downward N_2O and CO_2 fluxes in ploughed or no-tilled soils in relation to water-filled pore space, compaction and crop presence. *Soil Tillage Res.* **2008**, *101*, 20–30. [CrossRef]
63. Linn, D.M.; Doran, J.W. Aerobic and anaerobic microbial populations in no-till and plowed soils. *Soil Sci. Soc. Am. J.* **1984**, *48*, 794–799. [CrossRef]
64. Clagnan, E.; Thornton, S.F.; Rolfe, S.A.; Wells, N.S.; Knoeller, K.; Murphy, J.; Tuohy, P.; Daly, K.; Healy, M.G.; Ezzati, G.; et al. An integrated assessment of nitrogen source, transformation and fate within an intensive dairy system to inform management change. *PLoS ONE* **2019**, *14*, e0219479. [CrossRef]
65. Liu, X.J.; Mosier, A.R.; Halvorson, A.D.; Reule, C.A.; Zhang, F.S. Dinitrogen and N2O emissions in arable soils: Effect of tillage, N source and soil moisture. *Soil Biol. Biochem.* **2007**, *39*, 2362–2370. [CrossRef]
66. Dobbie, K.E.; Smith, K.A. The effects of temperature, water-filled pore space and land use on N_2O emissions from an imperfectly drained gleysol. *Eur. J. Soil Sci.* **2001**, *52*, 667–673. [CrossRef]
67. Daims, H.; Lebedeva, E.V.; Pjevac, P.; Han, P.; Herbold, C.; Albertsen, M.; Jehmlich, N.; Palatinszky, M.; Vierheilig, J.; Bulaev, A.; et al. Complete nitrification by nitrospira bacteria. *Nature* **2015**, *528*, 504–509. [CrossRef] [PubMed]
68. Carey, C.J.; Dove, N.C.; Beman, J.M.; Hart, S.C.; Aronson, E.L. Meta-analysis reveals ammonia-oxidizing bacteria respond more strongly to nitrogen addition than ammonia-oxidizing archaea. *Soil Biol. Biochem.* **2016**, *99*, 158–166. [CrossRef]
69. Gao, S.; Zhou, G.; Rees, R.M.; Cao, W. Green manuring inhibits nitrification in a typical paddy soil by changing the contributions of ammonia-oxidizing archaea and bacteria. *Appl. Soil Ecol.* **2020**, *156*, 103698. [CrossRef]
70. Shen, J.; Zhang, L.; Zhu, Y.; Zhang, J.; He, J. Abundance and composition of ammonia-oxidizing bacteria and ammonia-oxidizing archaea communities of an alkaline sandy loam. *Environ. Microbiol.* **2008**, *10*, 1601–1611. [CrossRef] [PubMed]

71. Shi, W.; Norton, J.M. Microbial control of nitrate concentrations in an agricultural soil treated with dairy waste compost or ammonium fertilizer. *Soil Biol. Biochem.* **2000**, *32*, 1453–1457. [CrossRef]
72. Yin, Z.; Bi, X.; Xu, C. Ammonia-oxidizing archaea (AOA) play with ammonia-oxidizing bacteria (AOB) in nitrogen removal from wastewater. *Archaea* **2018**, *2018*, 8429145. [CrossRef]
73. Hungria, M.; Kaschuk, G. Regulation of N_2 fixation and NO_3^-/NH_4 assimilation in nodulated and N-fertilized phaseolus vulgaris L. exposed to high temperature stress. *Environ. Exp. Bot.* **2014**, *98*, 32–39. [CrossRef]
74. Enrico, J.M.; Piccinetti, C.F.; Barraco, M.R.; Agosti, M.B.; Eclesia, R.P.; Salvagiotti, F. Biological nitrogen fixation in field pea and vetch: Response to inoculation and residual effect on maize in the pampean region. *Eur. J. Agron.* **2020**, *115*, 126016. [CrossRef]
75. Nezomba, H.; Tauro, T.P.; Mtambanengwe, F.; Mapfumo, P. Nitrogen fixation and biomass productivity of indigenous legumes for fertility restoration of abandoned soils in smallholder farming systems. *S. Afr. J. Plant Soil* **2008**, *25*, 161–171. [CrossRef]
76. Allen, O.N.; Allen, E.K. *The Leguminosae: A Source Book of Characteristics, Uses, and Nodulation*; University of Wisconsin Press: Madison, WI, USA, 1981.
77. Li, T.; Gao, J.; Bai, L.; Wang, Y.; Huang, J.; Kumar, M.; Zeng, X. Influence of green manure and rice straw management on soil organic carbon, enzyme activities, and rice yield in red paddy soil. *Soil Tillage Res.* **2019**, *195*, 104428. [CrossRef]
78. Chung, H.; Zak, D.R.; Reich, P.B.; Ellsworth, D.S. Plant species richness, elevated CO2, and atmospheric nitrogen deposition alter soil microbial community composition and function. *Global Change Biology.* **2007**, *13*, 980–989. [CrossRef]
79. Maltais-Landry, G.; Scow, K.; Brennan, E. Soil phosphorus mobilization in the rhizosphere of cover crops has little effect on phosphorus cycling in california agricultural soils. *Soil Biol. Biochem.* **2014**, *78*, 255–262. [CrossRef]
80. Henry, S.; Texier, S.; Hallet, S.; Bru, D.; Dambreville, C.; Chèneby, D.; Bizouard, F.; Germon, J.C.; Philippot, L. Disentangling the rhizosphere effect on nitrate reducers and denitrifiers: Insight into the role of root exudates. *Environ. Microbiol.* **2008**, *10*, 3082–3092. [CrossRef]

Article

The Effects of Agricultural Conservation Practices on the Small Water Cycle: From the Farm- to the Management-Scale

Nina Noreika *, Tailin Li, Julie Winterova, Josef Krasa and Tomas Dostal

Department of Landscape Water Conservation, Faculty of Civil Engineering, Czech Technical University in Prague, 12000 Prague, Czech Republic; tailin.li@cvut.cz (T.L.); julie.winterova@fsv.cvut.cz (J.W.); josef.krasa@cvut.cz (J.K.); dostal@fsv.cvut.cz (T.D.)
* Correspondence: nina.noreika@fsv.cvut.cz

Abstract: Reinforcing the small water cycle is considered to be a holistic approach to both water resource and landscape management. In an agricultural landscape, this can be accomplished by incorporating agricultural conservation practices; their incorporation can reduce surface runoff, increase infiltration, and increase the water holding capacity of a soil. Some typical agricultural conservation practices include: conservation tillage, contour farming, residue incorporation, and reducing field sizes; these efforts aim to keep both water and soil in the landscape. The incorporation of such practices has been extensively studied over the last 40 years. The Soil and Water Assessment Tool (SWAT) was used to model two basins in the Czech Republic (one at the farm-scale and a second at the management-scale) to determine the effects of agriculture conservation practice adoption at each scale. We found that at the farm-scale, contour farming was the most effective practice at reinforcing the small water cycle, followed by residue incorporation. At the management-scale, we found that the widespread incorporation of agricultural conservation practices significantly reinforced the small water cycle, but the relative scale and spatial distribution of their incorporation were not reflected in the SWAT scenario analysis. Individual farmers should be incentivized to adopt agricultural conservation practices, as these practices can have great effects at the farm-scale. At the management-scale, the spatial distribution of agricultural conservation practice adoption was not significant in this study, implying that managers should incentivize any adoption of such practices and that the small water cycle would be reinforced regardless.

Keywords: small water cycle; agricultural conservation practices; BMPs; SWAT

Citation: Noreika, N.; Li, T.; Winterova, J.; Krasa, J.; Dostal, T. The Effects of Agricultural Conservation Practices on the Small Water Cycle: From the Farm- to the Management-Scale. *Land* **2022**, *11*, 683. https://doi.org/10.3390/land11050683

Academic Editors: Chiara Piccini and Rosa Francaviglia

Received: 23 March 2022
Accepted: 30 April 2022
Published: 4 May 2022

Publisher's Note: MDPI stays neutral with regard to jurisdictional claims in published maps and institutional affiliations.

Copyright: © 2022 by the authors. Licensee MDPI, Basel, Switzerland. This article is an open access article distributed under the terms and conditions of the Creative Commons Attribution (CC BY) license (https://creativecommons.org/licenses/by/4.0/).

1. Introduction

The small water cycle is the local cycling of water, wherein water should fall as rain in the same geographic area from which it evapo(transpi)rates. The small water cycle also greatly emphasizes a reduction in surface runoff generation in a landscape, and the cycle's reinforcement is considered to be a holistic approach to managing water resources at the catchment scale [1–3]. In an agricultural landscape, certain conservation techniques can greatly improve the water holding capacity of a soil and can, in turn, strongly reinforce the small water cycle, making an agricultural landscape more resilient in the face of climate change.

Agricultural conservation practices have been extensively studied over the last 40 years and have been shown to significantly improve a soil's infiltration capacity and, consequently, significantly decrease the surface runoff in a landscape [4–8]. The most common agricultural conservation practices in modern literature include reduced/no-tillage, mulch cover/crop residues and cover crops, and reduced application of herbicides. The goal of conservation agriculture is to make soils "self-sustainable" by: maintaining sources of organic matter above and below the soil's surface, recycling water and nutrients within the system, and ensuring that the infiltration rate of a soil is greater than the predicted rainfall rate [9]. To maximize the benefits of implementing agricultural conservation practices, managers must maintain year-round organic matter cover, minimize soil disturbance, and diversify crop rotations [9–11]. The transition from conventional or reduced tillage to no-tillage has been shown to reduce surface runoff by upwards of 20% at the plot-scale [12]. A no-tillage management scheme can increase the infiltration capacity of a soil in two ways: by minimizing soil disruption and by preserving the highest percentage of crop residue cover. No-tillage has also been shown to reduce soil loss, splash erosion, and surface runoff, while increasing direct infiltration [10,13,14]. Maintaining adequate plant cover year-round provides numerous benefits, including improving soil quality, controlling soil erosion, and increasing soil water availability [15]. Plant cover percentage has a significant, negative relationship with final runoff rate, indicating that the greater the plant cover percentage, the lower the expected hourly runoff [16]. While cover crops and crop residues provide year-round soil coverage, they also provide an even-coverage mulching, which has been found to be a more successful mulching strategy in real-life scenarios when compared to artificial mulching with wheat straw, grass clippings, wood chips, etc., [13].

The Intergovernmental Panel on Climate Change (IPCC) predicts that in the face of future climate change, Central Europe will encounter more frequent, intensive storm events, which will magnify landscape management issues in the Czech Republic [17]. The Czech Republic is a highly agricultural country, with nearly 40% of its land area being arable. Agricultural intensification in the Czech Republic began in the 1970s when the landscape was publicly managed. Large fields, subsurface tile drainage systems, and artificially lined and straightened streams were incorporated across the landscape in an effort to increase crop production [18]. Unfortunately, these practices resulted in increased soil loss and reduced deep percolation and groundwater recharge. Since privatization in 1991, some small Czech farms have begun incorporating agricultural conservation practices and IPA (integrated pest management for agriculture) guidelines; however, much of the Czech agricultural landscape is managed by large agricultural conglomerates driven by profit [1,18]. By working to reinforce the small water cycle through the incorporation of agricultural conservation practices, the effects of extreme precipitation events (e.g., huge spikes in surface runoff ratios as well as extreme soil loss events) may be mitigated at the basin-scale, which should incentivize their incorporation to land managers and farmers [17].

The two basins of interest have been monitored for a number of years. The farm-scale basin (Nučice) is equipped to monitor localized basin processes, and previous studies have primarily focused on rainfall–runoff mechanisms and temporarily variable soil properties [19–23]. Sediment transport and erosion have been extensively studied in Vrchlice (the basin utilized for management-scale analysis), especially regarding the sediment trap efficiencies of the nearly 140 reservoirs across the basin [24–27]. The Soil and Water Assessment Tool (SWAT) has been previously utilized at both basins to assess the effects of land use changes on in-basin water balance [26,27], but since the Czech Republic is likely to remain quite agricultural for the foreseeable future, it is of great interest to assess the impacts of agricultural conservation practice incorporation at each of these scales. While sometimes data intensive, hydrologic models are a relatively easy and non-invasive way to run scenario analyses in a landscape. SWAT is a semi-distributed, semi-physically based, basin-scale hydrologic model. SWAT divides a basin into smaller elements called hydrologic response units (HRUs) that are each comprised of the same soil type, slope class, and land use classification [28–31]. SWAT was selected for this study because of its flexibility and applicability to agricultural catchments. SWAT makes running scenario analyses simple, and there is significant precedent for its incorporation of agricultural conservation practices [32,33].

The purpose of this study is to investigate the following questions: (i) do the incorporation of agricultural conservation practices impact the small water cycle proportionally at various scales? (ii) Which practice is most effective at reinforcing the small water cycle at the farm-scale? (iii) Does the spatial distribution of agricultural conservation practices affect their impacts on the small water cycle at the management scale? (iv) What do these results imply regarding catchment management and incentivizing farmers to adopt these practices?

2. Materials and Methods

2.1. Study Watersheds

Both study watersheds are located in the Central Bohemia region of the Czech Republic (Figure 1). This region is characterized by a humid continental climate and receives approximately 600 mm of precipitation per year. The rainy season in this region occurs from May through August, and the driest month is usually February. These two basins were selected for this study because they are typical of an intensively agricultural Czech landscape. Nučice is a simply-shaped catchment and represents the farm-scale, containing three large fields, each with very similar crop rotations and management. Vrchlice represents the management-scale. It is much larger (~100 km^2), with a more diverse landscape, and its water resources are managed to meet municipal needs. It is valuable to land owners as well as basin managers to determine the effects of agricultural conservation practice adoption at each scale.

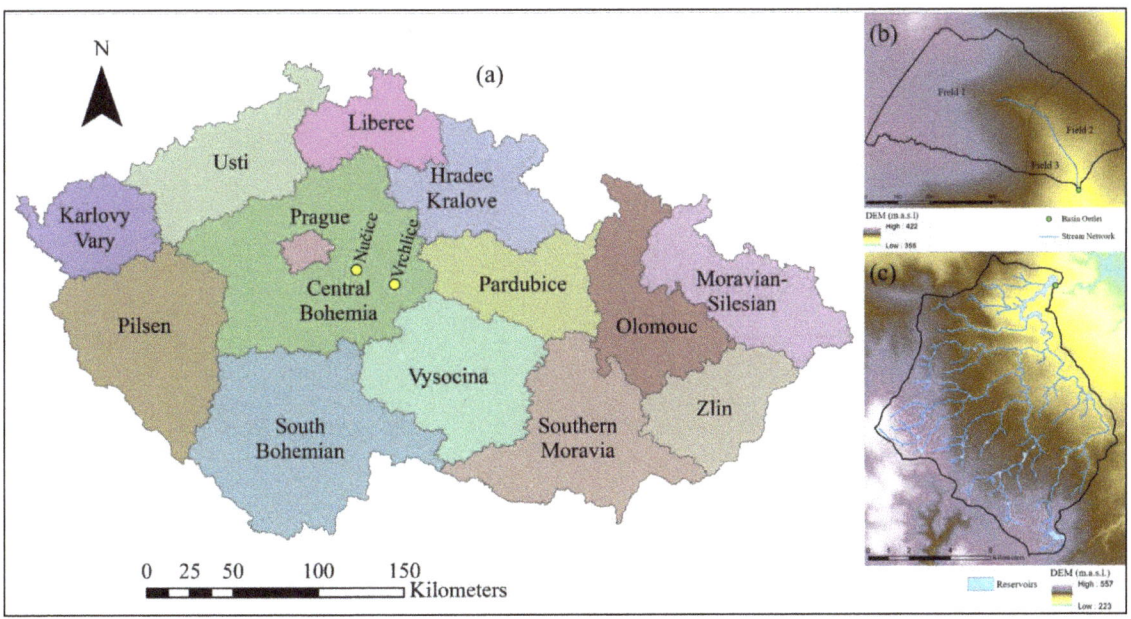

Figure 1. (a) A map of the Czech Republic. Prague is highlighted for reference as well as the outlet locations of the two study watersheds, Nučice (b) and Vrchlice (c).

The Nučice experimental catchment ("Nučice") has been monitored since 2011 by the Landscape Water Conservation Department (in the faculty of Civil Engineering) of Czech Technical University in Prague. It is a small watershed (~0.52 km^2) consisting of three fields that are managed by two farmers and is appropriate for modeling at the "farm-scale" in the Czech Republic. Its outlet is located at 49°57'49.230'' N, 14°52'13.242'' E (Figure 1b). The soils in Nučice are classified mainly as Luvisols and Cambisols overlaying siltstone and sandstone. The average slope in Nučice is 3.9% but ranges from 1 to 12%. Nučice is primarily cropland, with a very narrow riparian/brush zone around the stream; the basin is bisected horizontally by a 2-lane road (Table 1).

Table 1. Land use percent cover over the experimental basins.

Land Use	Nučice	Vrchlice
Impervious	2	3
Brush	2	4
Forest	-	25
Grassland	-	8
Cropland	95	54
Water	1	2
Gardens	-	4

The Vrchlice Basin ("Vrchlice") is much larger than Nučice, at ~97 km² (Figure 1c). Vrchlice also has a more diverse land use, with large areas of forested land as well as many townships (Table 1), but it is still primarily cropland. The Vrchlice Reservoir provides drinking water to the nearby town of Kutná Hora, serving approximately 40,000 inhabitants. Its outlet is located at 49°55′37.211″ N, 15°13′37.07″ E. The basin is covered in clayey soils classified as Cambisols overlaying a metamorphic bedrock [24]. Vrchlice contains a network of nearly 140 reservoirs, mostly small fish ponds, that serve cultural and hydrologic significance. The discharge at the outlet of the Vrchlice Reservoir has been monitored by the Elbe River Authority since 1979. The Vrchlice Basin is considered to be an appropriate size for modeling at the "management-scale" in the Czech Republic.

2.2. Soil and Water Assessment Tool (SWAT)

SWAT requires the following as its bare minimum regarding data requirements: soils, slopes, land uses, and daily weather data. The input data used for each of the models present in this study are outlined in Table 2.

Table 2. Input variables and their sources used for Soil and Water Assessment Tool (SWAT) modeling.

Input Data		Basin	Description	Source
Meteorological Data	Extreme Daily Temperatures	Nučice	2011–2019	On-site: 107 Temperature Probe (Campbell Sci., Logan, UT, USA)
		Vrchlice	1996–2019	Czech Hydrometeorological Institute
	Precipitation (Total Daily)	Nučice	2011–2019	On-site: MR3-01s Tipping Bucket (Meteo Servis, Vodnany, Czech Republic)
		Vrchlice	1996–2019	Czech Hydrometeorological Institute
Spatial Data	DEM	Nučice	3 m resolution	LiDAR Survey: Czech Institute of Geodesy and Cartography
		Vrchlice	5 m resolution	LiDAR Survey: Czech Institute of Geodesy and Cartography
	Soils	Nučice	1:5000 soil map	State Land Office of the Czech Republic
		Vrchlice	1:5000 soil map	Czech Research Institute of Soil Conservation & the State Land Office of the Czech Republic
	Land Use	Nučice	Digitized from detailed orthophoto	UAV Survey: Czech Technical University
		Vrchlice	1:10,000 land use map	ZABAGED (Fundamental Base of Geographic Data of the Czech Republic) & LPIS (Land Parcel Identification System)

The daily meteorological data for the Nučice SWAT model was obtained from on-site gauges (Table 2). The climate data for this model was obtained from the Climate Forecast System Reanalysis (CFSR) database; these data are used in case there are any gaps in the observed weather dataset. The digital elevation model (DEM) was obtained from the fifth generation of the digital relief model of the Czech Republic (DMR5G) and was point-cloud processed to obtain a 3 m spatial resolution. The SWAT model for Nučice was developed using the field boundary method [34]. In the field boundary method scheme, each field is defined as its own HRU by aggregating the primary soil type and elevation class for each field. This method was selected in order to incorporate reduced field sizes at the farm-scale and was accomplished through the use of soil dummy variables. The SWAT model for Nučice was run during the growing seasons (~April through October) from 2014 through 2019, using 2013 as a warmup period. Calibration and validation procedures followed those outlined in Noreika et al. 2020 [26].

The Vrchlice SWAT model was developed originally for Noreika et al. 2021 to study the effects of land use and management changes over time in the basin [27]. The model itself has not been edited further. This model was run at the monthly timestep from 2001 through 2019 with a 5-year warmup period (1996–2000). The monthly timestep was chosen to minimize daily effects due to reservoir processes that are not publicly available and therefore unable to be represented in SWAT. The model was calibrated (2001–2012) and validated (2013–2016) at the monthly timestep with discharge data from the basin's outlet. The basin boundaries and stream network were largely DEM-based, but ground-truthed to existing data. Vrchlice was divided into 63 sub-basins, containing 1058 HRUs that were defined by their unique combinations of land use, slope class, and soil type. For further detail, parameterization, and intricacies of the model setup, please refer to Noreika et al. 2021 [27].

2.3. Scenario Analysis

2.3.1. Literature Review

Contour farming results in a reduction of surface runoff by impounding water in small depressions, as well as a reduction of sheet and rill erosion by reducing the erosive power of surface runoff and preventing or minimizing the development of rills. This practice is represented by adjusting the Soil Conservation Service (SCS) curve number in SWAT. Residues are meant to slow down surface and peak runoff by increasing surface roughness. They also increase infiltration and reduce surface runoff by decreasing surface sealing and slowing down overland flow. Finally, residues reduce sheet and rill erosion by reducing surface flow volume. In SWAT, there is significant literature precedent to incorporate these practices; conservation tillage and residue management are typically represented by adjusting the curve number and Manning's roughness coefficient for overland flow, respectively. In order to incorporate these practices appropriately, a literature review was conducted using the following keywords: SWAT, best management practice, and conservation agriculture [35–67]. A total of 33 articles were downloaded and narrowed down to 25 based on relevance. The 25 remaining papers addressed the incorporation of conservation tillage operations, contour farming, and residue management into SWAT (Figure 2). Of the 25, 12 took place in the Midwest (of US and Canada), 1 in Texas, 6 in Europe, 1 in Africa, and 6 in Asia. Overwhelmingly, 17 of the 25 papers referenced Arabi et al. 2008 and Neitsch et al. 2011 publications [32,33], meaning that conservation practices were incorporated via the curve number (CN) method. Three publications introduced tillage operation changes and no CN edits (TO). Two introduced tillage operation changes along with the CN edits (CN + TO). Two modified the CN by a percent change, and two did not specify (NS) how the practices were incorporated into SWAT. We then conducted a scenario analysis at the Nučice basin to determine whether it is necessary to incorporate both CN shifts and tillage operation changes. We found no significant differences between water balance variable outputs (discharge at the basin's outlet, subsurface lateral flow, surface runoff, evapotranspiration, and soil water content, $p > 0.05$) when only the CN method was utilized versus shifting both the CN and the tillage practices. We concluded that the CN method is appropriate to incorporate agricultural conservation practices and is also more efficient in the modeling process.

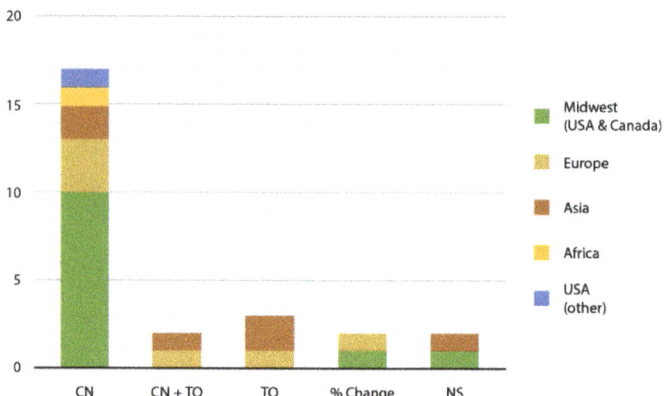

Figure 2. Literature review results of 25 articles outlining the incorporation of agricultural conservation practices into SWAT.

2.3.2. Scenarios Outlined

Five scenarios were run at the farm-scale for this study (Table 3). These scenarios incorporate contour farming, small residues (0.5–1 t/ha), large residues (1–9 t/ha), conservative tillage, and field size reductions at Nučice. To incorporate field size reductions, instead of three fields averaging 17 ha each, Nučice was divided into 52 fields averaging 1 ha each through the use of dummy soil variables. The other scenarios were incorporated as presented in Table 4.

Table 3. Outline of scenarios implemented in the Nučice and Vrchlice Basins.

Scale	Practice
Farm-Scale: Nučice	Conventional Tillage (Conv)
	Conservation Tillage * (Cons)
	Contour Farming (Cont)
	Small Residues (Res1)
	Large Residues (Res2)
	Field Reductions (SmFld)
Management-Scale: Vrchlice	Conventional Tillage *
	Full Adoption
	Lower Adoption
	Lower Extended
	Middle Adoption
	Upper Adoption
	Upper Extended
	Random Adoption

* denotes the original calibrated model for each

In Vrchlice, only the "General Measures" agricultural conservation scenario was adopted (Table 4) at various scales across the basin (Table 3, Figure 3). The "General Measures" outlined in Table 4 are considered to be "best case scenarios" to represent conditions if the practices were incorporated properly and if the landscape responds as expected, but it is likely that any real-world result would fall somewhere between the calibrated model without any conservation practices and the "General Measures" scenarios. Vrchlice was divided into three regions based on location in the basin and percent area cropland (Figure 3). Each area (Upper, Middle, Lower, Random) comprises approximately 1/3 of the cropland cover in the Vrchlice Basin. Additionally, the Upper Extended and Lower Extended scenarios encompass the Upper + Middle and Lower + Middle areas, respectively, to encompass approximately 2/3 of the cropland cover in the Vrchlice Basin.

A requirement for the Random scenario is that no selected sub-basins should be adjacent. The Random scenario controls for the effects of connectivity of agricultural conservation practices to determine if individual farm adoption is "enough" or if regional adoption is necessary to more greatly reinforce the small water cycle. These scenarios were outlined so that the individual impacts of agricultural conservation practice continuity and spatial adoption within the basin could be evaluated.

Table 4. Agricultural conservation measures applied to the Nučice Basin and how they are parameterized in SWAT. The Soil Conservation Service (SCS) Curve Number (CN) and Manning's Roughness values represent a relative change from the respective calibrated model [32,33].

Scenario	CN	USLE P	Manning's Roughness
Conventional Tillage *	+2	-	-
Contour Farming	−1	0.5	-
Small Residues	-	-	+0.07
Large Residues	-	-	+0.15
General Measures †	−3	0.5	+0.15

* Conventional tillage is present because the Nučice model was calibrated based on conservation tillage, and this is how the effects of conservation tillage will be compared to conventional tillage. † These general measures are applied to the Vrchlice scenarios at various levels of incorporation across the basin.

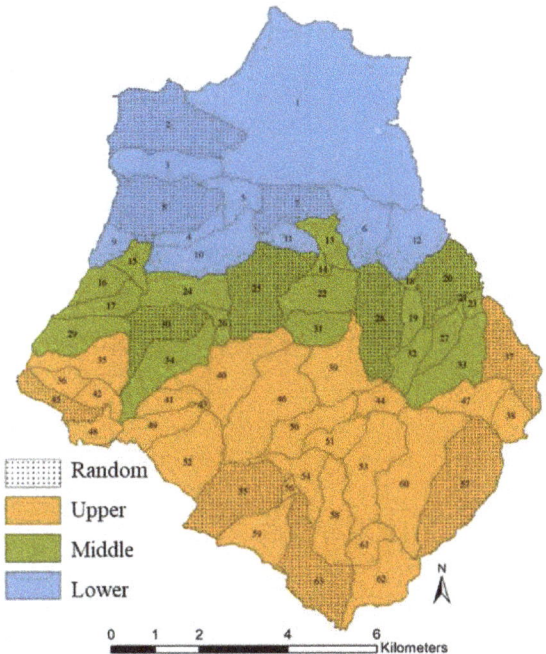

Figure 3. A map of the Vrchlice Basin, color coded to represent the various scenarios analyzed.

3. Results

According to the global sensitivity analysis that was conducted, three parameters significantly influenced the modeled discharge flowing out of the Nučice experimental basin (Table 5). RCHRG_DP is the deep aquifer percolation fraction; this value should fall between 0 and 1 as it is the fraction of percolation past the root zone which recharges the deep aquifer. Since this value is very close to 0, this indicates that a very small fraction of water entering the Nučice Basin recharges the deep aquifer. Saturated hydraulic conductivity (SOL_K) and the available water capacity of the soil (SOL_AWC) govern how water is infiltrated and retained in a soil, respectively, were also significantly sensitive parameters.

Table 5. Sensitive parameters and their calibrated (adjusted) values.

Parameter	Method	Calibration Values		
		Minimum	Adjusted	Maximum
RCHRG_DP	V	0.001	0.001	0.999
SOL_K	R	−0.5	−0.11	0.5
SOL_AWC	R	−0.90	0.88	0.90

V: replace, A: absolute, and R: relative.

Calibration (2016–2018) and validation (2019) for the Nučice basin were conducted with SWAT-Cup 2019, which is a semiautomatic calibration methodology [28]. Table 6 presents the selected model performance indicators during the calibration and validation periods for the Nučice SWAT model. Figure 4 presents a scatterplot, correlating the modeled discharge values with the observed discharge values at Nučice during the calibration and validation periods.

Table 6. Model performance indicators for the calibration and validation periods of the Nučice SWAT model.

Calibration	Performance Indicator	Validation
0.76	p-factor	0.80
0.46	r-factor	0.21
0.77	R^2	0.52
0.77	NSE	0.48
6.9	PBIAS	12.1
0.80	KGE	0.64

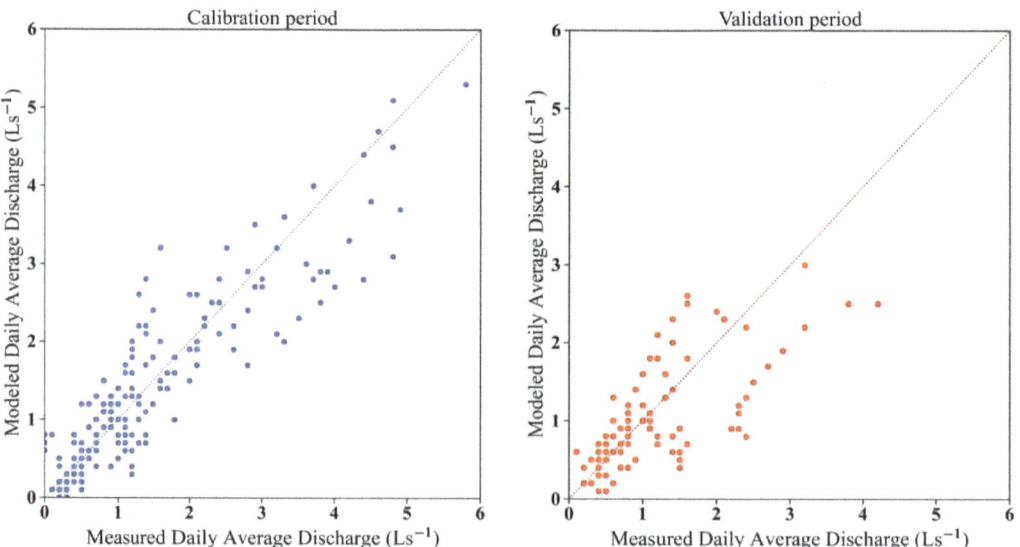

Figure 4. Correlation of modeled and observed discharge values at Nučice's outlet during the calibration and validation periods; a 1:1 line is included for reference.

There were significant shifts across water balance parameters with the incorporation of agricultural conservation practices at the Nučice scale (Figure 5). The incorporation of residues and contour farming reinforced all of the small water balance parameters when compared to the calibrated scenario, which included generalized conservation tillage. Resorting to conventional tillage from conservation tillage was consistently contradictory to the goal of reinforcing the small water cycle. Field size reductions resulted in the highest amount of streamflow contribution from subsurface lateral flow, but the model indicated that otherwise the adoption of smaller fields does not reinforce the small water cycle.

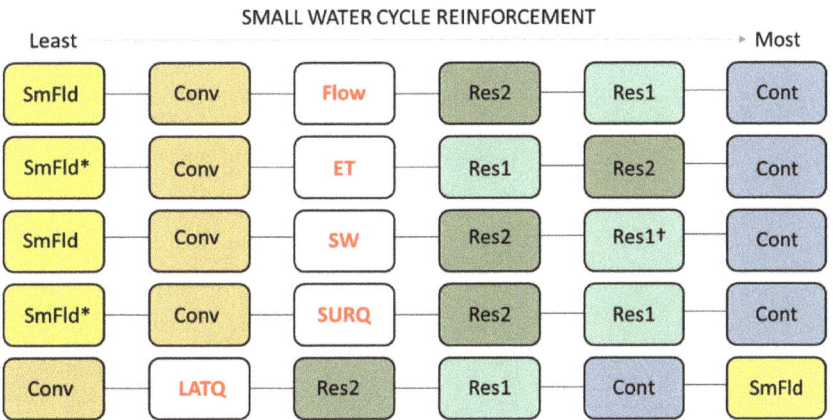

Figure 5. A ranking of each scenario (in the Nučice basin) according to its reinforcement of specific small water cycle parameters. All values are significantly different from the calibrated scenario (parameters in red, bold) unless indicated by *. † indicates a significant difference between Res1 and Res2.

All small water cycle parameters, except for discharge at Vrchlice's outlet, were significantly affected by the incorporation of agricultural conservation practices across the basin. Interestingly enough, neither the scale of adoption nor the spatial distribution of agricultural conservation practices significantly affected any small water cycle parameters at this scale; further figures presented compare only Vrchlice's conventional tillage (calibrated model) and the full adoption scenario. Both the available water content and evapotranspiration in the conventional tillage scenario are consistently lower than the full conservation adoption across the entire year (Figures 6 and 7). Both the surface runoff ratios and subsurface lateral flow were significantly higher throughout the year in the conventional tillage scenario when compared to the General Measures full adoption scenario (Figures 8 and 9). Generally surface runoff in the conventional tillage scenario is greater than 2× that of the conservation scenario (Figure 9).

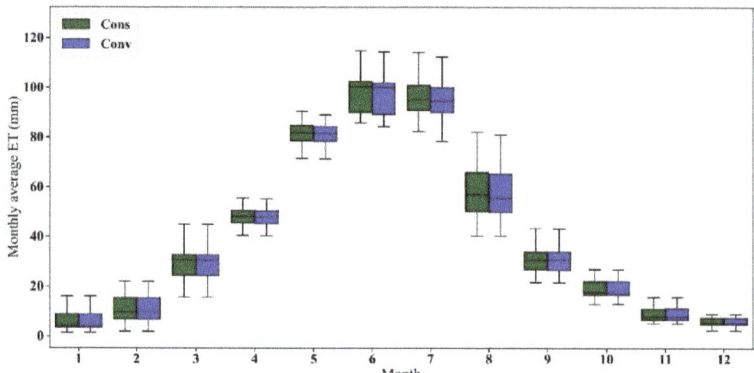

Figure 6. Average monthly evapotranspiration rates (mm) across the modeled time period in Vrchlice.

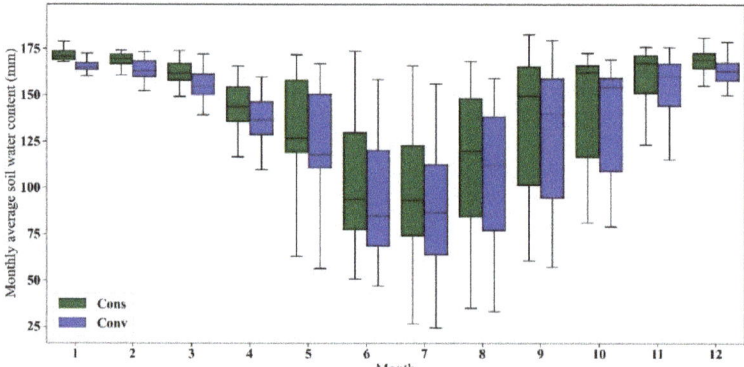

Figure 7. Average soil water content (mm) by month across the modeled time period in Vrchlice.

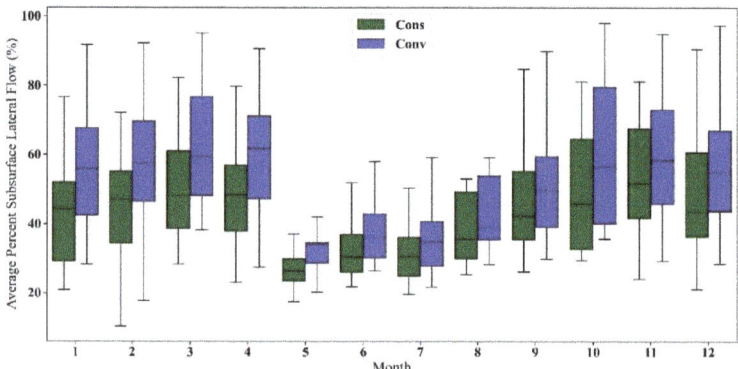

Figure 8. Average monthly percentage subsurface lateral flow contribution to streamflow across the modeling period in Vrchlice.

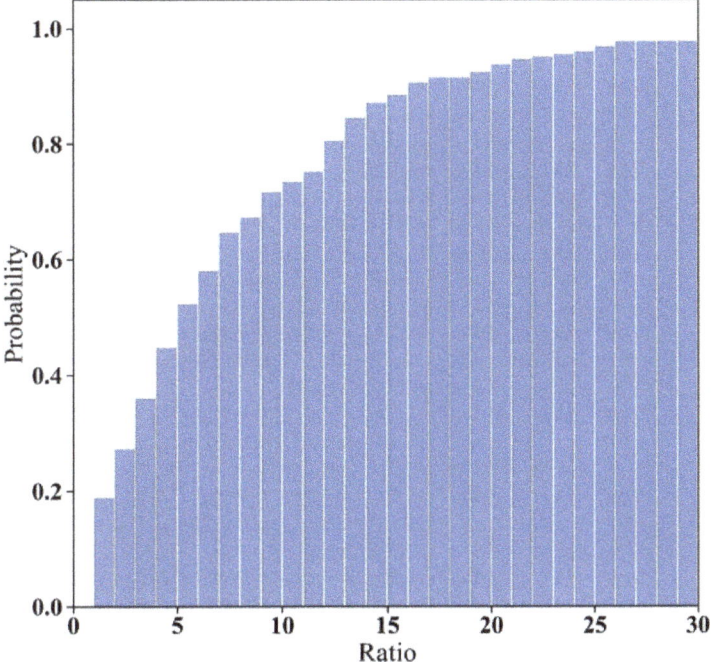

Figure 9. Surface runoff ratios between full agricultural conservation practice adoption and the default conventional tillage scenario. The histogram bars represent the cumulative probability that a value falls at or below the respective ratio.

4. Discussion

4.1. Hydrological Modeling with SWAT

There are several possible sources of error in any hydrologic model; the first is input parameter uncertainty, which is the largest possible source of error and also influences uncertainties associated with output data. Model parameterization and model structural uncertainties are additional possible sources of error [62,68]. Furthermore, since SWAT is neither fully physically based nor fully distributed, some processes may not be properly represented, such as temporal changes in topsoil hydraulic properties, preferential flow, or the influences of the spatial distribution of fine-scale land management [28,69,70]. While there are some drawbacks to the SWAT model (as stated above), it is a very useful tool for hydrologic modeling, especially regarding scenario analysis. Currently, Nučice is equipped to model generalized processes rather than more spatially distributed processes (piezometer clusters and a cosmic-ray neutron sensor are currently being installed). The soil data at this scale is fairly coarse and is nearing the lower spatial range of SWAT's modeling capabilities, but SWAT was still able to model Nučice effectively with "good" or "very good" performance across the selected indicators [71–73]. The uncertainties associated with the Vrchlice model primarily include generalized reservoir processes and crop rotations [27]. Vrchlice was able to be effectively modeled at the monthly timescale, also with "good" and "very good" performance indicators [71–73]. While SWAT was able to model significant shifts in water balance parameters with the incorporation of agricultural conservation practices in Vrchlice, it was unable to represent significant differences at varying scales and distributions of incorporation across Vrchlice. This could be due to the fact that Vrchlice, while primarily cropland, contains significant areas of forested areas and riparian zones, which may disguise the effects of agricultural conservation adoption. Additionally, since Vrchlice is of significant size and SWAT is not fully distributed, the effects of the scale of agricultural conservation practice adoption may be aggregated across the basin, leading to insignificant changes across the agricultural conservation practice adoption scenarios.

4.2. The Small Water Cycle at the Farm-Scale

At the farm-scale, SWAT was able to model significant differences in water balance parameters across agricultural conservation practice scenarios. According to SWAT, residue incorporation and contour farming were the most effective at reinforcing the small water cycle and should be prioritized by farmers to aid the holistic management of their land in the face of future climate change [17]. Although it was not in the scope of this study to investigate the effects of crop changes in addition to the incorporation of agricultural conservation practices, the previous SWAT study of the Nučice basin indicated that crop changes also have significant impacts on the small water cycle [26]. For instance, winter wheat reinforces the small water cycle to a greater degree than rapeseed in the Czech landscape. The incorporation of contour farming and crop residues may be able to mitigate water balance issues that arise from less-sustainable crop choices, and the interaction should be studied further.

SWAT was not able to effectively model the impacts of incorporating smaller field sizes at Nučice. This may be due to several factors: SWAT is not fully distributed and cannot model the spatial effects influenced by smaller field incorporation, crop changes were not incorporated across the smaller fields, and SWAT does not model true border effects between fields. To replicate this in future studies, a trap efficiency would need to be applied to each HRU to simulate flow disruption between fields. The field boundary HRU method may be more useful to identify "hotspot" fields that may be susceptible to erosive events due to their slopes, crops, and soil types [28,30,34,69,70].

4.3. The Small Water Cycle at the Management-Scale

The adoption of agricultural conservation practices in at least 33% of the cropland across Vrchlice had significant effects on the small water cycle within the basin. Neither the distribution nor the scale of adoption (anything above 33%) significantly affected the small water cycle variables at Vrchlice any further. While Vrchlice is a very agricultural basin (>50% cropland), there are also very large forested and riparian areas that may mask the effects of various intensities of agricultural conservation practice adoption. It may also be due to SWAT's model structure, being semi-distributed and semi-physically based, that some effects at this scale may be lost due to HRU aggregation or generalizations due to using the curve number method [34,74,75]. While SWAT models significant impacts on the small water cycle due to the adoption of agricultural conservation practices, SWAT cannot represent realistic effects when additional spatial distribution and connectivity scenarios are introduced; a fully distributed model would be necessary for this purpose. However, SWAT was able to model general trends and could represent significant differences between conventional agricultural practices and full conservation adoption.

When compared to the effects of land use changes at Vrchlice [27], average soil water content and subsurface lateral flow shifts fell in similar ranges to that of agricultural conservation practice incorporation scenarios. However, the modeled adoption of agricultural conservation practices reduced the proportion of surface runoff at the management scale by up to 30×, which greatly outweighs the effects of the land use change scenarios previously modeled [27]. These findings indicate that, at the management-scale, the incorporation of agricultural conservation practices can have similar effects to land use changes on the small water cycle and can greatly reduce the overall proportion of surface runoff contribution to streamflow.

4.4. Implications for Agricultural Conservation Practice Incorporation in the Czech Republic

The incorporation of agricultural conservation practices tend to reinforce the small water cycle regardless of scale of incorporation. These effects are more obvious at the farm-scale than at the management-scale, which should motivate individual farmers to adopt such practices. At the management-scale, the effects of agricultural conservation practices were still significant but the scale and the spatial distribution of adoption were not. This implies that managers should incentivize any willing famers/conglomerates within their management area to adopt such practices. In addition to agricultural conservation practices, other land and crop management factors can also have significant effects on the small water cycle and their interactions should be studied further [26,27]. While soil erosion and sediment transport were not explored in this study, agricultural conservation practices have also been shown to have positive effects concerning these issues and can lead to increased soil conservation [14–16,76,77].

5. Conclusions

This study reinforces SWAT's applicability to the Czech landscape at both the farm- and management-scales. SWAT is very effective in its ability to model various management, land use, and crop change scenarios. While likely exaggerated by the scale, agricultural conservation practice adoption at the farm-scale has significant effects on the small water cycle. The most effective practice modeled at this scale was the incorporation of contour farming. The effects of small field incorporation at the farm-scale tended to have significantly negative impacts on the small water cycle, but this result is likely an artifact due to the HRU processing in SWAT. At the management-scale in the Czech Republic, any degree of incorporation of agricultural conservation practices makes significant impacts on the small water cycle, according to the Vrchlice SWAT model. SWAT was able to model that the incorporation of agricultural conservation practices in a primarily agricultural landscape can have significant effects on the small water cycle, especially regarding surface runoff ratios. While SWAT is not fully distributed and real-world effects would likely vary, this study indicates that managers should encourage agricultural conservation practices, regardless of scale or spatial distribution. As this study only focuses on the effects of agricultural conservation practices on the small water cycle, further studies should be conducted to model their effects on erosion as well as the interactions between agricultural conservation practices and land use/management changes in the Czech landscape.

Author Contributions: Conceptualization, N.N. and T.D.; methodology, N.N.; modeling, N.N.; writing—original draft preparation, N.N.; writing—review and editing, N.N., T.L., J.W., J.K., and T.D.; figures, N.N., T.L., and J.W.; supervision, T.D. All authors have read and agreed to the published version of the manuscript.

Funding: The presented research has been performed within projects H2020 No. 773903 Shui, focused on water scarcity in European and Chinese cropping systems and the Grant Agency of Czech Technical University in Prague, No. SGS20/156/OHK1/3T/11.

Institutional Review Board Statement: Not applicable.

Informed Consent Statement: Not applicable.

Conflicts of Interest: The authors declare no conflict of interest.

References

1. Zelenakova, M.; Fialová, J.; Negm, A.M. *Assessment and Protection of Water Resources in the Czech Republic*; Springer: Cham, Switzerland, 2020. [CrossRef]
2. Marlow, D. Small Water Cycles: What They Are, Their Importance, Their Restoration. *Proc. R. Soc. Qld* **2019**, *127*.
3. Kravčík, M.; Pokorný, J.; Kohutiar, J.; Kováč, M.; Tóth, E. Water for the Recovery of the Climate-A New Water Paradigm. In Proceedings of the Joint Conference of APLU and ICA, Prague, Czech Republic, 23–26 June 2009.
4. Basch, G.; Friedrich, T.; Kassam, A.; Gonzalez-Sanchez, E. Conservation Agriculture in Europe. In *Conservation Agriculture*; Farooq, M., Siddique, K.H.M., Eds.; Springer: Cham, Switzerland, 2015; ISBN 978-3-319-11619-8.
5. Rockström, J.; Karlberg, L.; Wani, S.P.; Barron, J.; Hatibu, N.; Oweis, T.; Bruggeman, A.; Farahani, J.; Qiang, Z. Managing water in rainfed agriculture—The need for a paradigm shift. *Agric. Water Manag.* **2010**, *97*, 543–550. [CrossRef]
6. Stevenson, J.R.; Serraj, R.; Cassman, K.G. Evaluating conservation agriculture for small-scale farmers in Sub-Saharan Africa and South Asia. *Agric. Ecosyst. Environ.* **2014**, *187*, 1–10. [CrossRef]

7. van Wie, J.B.; Adam, J.C.; Ullman, J.L. Conservation tillage in dryland agriculture impacts watershed hydrology. *J. Hydrol.* **2013**, *483*, 26–38. [CrossRef]
8. Gómez Calero, J.A.; Krása, J.; Quinton, J.N.; Klik, A.; Fereres Castiel, E.; Intrigliolo, D.S.; Chen, L.; Strauss, P.; Yun, X.; Dostál, T. *Best Management Practices for Optimized Use of Soil and Water in Agriculture*; Spanish National Research Council (CSIC): Madrid, Spain, 2021; Available online: http://hdl.handle.net/10261/246622 (accessed on 22 February 2022).
9. Kassam, A.; Friedrich, T.; Shaxson, F.; Pretty, J. The spread of Conservation Agriculture: Justification, sustainability and uptake. *Int. J. Agric. Sustain.* **2009**, *7*, 292–320. [CrossRef]
10. Choudhary, M.A.; Lal, R.; Dick, W.A. Long-term tillage effects on runoff and soil erosion under simulated rainfall for a central Ohio soil. *Soil Tillage Res.* **1997**, *42*, 175–184. [CrossRef]
11. Chow, T.L.; Rees, H.W.; Monteith, J. Seasonal distribution of runoff and soil loss under four tillage treatments in the upper St. John River valley New Brunswick, Canada. *Can. J. Soil. Sci.* **2000**, *80*, 649–660. [CrossRef]
12. Sun, Y.; Zeng, Y.; Shi, Q.; Pan, X.; Huang, S. No-tillage controls on runoff: A meta-analysis. *Soil Tillage Res.* **2015**, *153*, 1–6. [CrossRef]
13. Hösl, R.; Strauss, P. Conservation tillage practices in the alpine forelands of Austria—Are they effective? *CATENA* **2016**, *137*, 44–51. [CrossRef]
14. Leys, A.; Govers, G.; Gillijns, K.; Berckmoes, E.; Takken, I. Scale effects on runoff and erosion losses from arable land under conservation and conventional tillage: The role of residue cover. *J. Hydrol.* **2010**, *390*, 143–154. [CrossRef]
15. Zuazo, V.H.D.; Pleguezuelo, C.R.R. Soil-erosion and runoff prevention by plant covers. A review. *Agron. Sustain. Dev.* **2008**, *28*, 65–86. [CrossRef]
16. Greene, R.S.B.; Kinnell, P.I.A.; Wood, J.T. Role of plant cover and stock trampling on runoff and soil-erosion from semi-arid wooded rangelands. *Soil Res.* **1994**, *32*, 953. [CrossRef]
17. Barros, V.R.; Field, C.B.; Dokken, D.J.; Mastrandrea, M.D.; Mach, K.J.; Bilir, T.E.; Chatterjee, M.; Ebi, K.L.; Estrada, Y.O.; Genova, R.C.; et al. *Climate Change 2014: Impacts, Adaptation and Vulnerability. Part B: Regional Aspects. Contribution of Working Group II to the Fifth Assessment Report of the Intergovernmental Panel on Climate Change*; Cambridge University Press: Cambridge, UK; New York, NY, USA, 2014.
18. Ministry of Agriculture of the Czech Republic. *We Support Traditions and Rural Development in the Czech Republic*; Ministry of Agriculture of the Czech Republic: Prague, Czech Republic, 2018; ISBN 978-80-7434-416-9.
19. Zumr, D.; Kubicek, J.; Dostal, T. Temporary Variable Soil Structure and Its Effect on Runoff Mechanism on Intensively Cultivated Land. European Geosciences Union. In *EGU General Assembly Conference Abstracts, Vienna, Austria, 7–12 April 2013*; Copernicus: Göttingen, Germany, EGU2013-9408.
20. Bauer, M.; Zumr, D.; Krása, J.; Dostál, T.; Jáchymová, B.; Rosendorf, P. Sediment and Phosphorus Fluxes-Monitoring and Modelling from Field to Regional Scale-Connectivity Implications. In *EGU General Assembly Conference Abstracts, Vienna, Austria, 12–17 April 2015*; Copernicus: Göttingen, Germany, EGU2015-11171.
21. Jeřábek, J.; Zumr, D.; Strouhal, L. Predominant Runoff Components During Heavy Rainfall Events on Cultivated Catchment. American Geosciences Union. In *AGU Fall Meeting Abstracts, Vienna, Austria, 14–18 December 2015*; Copernicus: Göttingen, Germany, AGU 2015 H43I-1664.
22. Zumr, D.; Strouhal, L.; Kavka, P. Runoff Generation and Flow Paths on an Inclined Cultivated Soil. In *EGU General Assembly Conference Abstracts, Vienna, Austria, 12–17 December 2015*; Copernicus: Göttingen, Germany, EGU2015-6718.
23. Zumr, D.; Vláčilová, M.; Dostál, T.; Jeřábek, J.; Sobotková, M.; Sněhota, M. Spatial Analysis of Subsoil Compaction on Cultivated Land by Means of Penetrometry, Electrical Resistence Tomography and X-ray Computed Tomography. In *EGU General Assembly Conference Abstracts, Vienna, Austria, 12–17 April 2015*; Copernicus: Göttingen, Germany, EGU2015-12926.
24. Krasa, J.; Dostal, T.; van Rompaey, A.; Vaska, J.; Vrana, K. Reservoirs' siltation measurements and sediment transport assessment in the Czech Republic, the Vrchlice catchment study. *CATENA* **2005**, *64*, 348–362. [CrossRef]
25. Krasa, J.; Dostal, T.; Jachymova, B.; Bauer, M.; Devaty, J. Soil erosion as a source of sediment and phosphorus in rivers and reservoirs-Watershed analyses using WaTEM/SEDEM. *Environ. Res.* **2019**, *171*, 470–483. [CrossRef]
26. Noreika, N.; Li, T.; Zumr, D.; Krasa, J.; Dostal, T.; Srinivasan, R. Farm-Scale Biofuel Crop Adoption and Its Effects on In-Basin Water Balance. *Sustainability* **2020**, *12*, 10596. [CrossRef]
27. Noreika, N.; Winterová, J.; Li, T.; Krása, J.; Dostál, T. The Small Water Cycle in the Czech Landscape: How Has It Been Affected by Land Management Changes Over Time? *Sustainability* **2021**, *13*, 13757. [CrossRef]
28. Arnold, J.G.; Moriasi, D.N.; Gassman, P.W.; Abbaspour, K.C.; White, M.J.; Srinivasan, R.; Santhi, C.; Harmel, R.D.; van Griensven, A.; van Liew, M.W.; et al. SWAT: Model Use, Calibration, and Validation. *Trans. ASABE* **2012**, *55*, 1491–1508. [CrossRef]
29. van Liew, M.W.; Veith, T.L.; Bosch, D.D.; Arnold, J.G. Suitability of SWAT for the Conservation Effects Assessment Project: Comparison on USDA Agricultural Research Service Watersheds. *J. Hydrol. Eng.* **2007**, *12*, 173–189. [CrossRef]
30. Arnold, J.G.; Srinivasan, R.; Muttiah, R.S.; Williams, J.R. LARGE AREA HYDROLOGIC MODELING AND ASSESSMENT PART I: MODEL DEVELOPMENT. *J. Am. Water Resour. Assoc.* **1998**, *34*, 73–89. [CrossRef]
31. Moriasi, D.N.; Arnold, J.G.; van Liew, M.W.; Bingner, R.L.; Harmel, R.D.; Veith, T.L. Model Evaluation Guidelines for Systematic Quantification of Accuracy in Watershed Simulations. *Trans. ASABE* **2007**, *50*, 885–900. [CrossRef]
32. Arabi, M.; Frankenberger, J.R.; Engel, B.A.; Arnold, J.G. Representation of agricultural conservation practices with SWAT. *Hydrol. Process.* **2008**, *22*, 3042–3055. [CrossRef]

33. Neitsch, S.L.; Arnold, J.G.; Kiniry, J.R.; Williams, J.R. *Soil and Water Assessment Tool Theoretical Documentation*; Technical Report for Texas Water Resources Institute: College Station, TX, USA, 2011.
34. Daggupati, P.; Douglas-Mankin, K.R.; Sheshukov, A.Y.; Barnes, P.L.; Devlin, D.L. Field-Level Targeting Using SWAT: Mapping Output from HRUs to Fields and Assessing Limitations of GIS Input Data. *Trans. ASABE* **2011**, *54*, 501–514. [CrossRef]
35. Abouabdillah, A.; White, M.; Arnold, J.G.; De Girolamo, A.M.; Oueslati, O.; Maataoui, A.; Lo Porto, A. Evaluation of soil and water conservation measures in a semi-arid river basin in Tunisia using SWAT. *Soil Use Manage.* **2014**, *30*, 539–549. [CrossRef]
36. Bosch, N.S.; Allan, J.D.; Selegean, J.P.; Scavia, D. Scenario-testing of agricultural best management practices in Lake Erie watersheds. *J. Great Lakes Res.* **2013**, *39*, 429–436. [CrossRef]
37. Bosch, N.S.; Evans, M.A.; Scavia, D.; Allan, J.D. Interacting effects of climate change and agricultural BMPs on nutrient runoff entering Lake Erie. *J. Great Lakes Res.* **2014**, *40*, 581–589. [CrossRef]
38. Briak, H.; Mrabet, R.; Moussadek, R.; Aboumaria, K. Use of a calibrated SWAT model to evaluate the effects of agricultural BMPs on sediments of the Kalaya river basin (North of Morocco). *Int. Soil Water Conserv. Res.* **2019**, *7*, 176–183. [CrossRef]
39. Chen, Y.; Marek, G.W.; Marek, T.H.; Porter, D.O.; Brauer, D.K.; Srinivasan, R. Simulating the effects of agricultural production practices on water conservation and crop yields using an improved SWAT model in the Texas High Plains, USA. *Agric. Water Manag.* **2021**, *244*, 106574. [CrossRef]
40. Daloğlu, I.; Nassauer, J.I.; Riolo, R.; Scavia, D. An integrated social and ecological modeling framework—Impacts of agricultural conservation practices on water quality. *Ecol. Soc.* **2014**, *19*, 12. [CrossRef]
41. Daloğlu, I.; Nassauer, J.I.; Riolo, R.L.; Scavia, D. Development of a farmer typology of agricultural conservation behavior in the American Corn Belt. *Agric. Syst.* **2014**, *129*, 93–102. [CrossRef]
42. Dechmi, F.; Skhiri, A. Evaluation of best management practices under intensive irrigation using SWAT model. *Agric. Water Manag.* **2013**, *123*, 55–64. [CrossRef]
43. Elçi, A. Evaluation of nutrient retention in vegetated filter strips using the SWAT model. *Water Sci. Technol.* **2017**, *76*, 2742–2752. [CrossRef] [PubMed]
44. Engebretsen, A.; Vogt, R.D.; Bechmann, M. SWAT model uncertainties and cumulative probability for decreased phosphorus loading by agricultural Best Management Practices. *CATENA* **2019**, *175*, 154–166. [CrossRef]
45. Gitau, M.W.; Gburek, W.J.; Bishop, P.L. Use of the SWAT Model to Quantify Water Quality Effects of Agricultural BMPs at the Farm-Scale Level. *Trans. ASABE* **2008**, *51*, 1925–1936. [CrossRef]
46. Gitau, M.W.; Veith, T.L.; Gburek, W.J. Farm level optimization of bmp placement for cost–effective pollution reduction. *Trans. ASAE* **2004**, *47*, 1923–1931. [CrossRef]
47. Himanshu, S.K.; Pandey, A.; Yadav, B.; Gupta, A. Evaluation of best management practices for sediment and nutrient loss control using SWAT model. *Soil Tillage Res.* **2019**, *192*, 42–58. [CrossRef]
48. Jang, S.S.; Ahn, S.R.; Kim, S.J. Evaluation of executable best management practices in Haean highland agricultural catchment of South Korea using SWAT. *Agric. Water Manag.* **2017**, *180*, 224–234. [CrossRef]
49. Kalcic, M.M.; Frankenberger, J.; Chaubey, I. Spatial Optimization of Six Conservation Practices Using Swat in Tile-Drained Agricultural Watersheds. *J. Am. Water Resour. Assoc.* **2015**, *51*, 956–972. [CrossRef]
50. Lamba, J.; Thompson, A.M.; Karthikeyan, K.G.; Panuska, J.C.; Good, L.W. Effect of best management practice implementation on sediment and phosphorus load reductions at subwatershed and watershed scale using SWAT model. *Int. J. Sediment Res.* **2016**, *31*, 386–394. [CrossRef]
51. Liu, R.; Zhang, P.; Wang, X.; Wang, J.; Yu, W.; Shen, Z. Cost-effectiveness and cost-benefit analysis of BMPs in controlling agricultural nonpoint source pollution in China based on the SWAT model. *Environ. Monit. Assess.* **2014**, *186*, 9011–9022. [CrossRef] [PubMed]
52. Liu, Y.; Guo, T.; Wang, R.; Engel, B.A.; Flanagan, D.C.; Li, S.; Pijanowski, B.C.; Collingsworth, P.D.; Lee, J.G.; Wallace, C.W. A SWAT-based optimization tool for obtaining cost-effective strategies for agricultural conservation practice implementation at watershed scales. *Sci. Total Environ.* **2019**, *691*, 685–696. [CrossRef]
53. Liu, Y.; Wang, R.; Guo, T.; Engel, B.A.; Flanagan, D.C.; Lee, J.G.; Li, S.; Pijanowski, B.C.; Collingsworth, P.D.; Wallace, C.W. Evaluating efficiencies and cost-effectiveness of best management practices in improving agricultural water quality using integrated SWAT and cost evaluation tool. *J. Hydrol.* **2019**, *577*, 123965. [CrossRef]
54. López-Ballesteros, A.; Senent-Aparicio, J.; Srinivasan, R.; Pérez-Sánchez, J. Assessing the Impact of Best Management Practices in a Highly Anthropogenic and Ungauged Watershed Using the SWAT Model: A Case Study in the El Beal Watershed (Southeast Spain). *Agronomy* **2019**, *9*, 576. [CrossRef]
55. Merriman, K.; Daggupati, P.; Srinivasan, R.; Toussant, C.; Russell, A.; Hayhurst, B. Assessing the Impact of Site-Specific BMPs Using a Spatially Explicit, Field-Scale SWAT Model with Edge-of-Field and Tile Hydrology and Water-Quality Data in the Eagle Creek Watershed, Ohio. *Water* **2018**, *10*, 1299. [CrossRef]
56. Merriman, K.; Russell, A.; Rachol, C.; Daggupati, P.; Srinivasan, R.; Hayhurst, B.; Stuntebeck, T. Calibration of a Field-Scale Soil and Water Assessment Tool (SWAT) Model with Field Placement of Best Management Practices in Alger Creek, Michigan. *Sustainability* **2018**, *10*, 851. [CrossRef]
57. Park, J.-Y.; Yu, Y.-S.; Hwang, S.-J.; Kim, C.; Kim, S.-J. SWAT modeling of best management practices for Chungju dam watershed in South Korea under future climate change scenarios. *Paddy Water Environ.* **2014**, *12*, 65–75. [CrossRef]

58. Phomcha, P.; Wirojanagud, P.; Vangpaisal, T.; Thaveevouthti, T. Modeling the impacts of alternative soil conservation practices for an agricultural watershed with the SWAT model. *Procedia Eng.* **2012**, *32*, 1205–1213. [CrossRef]
59. Ricci, G.F.; Jeong, J.; De Girolamo, A.M.; Gentile, F. Effectiveness and feasibility of different management practices to reduce soil erosion in an agricultural watershed. *Land Use Policy* **2020**, *90*, 104306. [CrossRef]
60. Rocha, J.; Roebeling, P.; Rial-Rivas, M.E. Assessing the impacts of sustainable agricultural practices for water quality improvements in the Vouga catchment (Portugal) using the SWAT model. *Sci. Total Environ.* **2015**, *536*, 48–58. [CrossRef]
61. Tripathi, M.P.; Panda, R.K.; Raghuwanshi, N.S. Development of effective management plan for critical subwatersheds using SWAT model. *Hydrol. Process.* **2005**, *19*, 809–826. [CrossRef]
62. Tuppad, P.; Kannan, N.; Srinivasan, R.; Rossi, C.G.; Arnold, J.G. Simulation of Agricultural Management Alternatives for Watershed Protection. *Water Resour. Manage.* **2010**, *24*, 3115–3144. [CrossRef]
63. Ullrich, A.; Volk, M. Application of the Soil and Water Assessment Tool (SWAT) to predict the impact of alternative management practices on water quality and quantity. *Agric. Water Manag.* **2009**, *96*, 1207–1217. [CrossRef]
64. Uniyal, B.; Jha, M.K.; Verma, A.K.; Anebagilu, P.K. Identification of critical areas and evaluation of best management practices using SWAT for sustainable watershed management. *Sci. Total Environ.* **2020**, *744*, 140737. [CrossRef] [PubMed]
65. Wang, W.; Xie, Y.; Bi, M.; Wang, X.; Lu, Y.; Fan, Z. Effects of best management practices on nitrogen load reduction in tea fields with different slope gradients using the SWAT model. *Appl. Geogr.* **2018**, *90*, 200–213. [CrossRef]
66. Yang, W.; Liu, Y.; Simmons, J.; Oginskyy, A.; McKague, K. SWAT Modelling of Agricultural BMPs and Analysis of BMP Cost Effectiveness in the Gully Creek Watershed. Available online: http://www.abca.on.ca/downloads/wbbe-huron-swat-modelling-2013-08-21.pdf (accessed on 5 February 2022).
67. Zhang, X.; Zhang, M. Modeling effectiveness of agricultural BMPs to reduce sediment load and organophosphate pesticides in surface runoff. *Sci. Total Environ.* **2011**, *409*, 1949–1958. [CrossRef] [PubMed]
68. Nyeko, M. Hydrologic Modelling of Data Scarce Basin with SWAT Model: Capabilities and Limitations. *Water Resour. Manage.* **2015**, *29*, 81–94. [CrossRef]
69. Beven, K. How far can we go in distributed hydrological modelling? *Hydrol. Earth SYstem Sci. Discuss. Eur. Geosci. Union* **2001**, *5*, 1–12. [CrossRef]
70. Martínez-Retureta, R.; Aguayo, M.; Stehr, A.; Sauvage, S.; Echeverría, C.; Sánchez-Pérez, J.-M. Effect of Land Use/Cover Change on the Hydrological Response of a Southern Center Basin of Chile. *Water* **2020**, *12*, 302. [CrossRef]
71. Qi, J.; Li, S.; Bourque, C.P.-A.; Xing, Z.; Meng, F.-R. Developing a decision support tool for assessing land use change and BMPs in ungauged watersheds based on decision rules provided by SWAT simulation. *Hydrol. Earth Syst. Sci.* **2018**, *22*, 3789–3806. [CrossRef]
72. Qi, J.; Zhang, X.; Yang, Q.; Srinivasan, R.; Arnold, J.G.; Li, J.; Waldholf, S.T.; Cole, J. SWAT ungauged: Water quality modeling in the Upper Mississippi River Basin. *J. Hydrol.* **2020**, *584*, 124601. [CrossRef]
73. Jodar-Abellan, A.; Valdes-Abellan, J.; Pla, C.; Gomariz-Castillo, F. Impact of land use changes on flash flood prediction using a sub-daily SWAT model in five Mediterranean ungauged watersheds (SE Spain). *Sci. Total Environ.* **2019**, *657*, 1578–1591. [CrossRef]
74. Chaplot, V. Impact of spatial input data resolution on hydrological and erosion modeling: Recommendations from a global assessment. *Phys. Chem. Earth Parts ABC* **2014**, *67–69*, 23–35. [CrossRef]
75. Geza, M.; McCray, J.E. Effects of soil data resolution on SWAT model stream flow and water quality predictions. *J. Environ. Manage.* **2008**, *88*, 393–406. [CrossRef] [PubMed]
76. Dickey, E.C.; Shelton, D.P.; Jasa, P.J.; Peterson, T. Tillage, Residue and Erosion on Moderately Sloping Soils. *Biol. Syst. Eng.* **1984**, *27*, 1093–1099.
77. Unger, P.W.; Vigil, M.F. Cover crop effects on soil water relationships. *J. Soil Water Conserv.* **1998**, *53*, 200–207.

Article

Comparison of Vegetation Types for Prevention of Erosion and Shallow Slope Failure on Steep Slopes in the Southeastern USA

Homayra Asima [†], Victoria Niedzinski [†,‡], Frances C. O'Donnell * and Jack Montgomery

Department of Civil and Environmental Engineering, Auburn University, Auburn, AL 36849, USA
* Correspondence: fco0002@auburn.edu; Tel.: +1-334-844-7168
† These authors contributed equally to this work.
‡ Current Address: School of Earth and Climate Sciences, University of Maine, Orono, ME 04469, USA.

Abstract: Shallow slope failures due to erosion are common occurrences along roadways. The use of deep-rooted vegetative covers is a potential solution to stabilize newly constructed slopes or repair shallow landslides. This study compared species that may provide slope stabilization for sites in the Piedmont region of the southeastern USA. Six species were tested on experimental plots under natural rainfall conditions, and vegetation health and establishment were monitored. Two methods were used to measure surface erosion, measurement of total suspended solids in collected runoff and erosion pins. While measurement uncertainty was high for both methods, differences were evident between species in the spatial distribution of surface erosion that was related to the quality of vegetation establishment. For three species that established well, soil cores were collected to measure root biomass at depths up to 40 cm. Vetiver grass (*Vetiveria zizaniodies*) had substantially higher mean root biomass (3.75 kg/m^3) than juniper shrubs (*Juniperus chinensis*; 0.45 kg/m^3) and fescue grass (*Lolium arundinaceum*; 1.28 kg/m^3), with the most pronounced difference in the deepest soil layers. Seeding with turf grass such as fescue is a common practice for erosion control in the region but replacing this with vetiver on steep slopes may help prevent shallow landslides due to the additional root reinforcement. Additional work is needed to measure the magnitude of the strength gain.

Keywords: erosion; vegetative covers; root biomass; erosion pins; vetiver grass

1. Introduction

Shallow slope failures due to erosion are common occurrences along roadway slopes in regions where high intensity rainfall is prevalent [1,2]. These instabilities are usually relatively small in size, but the consequences can cause major economic and social disruption [3,4]. Sediment from erosion can also have significant environmental impacts [5]. Many factors can impact roadside erosion, such as rainfall characteristics, slope gradient, rutting caused by lawn maintenance equipment, roadside construction, soil type, and the presence and type of vegetation [6].

The establishment of vegetation on newly constructed slopes can prevent erosion and increase slope stability [2,7–10]. Well-developed aboveground vegetation prevents surface erosion by intercepting rainfall and wind, increasing surface roughness, binding loose soil particles, and creating a physical barrier to sediment movement [11–14]. There is a non-linear relationship between precipitation and sediment yield, with precipitation driving erosion while also increasing vegetation growth up to the point where vegetation is no longer water-limited [15]. Plant root systems provide belowground support and can prevent shallow slope failures by increasing soil strength through reinforcement [16–19]. Roots are strong in resisting tension forces while soil is strong in resisting compression forces [20], so root-permeated soil creates a mixed material that can withstand both forces [21]. Roots perpendicular to the soil surface reinforce the soil mass on the sheared surface while roots growing parallel increase in-plane tensile strength [22,23]. Additionally, plants remove

water from the soil through transpiration, which prevents slope failures by reducing the unit weight of the soil and increasing apparent cohesion from matric suction [3,24–26].

Plant species and functional types differ in the traits that determine their suitability for use in vegetative covers on steep slopes. Recent reviews of vegetation traits and their effects on slope stability have been published by [8] and [10]. Rapid growth after planting and abundant, evenly distributed aboveground biomass are key to preventing surface erosion [6,27]. Generally, herbaceous vegetation performs better than woody species during the establishment phase [28]. Some vegetation types have specific characteristics, such as grasses that can grow in hedgerows [29,30] and ferns that form rhizome mats [31], which make them particularly effective in binding soil and forming physical barriers to erosion. High root length density [32] and fine-root content [33] are important to soil strength, and these are typically associated with herbaceous vegetation. However, woody vegetation tends to have deeper roots to prevent slope failures [18], though deep rooted grass species do exist [34,35]. These traits must be weighed against practical concerns, such as suitability for local conditions, ease of planting, and maintenance requirements [6,27,36–38].

In the southeastern USA., turf grasses grown from seed are the most common vegetative erosion control [6]. These perform well on relatively flat terrain, where deep root structure is not needed for stabilization [39]. However, turf grasses require mowing. On steeper terrain, mowing can cause ruts that increase erosion (Figure 1) and may expand into shallow slope failures during rain events [4]. There is interest among transportation management agencies in finding alternatives to turf grass that would provide deeper soil stabilization while still establishing quickly and preventing surface erosion. The goal of this study is to evaluate the field performance of several candidate species in experimental plots in the Piedmont region of Alabama. This area is especially prone to shallow slope failures and erosion along slopes due to mowing activities [4,5] (Figure 1a).

(a)   (b)

Figure 1. Eroded slopes along (**a**) US-280 near Waverly, Alabama (photo from Google StreetView) and (**b**) Alabama Highway 69 near Tuscaloosa, Alabama (photo provided by Jacob Hodnett, ALDOT).

Previous experimental plot studies of erosion and slope stability have found substantial differences between vegetation types. In a study of very steep (42.5°) slopes in central China, a mix of grass and shrubs reduced runoff and surface erosion, and had deeper roots than grass alone [2]. Plots with any type of planted vegetation performed better than those that were allowed to revegetate naturally. Studies in a semi-arid region of Spain found differences in erosion rate between species that were mainly driven by quality of establishment [40,41]. A previous study in the southeastern USA found a plot planted with a mix of native grass seed had a lower sediment yield than an exotic seed mix [42], though the difference was not statistically significant. A study by [43] examined combining biochemical surface treatments with vegetation and found that using seeded biochemical treatments on slopes was effective for both short- and long-term stabilization against erosion. Recent work by [44] highlighted the potential for vetiver grass to support resilient transportation systems by mitigating slope failures and improving stormwater quality, but this study also highlighted the relative lack of use of vetiver in the United States.

To determine the efficacy of a species for slope stabilization, both prevention of surface erosion and stabilization of deeper soil layers must be assessed. Previous studies have measured sediment yield by collecting runoff and measuring total suspended solids (TSS) in the collected runoff [2,42]. While this method is well-established, constructing runoff collection infrastructure is costly and adequate replication for statistical analysis is difficult to achieve. The method is also prone to missing data [42], especially during periods of high intensity rainfall when most erosion occurs. Methods that directly measure changes in the elevation of the vegetated surface are an alternative that have the advantage of characterizing spatial patterns in surface erosion. Some techniques that are common for bare soil surfaces, such as total station surveys and lidar scanning, do not work well on vegetated surfaces [45,46]. Erosion pins offer a simple, low-cost alternative that provides point-based measurements of erosion or deposition through a manual measurement of surface height relative to the fixed reference point of the pin head. In this study, both runoff collection and erosion pins are used to assess surface erosion. For deeper soil stabilization, the measurement of root biomass and morphological characteristics in soil core samples is an accepted and established method [2,47].

This study has the following objectives:

- Select vegetation types that may provide erosion control and slope stabilization for priority sites and compare them to the current management practice of planting turf grass from seed.
- Determine which species establish and grow well on moderately steep roadside slopes in the Piedmont region of the southeastern USA using experimental plots. This includes vetiver grass, which has not previously been used in this region.
- Compare surface erosion rates from the experimental plots based on whole-plot sediment yield determined from runoff collection and point-based erosion or deposition measured with erosion pins.
- Estimate the contribution of each species to increased slope stability by measuring root biomass and diameter distribution in soil cores collected from the experimental plots.

This study addresses processes on small (<50 m) constructed slopes in humid environments over short time scales (<2 years) post-construction. Recent research on vegetation-sediment interaction has emphasized the importance of orographic effects on precipitation that occur over large elevation gradients [48] and ecogeomorphic coevolution of landforms that occurs over centennial scales in arid and semi-arid environments [49], and these are outside the scope of the current research. Based on erosion control strategies that were successful in previous studies [40,41], we focus on planted vegetation rather than allowing for vegetation to establish naturally after disturbance. Therefore, the variation in species prevalence associated with slope, aspect, soil type, and other factors is not considered.

2. Materials and Methods

The materials and methods for this study are summarized in the flow chart shown in Figure 2.

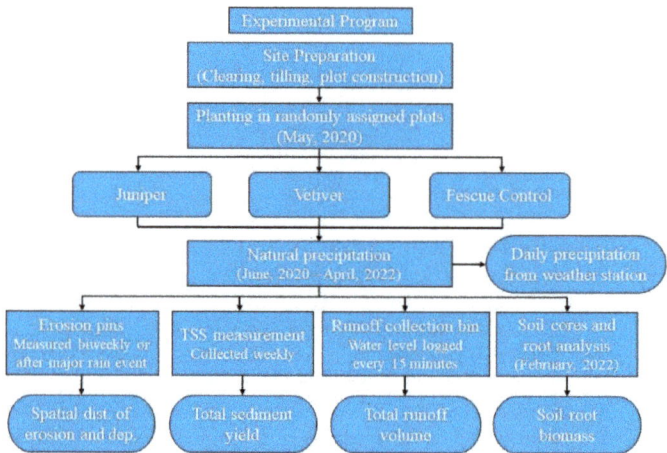

Figure 2. Flow chart illustrating the materials and methods for this study.

2.1. Study Site

This study focuses on the Piedmont region of the southeastern USA, and particularly the state of Alabama. Shallow slope failures and erosion on highway cut slopes are a common occurrence in this region due to the prevalence of high-intensity rainfall [50] and hilly terrain. The Piedmont region is one of the fastest growing areas of the United States in terms of population and land-use change [51] and there is a need to identify sustainable solutions to managing soil erosion in this region. This area has a humid subtropical climate with mean annual rainfall of 132 cm and mean annual temperature of 17.9 °C. Most rainfall occurs as localized convective thunderstorms occurring in the summer and as widespread frontal precipitation, including tropical storms that occur in the fall through spring. Class A annual pan evaporation is 122 cm [52]. While vegetation growth is generally not water-limited, droughts do occur, particularly in the fall.

The experimental plots were established on a roadside slope along the National Center for Asphalt Technology (NCAT) Test Track in Opelika, AL, USA (32.595390°, −85.296363°). The plots have a 25–30° slope. Particle size analysis of study area soil was performed with the Integral Suspension Pressure method [53] using a Pario device (Meter Environment, Pullman, WA). The surface soil layer (0–25 cm) is clay loam (29% sand, 40% silt and 31% clay) and deeper layers (>25 cm) are silt loam (23% sand, 65% silt and 12% clay). The plots are on a north-facing slope, so conditions are slightly cooler with less radiation than the local average. Daily precipitation data were collected onsite by NCAT using an automated weather station.

2.2. Experimental Plot Design

The experimental plot design was based on [2] and [42]. The plots were prepared, built and planted in May 2020. Existing vegetation was treated with Round-Up herbicide (Bayer, Germany) and removed with a small excavator. Each plot consisted of a 1.5 m × 3 m wooden frame built from pressure-treated lumber (Figure 3). The outlet of each plot was tapered to a 45 cm exit to create a total surface area of 5.23 m². The four corners of all the frames had rebar installed to keep the plots stable on the slope and maintain the shape of the 45 cm outlet. The planks at the end of all the frames were wrapped with plastic sheeting and the ground between them had sheeting shingled under the earth to create a smooth path. A trench was dug 2 m above the frames, creating a berm directly above the plot to divert water coming from the slope above the plots. An erosion fence was erected at the top of the frames and below the berm to prevent sediment movement from the slope above the plots.

Figure 3. Completed experimental plot after installation and planting with plots 1–5 shown left to right (**a**) and plot dimensions (**b**).

2.3. Vegetation Planting and Maintenance

Prior to planting, the area was tilled with a mechanical rototiller. Vegetation and seeds were planted according to the providers' instructions. Seeding straw was spread over plots where plants were grown from seed and in bare areas of plots with potted plants. Plots were watered regularly for three weeks after planting. Six vegetation types were selected for testing based on previous literature on vegetative covers for erosion control:

- Grass Control: One plot was planted with turf grass typical of what would be used for erosion control under current management practices in Alabama [6]. The Kentucky-31 cultivar of fescue grass (*Lolium arundinaceum*) was planted from seed. This plot was used to compare the other species with the status quo.
- Deep-rooted Grass: Vetiver grass (*Vetiveria zizaniodies*) is a deep-rooted grass native to southeast Asia that has been used for decades to improve slope stability, improve streambank establishment, and decrease sediment run-off in agricultural areas [44,54–56]. The grass is planted as a slip rather than as a seed and is generally sold sterile so it will not flower and be invasive to the surrounding native flora [30]. Slips were planted in four hedgerows parallel to the slope.
- Woody Shrub: Juniper is a deep-rooted, drought tolerant woody shrub that grows well on steep slopes [56]. It was grown from potted plants. Either *Juniperus chinensis* or *Juniperus horizontalis* was planted based on availability. The two species have similar characteristics.
- Perennial Legume: Hairy vetch (*Vivica villosa*) is a winter-active legume species used for erosion control and as an agricultural cover crop due to its ability to fix atmospheric nitrogen. It has a fast above- and belowground growth rate and high transpiration rate [57]. Hairy vetch grows best when planted in the fall so it can be beneficial for fall/winter construction projects when other species are typically dormant [56]. It was planted from seed in this study.
- Fern: Ferns are useful in erosion control practices as they create dense, long-lasting ground cover and naturally grow in disturbed areas with low nutrient and moisture access [58]. Maidenhair fern (*Adiantum pedatum*) is native to the southeast and is able to grow on near-vertical faces [31]. They were planted from potted plants.
- Native Prairie Grass: Native species are generally preferred in landscaping because they are adapted to local conditions and can enhance native biodiversity [59–61]. Unlike non-native deep-rooted grasses, they can be grown from seed. Switchgrass (*Panicum virgatum*), which has a deep and fibrous root system and is climatically adapted throughout the USA [27], was planted from seed.

Initially, each species, except switchgrass, was planted in one randomly assigned plot. There were problems with the establishment of maidenhair fern and hairy vetch. One was replaced with switchgrass and the other with a mix of juniper and fescue grass after the first year of the study. The species used and planting dates are summarized in Table 1. The vetiver grass was trimmed from a height of 2 m to a height of 1 m in April 2021. Weeds were a persistent problem. A broad-spectrum herbicide (Spectracide, United Industries, St. Louis, MO, USA) was applied and weeds were manually removed from all plots in July 2020 and late June and early July of 2021.

Table 1. Species planting dates in experimental plots. Plot numbers are described in Figure 3.

Species	Plot	Planting Date	Termination Date
Juniper [1]	1	May 2020	August 2022
Vetiver	2	May 2020	August 2022
Fescue	3	May 2020	August 2022
Maidenhair Fern	4	May 2020	July 2021
Hairy Vetch	5	May 2020	April 2021
Juniper [2] and Fescue	4	July 2021	August 2022
Switchgrass	5	April 2021	August 2022

[1] *Juniperus chinensi*; [2] *Juniperus horizontalis*.

2.4. Runoff and Erosion

Two methods were used to estimate erosion: runoff collection with TSS measurements and erosion pins. The runoff collection method was applied from June 2020 to March 2021, and erosion pins were applied from August 2021 to April 2022. Previous studies have traditionally only used one of these methods [2,16,62], but we found them to be complimentary to obtain both the spatial distribution of erosion and deposition and the total settlement yield.

2.4.1. Runoff Collection with TSS

A hole was dug at the base of each plot, and a 68-L plastic bin was placed in the hole. The bins were positioned with the plastic sheeting at the base of each plot flowing into the bins. Bins were partially covered with lids to minimize water loss due to evaporation. In each bin, a U20L HOBO water level logger (Onset, Bourne, MA, USA) was suspended from the lid with wire and submerged in water. The loggers were deployed on October 7 2020 and were set to record pressure and temperature every 15 minutes. Prior to this date, water depths were measured manually once per week. An additional logger was placed outside of the bins to record atmospheric pressure. Measured pressures were converted to change in water level at 15-min intervals using software HOBOware V3.7 analysis software (Onset, Bourne, MA, USA). Change in water level was converted to change in volume using the dimensions of the bin.

Once per week, a well-mixed sample of water from each bin was collected. TSS in each sample was measured by filtration following US Environmental Protection Agency method 160.2 [63]. After sampling, the bins were emptied and cleaned and filled with a known volume of clean water such that the water was above the measurement threshold of the water level logger. Assuming TSS of the clean water is negligible, sediment yield (SY) in g for each one-week collection period was determined by multiplying the measured TSS (g/L) by the total volume in the bin (L) at the end of the week. The volume was divided by the surface area of the plot (5.23 m^2) to give runoff depth (mm, after unit conversion).

2.4.2. Erosion Pins

Erosion pins estimate erosion and deposition at point locations by indicating the change in height of the land surface relative to a fixed reference [64]. EasyFlex 8-inch (20 cm) nylon anchoring spikes (Dimex, Marietta, OH, USA) were used as erosion pins and were installed perpendicular to the ground. Three erosion pins were installed in each plot on 12 April 2021. Pins were installed 0.3 m inside the left boundary of the plots. This distance minimizes edge effects while making it possible to measure the pin height without requiring foot traffic in the plot. The pins were placed 0.6, 1.5, and 2.4 m away from the upper boundary of the plot. These pins are considered upslope, midslope, and downslope, respectively, for analysis. After the first erosion pins produced reasonable data, four additional pins were added to each plot to increase statistical power. Three were added inside the right boundary of the plot in the same configuration as the previous pins. One additional upslope pin was added in the middle of the plot 0.3 m from the top boundary.

A ruler with 1 mm gradations was used to measure the visible height of the erosion pins above the ground. Values that are greater than the baseline value indicate erosion is occurring at the point, while values less than baseline indicate that deposition is occurring. The erosion pins were monitored and measured biweekly or after any large rain event from 12 April 2021 to 15 April 2022. Due to soil disturbance from installation of the erosion pins, data were not collected during a one-month stabilization period after installation [65]. The first measurement after this period was used as the baseline height for the study. Thus, the first set of pins was analyzed from 25 May 2021 to 15 April 2022 and the second set of pins was analyzed from 12 October 2021 to 15 April 2022.

Some studies have suggested that the absolute value of change in erosion pin height is a better indicator of erosion when multiple pins are considered, because it differentiates plots with both erosion and deposition [65]. In this study, we are interested in overall sediment yield from the plots, so actual change in pin height is used for analysis. Linear regression analysis with either time in days or cumulative rainfall based on daily measurements is the independent (X) variable and change in erosion pin height is the dependent (Y) variable. Time and cumulative rainfall are strongly correlated, so they were considered in separate single-variable regression models rather than a multivariate model. The measurement on 12 October 2021 was used as the baseline because all pins were installed and stabilized by that date. A statistically significant positive slope indicates erosion is occurring at the pin location while a negative slope indicates deposition. One-way ANOVA was used to determine if change in erosion pin height from 12 October 2021 to 15 April 2022 was significantly different among plots. An unpaired t-test was used to determine if the mean change in erosion pin height over the same time period was different for upslope and downslope pins. A significance level of 10% was used for statistical analyses due to the inherently high variability in erosion data. All statistical analyses were performed in Microsoft Excel V16.

2.5. Root Biomass Analysis

Samples for root biomass testing were collected from the three plots that were planted at the beginning of the study and had established well: fescue, vetiver, and juniper. Samples were collected in February 2022 after nearly two years of growth. Sampling was carried out when the soil was moist, for best results [66]. A fixed-volume soil core sampler (AMS, American Falls, ID, USA) was used to collect the samples. Cores were collected with a slide hammer until the point of resistance, which was reached at 40 cm depth. One upslope and one downslope sample was collected 0.6 m and 1.8 m from the upper plot boundary.

The soil cores were cut into 5 cm sections to determine the distribution of root biomass with depth. Due to dry soil near the bottom of the soil profile, the core could not be sectioned below 30 cm depth, so 30–40 cm depth is analyzed together. Methods from [67] were used to determine root biomass. The samples were dried at 110 °C for 24 h and weighed before and after drying to determine moisture content. After drying, samples were soaked in tap water for 30 min to break down soil aggregates. Roots were collected by washing the samples through a 2 mm (#10) sieve followed by a 600 μm (#30) sieve under running tap water. The roots collected on the sieves were dried at 60°C for 24 h. Roots were weighed after drying to determine dry root biomass in each sample (g) and converted to a soil root biomass (g/m^3) by dividing by the volume of the core section.

3. Results

3.1. Vegetation Growth and Establishment

After the initial planting, all five of the plant species were able to grow and establish successfully. However, native weeds and plants in the surrounding area began to encroach and quickly took over the plots, and weeding was required. After weeding, the two plots containing the hairy vetch and maidenhair fern were overwhelmed by the disturbance and showed little new growth. The plots were overtaken by fescue grass from the adjacent plot, as well as white clover (*Trifolium repens*) and ground-ivy (*Glechoma hederacea*). Fescue grass grew well if it had direct sun. The small amount of shading caused by the silt fence resulted in poor establishment at the top of the plot. The height of the fence was reduced after the first year of the study.

During the first year of the study, vetiver and juniper were the most successful at general establishment. The area surrounding the junipers was overgrown by similar weeds as the other plots, but this did not impact the growth of the juniper. A mix of juniper and fescue grass was planted at the start of the second year (Table 1) as a potential strategy to improve the weed resistance of the area surrounding the juniper. However, weeds were still an issue. By September 2020, the vetiver was nearly at its full height (2 m) and the hedgerows developed an almost impenetrable layer that was resistant to weeds. In November 2020, the vetiver reverted to a dormant stage but remained healthy and regrew the next year. The switchgrass that was planted in the second year of the study did not grow well as it was planted off season due to issues with obtaining seeds. Thus, switchgrass is excluded from further analysis. Juniper, vetiver, and fescue are compared in the subsequent analyses due to their good establishment and consistent growth throughout the study (Figure 4).

Figure 4. Final condition of the three plots where vegetation established well, photographed in April 2022. Shown from left to right are juniper, vetiver, and fescue.

3.2. Runoff and Erosion

3.2.1. Runoff Collection Method

The time series of runoff for the three plots is shown in Figure 5 for two representative rainy periods, one during the first year after planting and one almost one year after planting. The juniper had the lowest runoff volumes while vetiver and fescue had similar values during most of each rain event. While subtle differences between plots in slope or underlying soils cannot be ruled out, this difference was consistent across rain events.

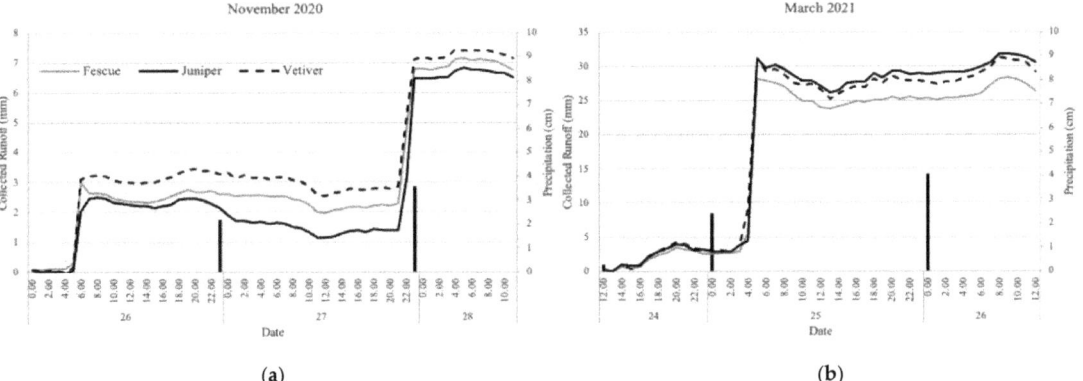

Figure 5. Daily precipitation and hourly change in runoff volume from three erosion plots during rainy periods: (a) six months after planting; and (b) ten months after planting.

Based on the runoff collection and TSS method, the plot with fescue grass had the highest sediment yield over the first nine months of the study while the plot with the juniper shrubs had the lowest (Figure 6). The initial spike in sediment yield in June was collected one month after planting and shows that the grass provided the least amount of initial surface-soil stabilization. All three species showed similar spikes in sediment during rain events at the end of November and the end of March.

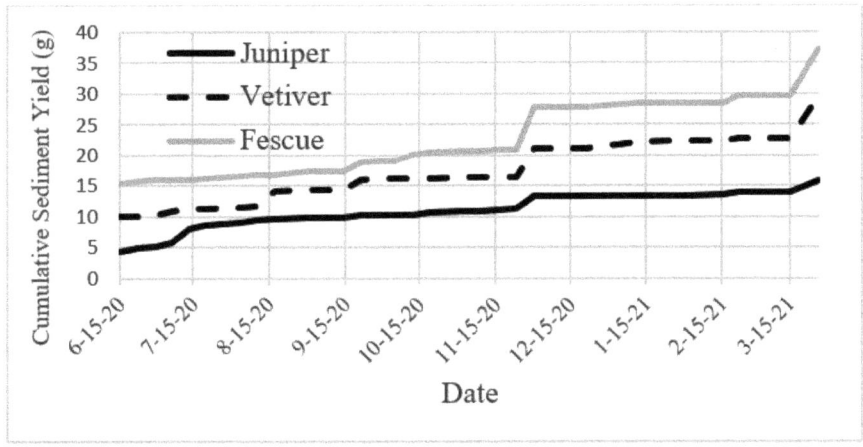

Figure 6. Cumulative sediment yield during the first year of the study measured using the runoff collection with TSS method.

3.2.2. Erosion Pins

As previously discussed, a single-variable regression model was used to assess changes in erosion pin height during the study period. The linear regression models with time and cumulative rainfall as the independent variables produced similar results in terms of which subsets of the data had the best model performance. However, R^2 values were consistently higher for models with time as the dependent variable, so this is considered for analysis. Linear regression between time and erosion pin height showed spatial variability in erosion and deposition among the species tested. For juniper (Figure 7) the slopes for upslope, midslope and downslope positions are 0.018, 0.006 and −0.02 respectively. This indicates erosion from the top pins, almost no erosion at the middle pins, and deposition at the bottom pins. The juniper showed more growth at the base of the plot, which may have slowed water flow leading to deposition. For fescue grass (Figure 8), the slopes for upslope, midslope and downslope positions are 0.011, −0.007 and 0.013, respectively. There is deposition and erosion evident at the midslope and downslope, respectively. While the slope was positive for the upslope pins, the high variability in the data for this area makes it difficult to draw conclusions about the dynamics. For vetiver (Figure 9), the slopes for upslope, midslope and downslope positions are 0.001, 0.011 and 0.018, respectively. The vetiver established uniformly across the plot but was not present near the plot outlet because it was not possible to plant slips in this small area. For a uniform surface, the flow velocity is expected to be highest at the bottom of the plot because of the accumulation of rainfall over the plot. This could be why the vetiver shows little to no erosion at the upslope and midslope and erosion at the downslope. The denser growth of the vetiver may also change the overland flow patterns and velocities [68], but these flow patterns were not measured in this study.

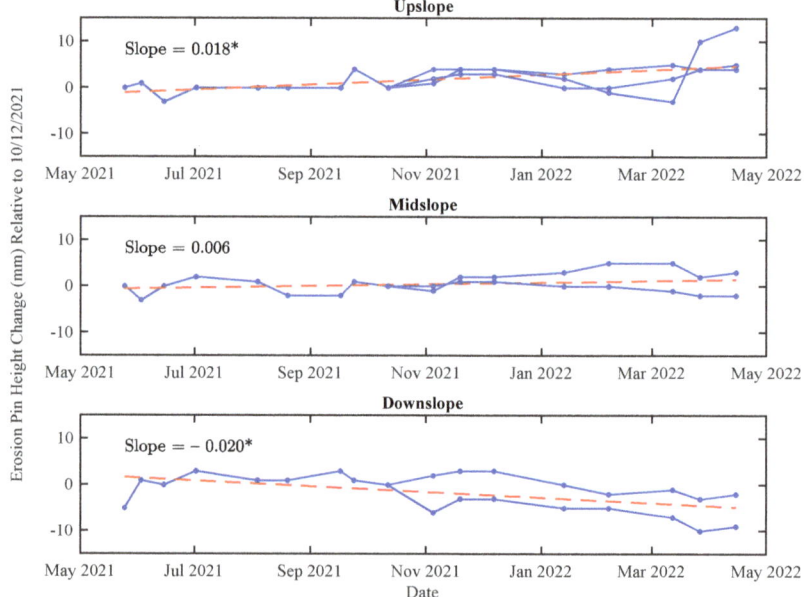

Figure 7. Linear regression for the plot with juniper between time in days and erosion pin height relative to the measurement on 12 October 2021, when all erosion pins were installed and stabilized. Solid blue lines and markers show the trajectory of measurements for each erosion pin and dashed red lines show the regression line. Regression slope is given in each plot with * indicating a regression p-value less than 0.1. Regression lines are calculated separately for upslope (**top**), midslope (**middle**), and downslope (**bottom**).

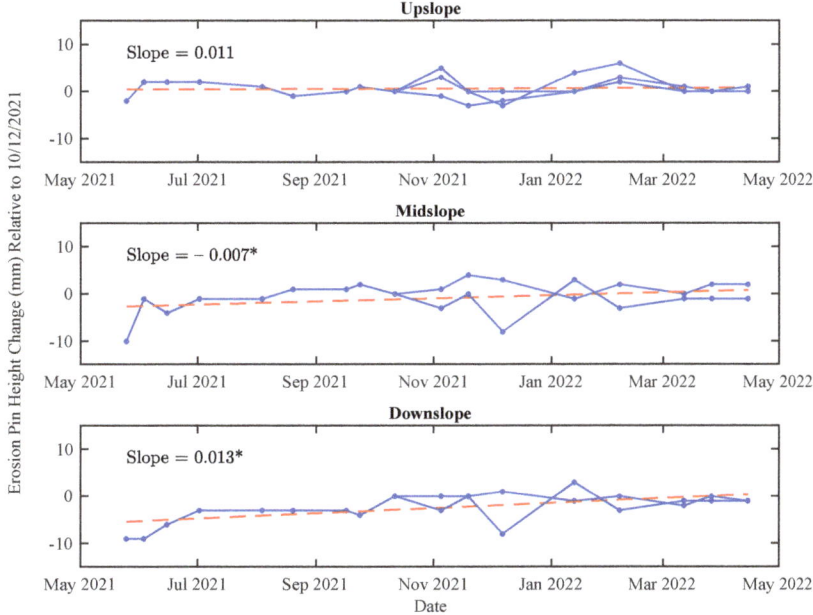

Figure 8. Linear regression for the plot with vetiver between time in days and erosion pin height relative to the measurement on 12 October 2021, when all erosion pins were installed and stabilized. Solid blue lines and markers show the trajectory of measurements for each erosion pin and dashed red lines show the regression line. Regression slope is given in each plot with * indicating a regression p-value less than 0.1. Regression lines are calculated separately for upslope (**top**), midslope (**middle**), and downslope (**bottom**).

The regression analysis summary (Table 2) shows that the R^2 value for every species is less than 0.5, indicating that the independent variable (time) is explaining only a small amount of the variation in the dependent variable (erosion pin height). Other potential sources of variation include spatial variability in erosion patterns and measurement errors. The species differed in which positions showed significant erosion or deposition. However, where erosion or deposition was occurring, the rates were similar, as indicated by overlapping 90% confidence intervals of the slope values. An exception to this is between the vetiver and juniper. These do not overlap at the upslope or downslope locations, though it should be noted that the R^2 and p values for vetiver upslope indicate that time was not a significant predictor of change in erosion pin height (Table 2).

One-way ANOVA analysis of the effect of species on change in erosion pin height did not show a significant effect ($F = 0.55$, $p = 0.58$), indicating similar mean change among plots (Figure 10a). Thus, the differences between plots were primarily in the spatial distribution of erosion and deposition due to differences in the uniformity of vegetation establishment. The influence of slope position indicated a clear pattern of positive change in pin height for upslope pins, indicating erosion negative values, and deposition in downslope pins (Figure 9b). A t-test demonstrated a significant difference between upslope and downslope pins ($t = 1.48$, $p = 0.08$).

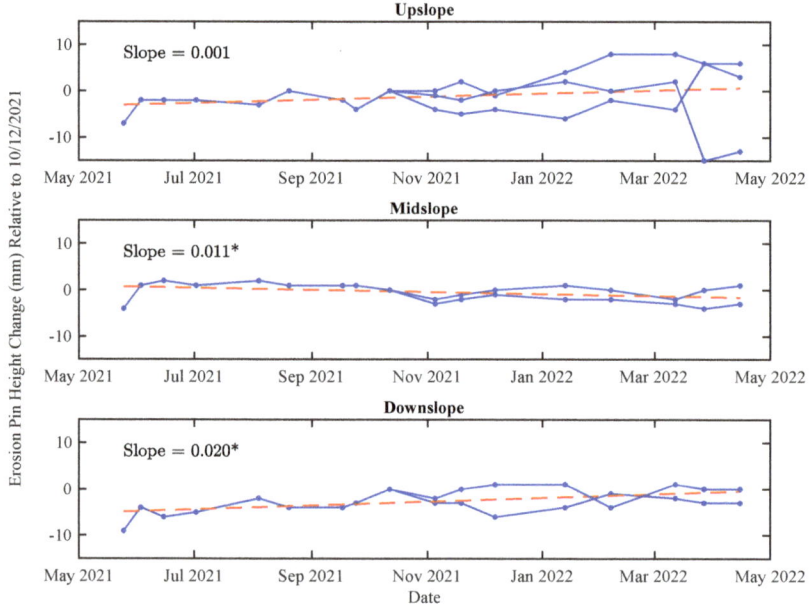

Figure 9. Linear regression for the plot with fescue grass between time in days and erosion pin height relative to the measurement on 12 October 2021, when all erosion pins were installed and stabilized. Solid blue lines and markers show the trajectory of measurements for each erosion pin and dashed red lines show the regression line. Regression slope is given in each plot with * indicating a regression p-value less than 0.1. Regression lines are calculated separately for upslope (top), midslope (**middle**), and downslope (**bottom**).

Table 2. Slope of the regression line with 90% confidence interval between time in days and change in erosion pin height in mm and regression statistics for each plot and slope position.

Species	Position	Slope (90% CI)	R^2	p-Value
Juniper	Upslope [1]	0.018 (0.009, 0.026)	0.28	<0.01
Juniper	Midslope	0.006 (−0.001, 0.013)	0.09	0.13
Juniper	Downslope [2]	−0.020 (−0.031, −0.009)	0.29	<0.01
Fescue	Upslope	0.011 (−0.004, 0.026)	0.05	0.22
Fescue	Midslope [2]	−0.007 (−0.013, −0.001)	0.16	0.04
Fescue	Downslope [1]	0.013 (0.006, 0.021)	0.30	<0.01
Vetiver	Upslope	0.001 (−0.005, 0.007)	0.00	0.76
Vetiver	Midslope [1]	0.011 (0.000, 0.021)	0.11	0.09
Vetiver	Downslope [1]	0.020 (0.014, 0.027)	0.36	<0.01

[1] Significant erosion ($p < 0.10$); [2] Significant deposition ($p < 0.10$).

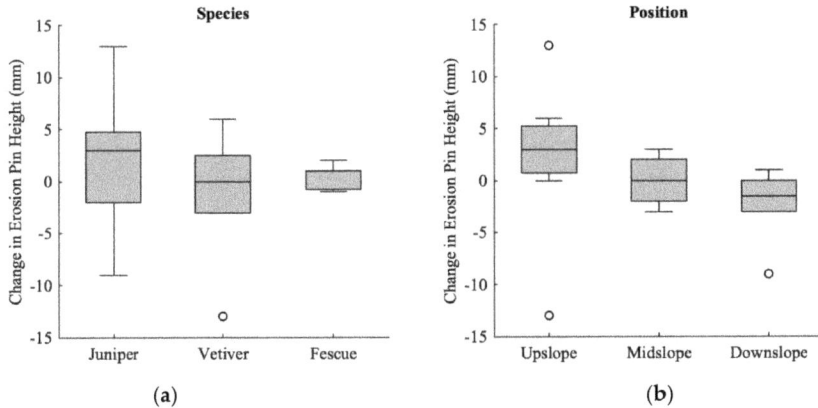

Figure 10. Boxplots showing the change in erosion pin height between 12 October 2021 and 15 April 2022, as grouped by (**a**) species planted on the experimental plot (data from all slope positions included); and (**b**) slope position of the erosion pin (data from all plots included).

3.3. Root Biomass

Vetiver had the highest overall root biomass in the top 40 cm of soil (3.75 kg/m^3), followed by fescue (1.28 kg/m^3) and juniper (0.45 kg/m^3). In the upper layers of the soil, the amount of root biomass is very similar among the species (Figure 11). However, the root biomass of vetiver increases with depth and shows higher amounts of root biomass than the other species in the deeper soil layers. Total root biomass was substantially higher for vetiver while juniper was lower than the other species (Table 3). Root biomass was generally higher in the upslope core, though this was most pronounced for juniper. It should be noted that roots could not be identified by species, so some of the roots sampled from the juniper and fescue plots are likely from weeds. Very few weeds were present on the vetiver plots.

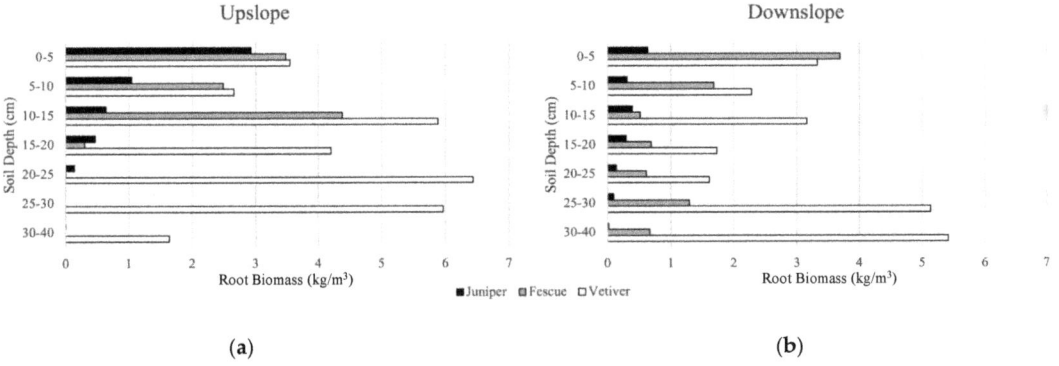

Figure 11. Root biomass by layer for three species in the (**a**) upslope core; and (**b**) downslope core.

Table 3. Root biomass by species in the top 40 cm of soil for the upslope core, the downslope core, and the mean of the two cores.

Species	Upslope (kg/m^3)	Downslope (kg/m^3)	Mean (kg/m^3)
Juniper	0.65	0.24	0.45
Fescue	1.33	1.23	1.28
Vetiver	4.00	3.51	3.75

Vetiver produced roots with a larger diameter than both the fescue and juniper (Figure 12). Root tensile strength, which is assessed per unit area, is inversely proportional to diameter [20,69], so a dense, fibrous root system with many small diameter roots is better for slope stability [18,70,71]. Given the similar biomass abundance in upper soil layers and prevalence of small diameter roots, fescue may be a better choice than juniper if only surface stabilization is needed. However, the root biomass analysis (Figure 11) demonstrated that vetiver is clearly better for deeper slope stabilization due to the greater abundance of deep root biomass. Additional work is needed to directly measure the impact of these roots on the strength of the slopes.

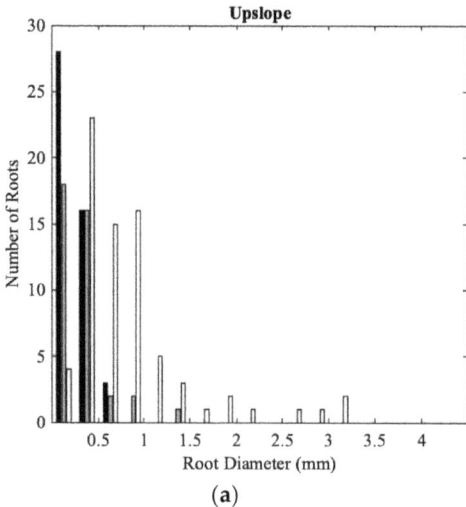
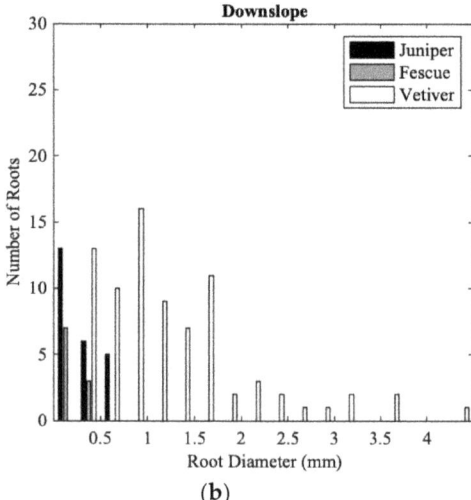

Figure 12. Histograms of root diameter measured at the midpoint of roots extracted from the (a) upslope core; and (b) downslope core.

3.4. Results Summary

This study addressed four research objectives. The first objective was to select vegetation types that may provide erosion control and slope stabilization for priority sites and compare them to the current management practice of planting turf grass from seed. Five vegetation types were tested—deep-rooted (vetiver) grass, woody shrubs, perennial legume, fern, and native prairie grass—and were compared to fescue grass grown from seed. The second objective was to determine which species establish and grow well on moderately steep roadside slopes in the Piedmont region of the southeast USA. Of the species tested, vetiver grass and juniper shrubs grew well (Figure 4). The third objective was to compare surface erosion rates from the experimental plots. Juniper and vetiver both had slightly lower sediment yield than fescue grass when sediment yield from the whole plot was considered (Figure 6). Erosion pins indicated that there was more spatial variability in erosion and deposition within the juniper and fescue plots due to uneven vegetation establishment (Figures 7–9). The final objective was to estimate the contribution of each species to increased slope stability by measuring root biomass and diameter in soil cores. Vetiver grass had substantially higher root biomass than the other species, particularly in deeper soil layers (Figure 11).

4. Discussion

Vetiver grass and juniper shrubs both established well on the experimental slopes and present possible alternatives to seeding with turf grasses such as fescue on high-priority sites. The failure of the perennial legume (hairy vetch) and maidenhair fern demonstrate the importance of weed resistance in species used for erosion control and slope stabilization in the study region. Vetiver showed the best weed resistance of the species tested while juniper may be suitable if a weed resistant species is planted in the interspace between plants.

The plot containing juniper had the lowest runoff amounts and sediment yield based on the runoff collection with TSS method (Figures 5 and 6). The erosion pin data suggest that this is because deposition is occurring at the base of the plot, as measured by the downslope pins (Figures 7–9), where aboveground vegetation establishment was strong (Figure 4). This suggests that the vegetation near the bottom of the plot was slowing runoff and allowing for deposition. If runoff was ponding as it slowed, it could increase infiltration [14], causing the lower runoff volumes observed. The highest rates of erosion were observed on the upslope pins in the juniper plot (Figure 10a), but the plot had the lowest total sediment yield (Figure 6). This demonstrates the high potential of juniper to create a barrier to sediment movement with good establishment.

In the other plots, with more even vegetation establishment, erosion and deposition rates were more consistent across the plot. This agrees with previous studies that emphasized the importance of good vegetation establishment in preventing erosion [41,42]. This study used two methods to compare surface erosion: runoff collection with TSS, which measures total sediment yield, and erosion pins, which measure the spatial distribution of erosion and deposition. The methods proved to be complimentary, as information on the spatial pattern of erosion was helpful in relating vegetation establishment to observed runoff and sediment yield. We did not directly quantify the effects of vegetation density on overland flow velocities or patterns and this remains an important topic for future research [68,72–75].

Despite the issues with establishment, the overall erosion rates observed in this study were within the limit of 2–7 mm/yr. This is much lower than a previous study conducted on a completely bare steep slope which had erosion rates of 20 mm/yr [62]. This study, which was conducted over a 10-year period, also recommended longer monitoring duration than was possible in the current study for erosion pins. Overall, the changes observed in the erosion pin heights were very small relative to the measurement precision ($\pm$ 1 mm) of a manual ruler. Longer monitoring until higher erosion or deposition values are observed may allow for more robust statistical analysis. Another difference with studies on bare slopes was in the pattern of erosion. On bare slopes, higher erosion rates are typically observed near the bottom of the slope, because that is where sheet flow velocities are highest [76]. The vetiver plot showed this same pattern. The pattern of higher erosion rates in the upslope pins observed for this study in the juniper and fescue (Figure 10b) was also found in a large study of vegetated streambanks slopes using erosion pins [77].

Slope stability also depends on root biomass, diameter distribution, and architecture [8]. Vetiver grass added substantially more belowground biomass than the other species tested (Figure 11). In general, fine roots (roots < 3 mm in diameter) are considered more important to soil stabilization than coarse roots [18]. Most of the biomass sampled from the plots, including vetiver, would be considered fine roots. While juniper and fescue had a slightly higher abundance of small diameter roots in upper soil layers, the deeper biomass of vetiver is key for increasing soil strength [20,24]. A previous study of vetiver grass in Brazil found similarly high levels of root biomass [78]. Based on the outcomes of this study, vetiver grass is a promising alternative to the current practice of seeding with turf grass that should be explored by managers. This study is one of the first to examine the use of vetiver grass in Alabama and the Piedmont region of the southern USA. and so this strong establishment and growth should be encouraging to those considering using vetiver. Future research should also measure additional factors that can influence slope failures, such as soil moisture and soil organic matter content [79–81]. Given the weak predictive power of daily rainfall as a predictor of change in erosion pin height, future studies should also measure sub-daily rainfall for the calculation of rainfall intensity as this is a better predictor of erosion than rainfall depth [82].

The performance of a vegetative cover must be weighed against practical concerns, such as ease and cost of planting and maintenance requirements. Vetiver is a non-native species that must be planted from slips that have been sterilized so they will not produce seeds. This is more costly and labor intensive than planting from seed. Future work should consider other native grasses, such as switchgrass, that are deep-rooted and can be planted from seed. However, to grow vegetations from seed, seeds need to be planted during the season recommended and do require some care, such as watering and reseeding of bare patches. Future studies could also consider applying seeded biochemical solutions such as those used by [43] to combine temporary biochemical surface stabilization with the benefits of native grasses.

5. Conclusions

This study compared several vegetation types for erosion control and slope stabilization on roadside slopes in the Piedmont region of Alabama. The focus was on deep-rooted species that could be an alternative to planting turf grass from seed, the prevailing slope management practice in the region. The test plots were established on a relatively steep cut slope (25–30°) at the NCAT Test Track in Opelika, AL, USA. The response of the plots was monitored for over two years using erosion pins and TSS measurements.

Vetiver grass and juniper shrubs established well. Juniper and vetiver both had slightly lower sediment yield than fescue grass when sediment yield from the whole plot was considered. Erosion pins indicated that there was more spatial variability in erosion and deposition within the juniper and fescue plots, likely due to uneven vegetation establishment. Selecting a species with strong and even establishment is important to preventing surface erosion. Vetiver grass had more abundant and deeper root biomass than the other species in the study, suggesting it will be best for slope stabilization. This was the first study to test vetiver grass in the Piedmont region of the southeast and demonstrates that it is a promising option for slope stabilization. Future work is needed to directly measure the impact of vetiver roots on slope stability and to investigate effects of vegetation density and planting arrangement.

Supplementary Materials: The following supporting information can be downloaded at: https://www.mdpi.com/article/10.3390/land11101739/s1, Table S1: Water Level Logger Dataset; Table S2: Runoff Collection and TSS Dataset; Table S3: Erosion Pin Dataset; Table S4: Root Analysis Dataset.

Author Contributions: Conceptualization, F.C.O. and J.M.; methodology, H.A., V.N., F.C.O. and J.M.; formal analysis, H.A. and V.N.; investigation, H.A. and V.N.; resources, F.C.O. and J.M.; data curation, H.A. and V.N.; writing—original draft preparation, H.A., V.N. and F.C.O.; writing—review and editing, H.A., V.N. and J.M.; visualization, H.A. and V.N.; supervision, F.C.O.; project administration, F.C.O. and J.M.; funding acquisition, F.C.O. and J.M. All authors have read and agreed to the published version of the manuscript.

Funding: Financial support for this work was provided by the Auburn University Highway Research Center.

Institutional Review Board Statement: Not applicable.

Informed Consent Statement: Not applicable.

Data Availability Statement: The data presented in this study are available in the Supplementary Materials as Tables S1–S4.

Acknowledgments: The authors thank M. Kiernan, J.R. Ellis, H. Xiao, Z. Hilliard-Shepherd, J. Anderson, C. Barrie, L. Rahimikhameneh and V. Aguilar for assistance with field and lab work. Michael Perez, Guy Savage, and the staff of the Auburn University Erosion and Sediment Control Test Facility assisted with site selection, preparation, and maintenance. Jason Nelson and the National Center for Asphalt Technology provided site access and coordination.

Conflicts of Interest: The authors declare no conflict of interest.

References

1. Knights, M.J.; Montgomery, J.; Carcamo, P.S. Development of a Slope Failure Database for Alabama Highways. *Bull. Eng. Geol. Environ.* **2020**, *79*, 423–438. [CrossRef]
2. Liu, Y.J.; Wang, T.W.; Cai, C.F.; Li, Z.X.; Cheng, D.B. Effects of Vegetation on Runoff Generation, Sediment Yield and Soil Shear Strength on Road-Side Slopes under a Simulation Rainfall Test in the Three Gorges Reservoir Area, China. *Sci. Total Environ.* **2014**, *485*, 93–102. [CrossRef] [PubMed]
3. Morgan, R.P.; Rickson, R.J. *Slope Stabilization and Erosion Control: A Bioengineering Approach*; Taylor & Francis: Oxfordshire, UK, 2003.
4. Montgomery, J.; Knights, M.; Xuan, M.; Carcamo, P. *Evaluation of Landslides along Alabama Highways*; Report to the Alabama Department of Transportation, Highway Research Center: Auburn, AL, USA, 2019.
5. National Research Council. *Assessing and Managing the Ecological Impacts of Paved Roads*; National Academies Press: Washington, DC, USA, 2005.
6. Alabama Soil and Water Conservation Committee Alabama Handbook for Erosion Control, Sediment Control and Stormwater Management on Construction Sites and Urban Areas, Alabama Soil and Water Conservation Committee, Montgomery, AL, USA. 2018. Available online: https://alabamasoilandwater.gov/wp-content/uploads/2021/03/2018-Handbook-Vol-1.pdf (accessed on 21 September 2022).
7. Stokes, A.; Spanos, I.; Norris, J.E.; Cammeraat, E. *Eco- and Ground Bio-Engineering: The Use of Vegetation to Improve Slope Stability: Proceedings of the First International Conference on Eco-Engineering 13–17 September 2004*; Springer Science & Business Media: Berlin/Heidelberg, Germany, 2007; ISBN 978-1-4020-5593-5.
8. Bordoloi, S.; Ng, C.W.W. The Effects of Vegetation Traits and Their Stability Functions in Bio-Engineered Slopes: A Perspective Review. *Eng. Geol.* **2020**, *275*, 105742. [CrossRef]
9. Liu, X.; Lan, H.; Li, L.; Cui, P. An Ecological Indicator System for Shallow Landslide Analysis. *CATENA* **2022**, *214*, 106211. [CrossRef]
10. Löbmann, M.T.; Geitner, C.; Wellstein, C.; Zerbe, S. The Influence of Herbaceous Vegetation on Slope Stability–A Review. *Earth-Sci. Rev.* **2020**, *209*, 103328. [CrossRef]
11. He, J.J.; Cai, Q.G.; Tang, Z.J. Wind Tunnel Experimental Study on the Effect of PAM on Soil Wind Erosion Control. *Environ. Monit. Assess.* **2008**, *145*, 185–193. [CrossRef]
12. Roundy, D. How Do Plants Help Prevent Erosion Control? Granite Seed: Lehi, Utah, 2019.
13. Rousseva, S.; Torri, D.; Pagliai, M. Effect of Rain on the Macroporosity at the Soil Surface. *Eur. J. Soil Sci.* **2002**, *53*, 83–93. [CrossRef]
14. Greenway, D.R. Vegetation and Slope Stability. In *Slope Stability: Geotechnical Engineering and Geomorphology*; Chichester: West Sussex, UK, 1987; pp. 187–230.
15. Langbein, W.B.; Schumm, S.A. Yield of Sediment in Relation to Mean Annual Precipitation. *Eos Trans. Am. Geophys. Union* **1958**, *39*, 1076–1084. [CrossRef]
16. Chen, H.; Zhang, X.; Abla, M.; Lü, D.; Yan, R.; Ren, Q.; Ren, Z.; Yang, Y.; Zhao, W.; Lin, P. Effects of Vegetation and Rainfall Types on Surface Runoff and Soil Erosion on Steep Slopes on the Loess Plateau, China. *Catena* **2018**, *170*, 141–149. [CrossRef]
17. Ranjan, V.; Sen, P.; Kumar, D.; Sarsawat, A. A Review on Dump Slope Stabilization by Revegetation with Reference to Indigenous Plant. *Ecol. Process.* **2015**, *14*, 1–11. [CrossRef]
18. Reubens, B.; Poesen, J.; Danjon, F.; Geudens, G.; Muys, B. The Role of Fine and Coarse Roots in Shallow Slope Stability and Soil Erosion Control with a Focus on Root System Architecture: A Review. *Trees* **2007**, *21*, 385–402. [CrossRef]
19. Masi, E.B.; Segoni, S.; Tofani, V. Root Reinforcement in Slope Stability Models: A Review. *Geosciences* **2021**, *11*, 212. [CrossRef]
20. De Baets, S.; Poesen, J.; Reubens, B.; Wemans, K.; Baerdemaeker, J.D.; Muys, B. Root Tensile Strength and Root Distribution of Typical Mediterranean Plant Species and Their Contribution to Soil Shear Strength. *Plant Soil* **2008**, *305*, 207–226. [CrossRef]
21. Bennett, S.J.; Simon, A. *Riparian Vegetation and Fluvial Geomorphology*; American Geophysical Union: Washington, DC, USA, 2004; Volume 8.
22. Zhou, Y.; Watts, D.; Li, Y.; Cheng, X. A Case Study of Effect of Lateral Roots of Pinus Yunnanensis on Shallow Soil Reinforcement. *For. Ecol. Manag.* **1998**, *103*, 107–120. [CrossRef]
23. Stokes, A.; Atger, C.; Bengough, A.G.; Fourcaud, T.; Sidle, R.C. Desirable Plant Root Traits for Protecting Natural and Engineered Slopes against Landslides. *Plant Soil* **2009**, *324*, 1–30. [CrossRef]
24. Simon, A.; Collison, A.J.C. Quantifying the Mechanical and Hydrologic Effects of Riparian Vegetation on Streambank Stability. *Earth Surf. Process. Landf.* **2002**, *27*, 527–546. [CrossRef]

25. Nelson, S.A. Slope Stability, Triggering Events, Mass Movement Hazards. *Nat. Disasters* **2013**. Tulane University, New Orleans, LA, USA. Available online: https://www2.tulane.edu/~{}sanelson/Natural_Disasters/slopestability.pdf (accessed on 21 September 2022).
26. Patil, U.D.; Shelton III, A.J.; Aquino, E. Bioengineering Solution to Prevent Rainfall-Induced Slope Failures in Tropical Soil. *Land* **2021**, *10*, 299. [CrossRef]
27. Cook, T. Tall Fescue Lolium Arundinaceum (Schreb.) Darbysh. Oregon State University, Corvallis, OR, USA. 2005. Available online: https://agsci.oregonstate.edu/sites/agscid7/files/horticulture/beaverturf/TallFescue-1-5-05V.pdf (accessed on 21 September 2022).
28. Burylo, M.; Hudek, C.; Rey, F. Soil Reinforcement by the Roots of Six Dominant Species on Eroded Mountainous Marly Slopes (Southern Alps, France). *Catena* **2011**, *84*, 70–78. [CrossRef]
29. Sharif, M. US Army Experience: Cold Tolerance and Seed Viability Characteristics of Vetiver. In Proceedings of the Second International Vetiver Conference, Thailand, 18–22 January 2000.
30. Truong, P. Vetiver Grass System: Potential Applications for Soil and Water Conservation in Northern California. In Proceedings of the Application of the Vetiver System for Phytoremediation of Mercury Pollution in the Lake and Yolo Counties, Northern California, 10 May 2000; Volume 9.
31. Knouse, J.A. Ferns as Conservation Agents. Available online: https://www.jaknouse.athens.oh.us/ferns/ferncons.html (accessed on 30 March 2021).
32. Hamidifar, H.; Keshavarzi, A.; Truong, P. Enhancement of River Bank Shear Strength Parameters Using Vetiver Grass Root System. *Arab. J. Geosci.* **2018**, *11*, 611. [CrossRef]
33. Saifuddin, M.; Osman, N. Evaluation of Hydro-Mechanical Properties and Root Architecture of Plants for Soil Reinforcement. *Curr. Sci.* **2014**, *107*, 845–852.
34. Ali, F.H.; Osman, N. Shear Strength of a Soil Containing Vegetation Roots. *Soils Found.* **2008**, *48*, 587–596. [CrossRef]
35. Kokutse, N.K.; Temgoua, A.G.T.; Kavazović, Z. Slope Stability and Vegetation: Conceptual and Numerical Investigation of Mechanical Effects. *Ecol. Eng.* **2016**, *86*, 146–153. [CrossRef]
36. Levine, E.R.; Knox, R.G.; Lawrence, W.T. Relationships between Soil Properties and Vegetation at the Northern Experimental Forest, Howland, Maine. *Remote Sens. Environ.* **1994**, *47*, 231–241. [CrossRef]
37. Raich, J.W.; Tufekciogul, A. Vegetation and Soil Respiration: Correlations and Controls. *Biogeochemistry* **2000**, *48*, 71–90. [CrossRef]
38. Lynne, K. Starting Seeds vs. Buying Plants: Which Is Better? New Life on A Homestead. *New Life Homestd.* **2013**. Available online: https://www.newlifeonahomestead.com/starting-seeds-vs-buying-plants/ (accessed on 21 September 2022).
39. Messer, R.C. Evaluating Maintenance Techniques for Long-Term Vegetation Establishment on Disturbed Slopes in Alabama. M.S. Thesis, Auburn University, Auburn, AL, USA, 2011.
40. Bochet, E.; García-Fayos, P.; Tormo, J. How Can We Control Erosion of Roadslopes in Semiarid Mediterranean Areas? Soil Improvement and Native Plant Establishment. *Land Degrad. Dev.* **2010**, *21*, 110–121. [CrossRef]
41. Bochet, E.; García-Fayos, P.; Poesen, J. Topographic Thresholds for Plant Colonization on Semi-Arid Eroded Slopes. *Earth Surf. Process. Landf. J. Br. Geomorphol. Res. Group* **2009**, *34*, 1758–1771. [CrossRef]
42. Grace, J.M. Effectiveness of Vegetation in Erosion Control from Forest Road Sideslopes. *Trans. ASAE* **2002**, *45*, 681.
43. Hodges, T.M.; Lingwall, B.N. Effects of Microbial Biomineralization Surface Erosion Control Treatments on Vegetation and Revegetation along Highways. *Transp. Res. Rec.* **2020**, *2674*, 1030–1040. [CrossRef]
44. Kim, K.; Riley, S.; Fischer, E.; Khan, S. Greening Roadway Infrastructure with Vetiver Grass to Support Transportation Resilience. *CivilEng* **2022**, *3*, 147–164. [CrossRef]
45. Myers, D.T.; Rediske, R.R.; McNair, J.N. Measuring Streambank Erosion: A Comparison of Erosion Pins, Total Station, and Terrestrial Laser Scanner. *Water* **2019**, *11*, 1846. [CrossRef]
46. Thomsen, L.M.; Baartman, J.E.M.; Barneveld, R.J.; Starkloff, T.; Stolte, J. Soil Surface Roughness: Comparing Old and New Measuring Methods and Application in a Soil Erosion Model. *Soil* **2015**, *1*, 399–410. [CrossRef]
47. Hunolt, A.E.; Brantley, E.F.; Howe, J.A.; Wright, A.N.; Wood, C.W. Comparison of Native Woody Species for Use as Live Stakes in Streambank Stabilization in the Southeastern United States. *J. Soil Water Conserv.* **2013**, *68*, 384–391. [CrossRef]
48. Srivastava, A.; Yetemen, O.; Saco, P.M.; Rodriguez, J.F.; Kumari, N.; Chun, K.P. Influence of Orographic Precipitation on Coevolving Landforms and Vegetation in Semi-Arid Ecosystems. *Earth Surf. Process. Landf.* **2022**, *47*, 2846–2862. [CrossRef]
49. Saco, P.M.; Moreno-de las Heras, M. Ecogeomorphic Coevolution of Semiarid Hillslopes: Emergence of Banded and Striped Vegetation Patterns through Interaction of Biotic and Abiotic Processes. *Water Resour. Res.* **2013**, *49*, 115–126. [CrossRef]
50. NRCS Information on Rainfall, Frequency, & Distributions NRCS NEH Part 630 Ch.4. Available online: https://www.nrcs.usda.gov/wps/portal/nrcs/detailfull/national/water/?cid=stelprdb1044959 (accessed on 27 May 2020).
51. Terando, A.J.; Costanza, J.; Belyea, C.; Dunn, R.R.; McKerrow, A.; Collazo, J.A. The Southern Megalopolis: Using the Past to Predict the Future of Urban Sprawl in the Southeast US. *PLoS ONE* **2014**, *9*, e102261. [CrossRef]
52. US Department of Agriculture. National Weather and Climate Service Interactive Map. Available online: https://www.nrcs.usda.gov/wps/portal/ (accessed on 21 September 2022).
53. Durner, W.; Iden, S.C. The Improved Integral Suspension Pressure Method (ISP+) for Precise Particle Size Analysis of Soil and Sedimentary Materials. *Soil Tillage Res.* **2021**, *213*, 105086. [CrossRef]

54. Dalton, P.A.; Smith, R.J.; Truong, P.N.V. Vetiver Grass Hedges for Erosion Control on a Cropped Flood Plain: Hedge Hydraulics. *Agric. Water Manag.* **1996**, *31*, 91–104. [CrossRef]
55. Kemper, W.D. Vetiver Grass: A Thin Green Line against Erosion. *J. Soil Water Conserv.* **1993**, *48*, 426–427.
56. US Department of Agriculture, The PLANTS Database: Plant List of Attributes, Names, Taxonomy, and Symbols. US Department of Agriculture, Natural Resource Conservation Service, National Plant Data Team, Baton Rouge, LA, USA. 2015. Available online: http://Plants.Usda.Gov (accessed on 21 September 2022).
57. Wiesmeier, M.; Lungu, M.; Hübner, R.; Cerbari, V. Remediation of Degraded Arable Steppe Soils in Moldova Using Vetch as Green Manure. *Solid Earth* **2015**, *6*, 609–620. [CrossRef]
58. Chau, N.L.; Chu, L.M. Fern Cover and the Importance of Plant Traits in Reducing Erosion on Steep Soil Slopes. *Catena* **2017**, *151*, 98–106. [CrossRef]
59. DuBois, L.; Latimer, J.G.; Appleton, B.L.; Close, D. America's Anniversary Garden. Native Plants. Publication 426-223. Virginia Cooperative Extension, Blacksburg, VA, USA. 2009. Available online: https://vtechworks.lib.vt.edu/bitstream/handle/10919/48226/426-223_pdf.pdf?sequence=1 (accessed on 21 September 2022).
60. Correll, D.L. Principles of Planning and Establishment of Buffer Zones. *Ecol. Eng.* **2005**, *24*, 433–439. [CrossRef]
61. Hoag, C.J. Vertical Bundles: A Streambank Bioengineering Treatment to Establish Willows and Dogwoods on Streambanks. In *Technical Note USDA-NRCS. Boise, Idaho-Salt Lake City, Utah. Plant Material*; USDA-NRCS (US Department of Agriculture Natural Resource Conservation Service): Boise, Idaho; Idaho-Salt Lake City, Utah, 2009; Volume 53, pp. 1–6.
62. Hart, E.A.; Mills, H.H.; Li, P. Measuring Erosion Rates on Exposed Limestone Residuum Using Erosion Pins: A 10-Year Record. *Phys. Geogr.* **2017**, *38*, 541–555. [CrossRef]
63. USEPA Method 160.2: Total Suspended Solids (TSS) (Gravimetric, Dried at 103–105° C) US Environmental Protection Agency, Washington, DC, USA. 1999. Available online: https://www.nemi.gov/methods/method_summary/5213/ (accessed on 21 September 2022).
64. Haigh, M.J. The Use of Erosion Pins in the Study of Slope Evolution. *Br. Geomorphol. Res. Group Tech. Bull.* **1977**, *18*, 31–49.
65. Kearney, S.P.; Fonte, S.J.; García, E.; Smukler, S.M. Improving the Utility of Erosion Pins: Absolute Value of Pin Height Change as an Indicator of Relative Erosion. *CATENA* **2018**, *163*, 427–432. [CrossRef]
66. Frasier, I.; Noellemeyer, E.; Fernández, R.; Quiroga, A. Direct Field Method for Root Biomass Quantification in Agroecosystems. *MethodsX* **2016**, *3*, 513–519. [CrossRef] [PubMed]
67. Franks, C.D.; Goings, K.A. Separating Roots from the Soil by Hand Sieving. Available online: https://www.nrcs.usda.gov/Internet/FSE_DOCUMENTS/nrcs142p2_050967.pdf (accessed on 21 September 2022).
68. Wang, L.; Zhang, Y.; Jia, J.; Zhen, Q.; Zhang, X. Effect of Vegetation on the Flow Pathways of Steep Hillslopes: Overland Flow Plot-Scale Experiments and Their Implications. *CATENA* **2021**, *204*, 105438. [CrossRef]
69. Pollen, N.; Simon, A. Estimating the Mechanical Effects of Riparian Vegetation on Stream Bank Stability Using a Fiber Bundle Model. *Water Resour. Res.* **2005**, *41*, W07025. [CrossRef]
70. Grimshaw, R.G.; Helfer, L. *Vetiver Grass for Soil and Water Conservation, Land Rehabilitation, and Embankment Stabilization: A Collection of Papers and Newsletters Compiled by the Vetiver Network*; The World Bank: Washington, DC, USA, 1995.
71. Yoon, P.K. Important Biological Considerations in Use of Vetiver Grass Hedgerows (VGHR) for Slope Protection and Stabilisation. In *Vegetation and Slopes*; Institution of Civil Engineers: London, UK, 1995; pp. 212–221.
72. Zhang, S.; Zhang, J.; Liu, Y.; Liu, Y.; Wang, Z. The Effects of Vegetation Distribution Pattern on Overland Flow. *Water Environ. J.* **2018**, *32*, 392–403. [CrossRef]
73. Tinoco, R.O.; San Juan, J.E.; Mullarney, J.C. Simplification Bias: Lessons from Laboratory and Field Experiments on Flow through Aquatic Vegetation. *Earth Surf. Process. Landf.* **2020**, *45*, 121–143. [CrossRef]
74. Nicosia, A.; Di Stefano, C.; Pampalone, V.; Palmeri, V.; Ferro, V.; Polyakov, V.; Nearing, M.A. Testing a Theoretical Resistance Law for Overland Flow under Simulated Rainfall with Different Types of Vegetation. *CATENA* **2020**, *189*, 104482. [CrossRef]
75. Li, C.; Pan, C. Overland Runoff Erosion Dynamics on Steep Slopes with Forages under Field Simulated Rainfall and Inflow. *Hydrol. Process.* **2020**, *34*, 1794–1809. [CrossRef]
76. Zhuang, X.; Wang, W.; Ma, Y.; Huang, X.; Lei, T. Spatial Distribution of Sheet Flow Velocity along Slope under Simulated Rainfall Conditions. *Geoderma* **2018**, *321*, 1–7. [CrossRef]
77. Zaimes, G.N.; Tamparopoulos, A.E.; Tufekcioglu, M.; Schultz, R.C. Understanding Stream Bank Erosion and Deposition in Iowa, USA: A Seven Year Study along Streams in Different Regions with Different Riparian Land-Uses. *J. Environ. Manag.* **2021**, *287*, 112352. [CrossRef] [PubMed]
78. Machado, L.; Holanda, F.S.R.; da Silva, V.S.; Maranduba, A.I.A.; Lino, J.B. Contribution of the Root System of Vetiver Grass towards Slope Stabilization of the São Francisco River. *Semin. Cienc. Agrar.* **2015**, *36*, 2453–2463. [CrossRef]
79. Fitzjohn, C.; Ternan, J.L.; Williams, A.G. Soil Moisture Variability in a Semi-Arid Gully Catchment: Implications for Runoff and Erosion Control. *Catena* **1998**, *32*, 55–70. [CrossRef]
80. Polyakov, V.; Lal, R. Modeling Soil Organic Matter Dynamics as Affected by Soil Water Erosion. *Environ. Int.* **2004**, *30*, 547–556. [CrossRef]

81. Polyakov, V.O.; Lal, R. Soil Organic Matter and CO_2 Emission as Affected by Water Erosion on Field Runoff Plots. *Geoderma* **2008**, *143*, 216–222. [CrossRef]
82. Williams, J.R. Sediment-Yield Prediction with Universal Equation Using Runoff Energy Factor. In *Present Prospect. Technol. Predict. Sediment Yield Sources 1975*; US Department of Agriculture, Agriculture Research Service: Washington DC, USA, 1975; pp. 244–252.

Article

Inversion Estimation of Soil Organic Matter in Songnen Plain Based on Multispectral Analysis

Siyu Tang, Chong Du * and Tangzhe Nie

School of Hydraulic and Electric Power, Heilongjiang University, Harbin 150080, China; 2201873@s.hlju.edu.cn (S.T.); 2019036@hlju.edu.cn (T.N.)
* Correspondence: duchong@hlju.edu.cn

Abstract: Sentinel-2A multi-spectral remote sensing image data underwent high-efficiency differential processing to extract spectral information, which was then matched to soil organic matter (SOM) laboratory test values from field samples. From this, multiple-linear stepwise regression (MLSR) and partial least square (PLSR) models were established based on a differential algorithm for surface SOM modeling. The original spectra were subjected to basic transformations with first- and second-derivative processing. MLSR and PLSR models were established based on these methods and the measured values, respectively. The results show that Sentinel-2A remote sensing imagery and SOM content correlated in some bands. The correlation between the spectral value and SOM content was significantly improved after mathematical transformation, especially square-root transformation. After differential processing, the multi-band model had better predictive ability (based on fitting accuracy) than single-band and unprocessed multi-band models. The MLSR and PLSR models of SOM had good prediction functionality. The reciprocal logarithm first-order differential MLSR regression model had the best prediction and inversion results (i.e., most consistent with the real-world data). The MLSR model is more stable and reliable for monitoring SOM content, and provides a feasible method and reference for SOM content-mapping of the study area.

Keywords: soil organic matter; Sentinel-2A; remote sensing; differential algorithm; multispectral modeling; PLSR

1. Introduction

Soil organic matter (SOM), as an extensive component of soil, is an important indicator to measure the fertility, status and degradation degree of cultivated soil [1]. It plays a role in increasing moisture retention and, consequently, the drought tolerance of crops [2,3]. SOM also constitutes a huge organic carbon pool in terrestrial ecosystems [4,5]. It is of great practical significance to estimate the soil organic pool by mastering the spatial distribution information of large-scale soil organic matter content instantaneously [6]. Estimating soil organic matter pools has a significant impact on ecology and sustainable land use in the long term. Precision agriculture and long-term regional land development are aided by timely monitoring of SOM data [7]. Traditional biochemical analysis methods are time-consuming and labor-intensive, in addition to being ineffective and unsuitable for gathering information such as soil organic matter content over a large expanse [8,9].

Soil-reflection spectroscopy has successfully enabled rapid and cost-effective SOM estimation, assisting in fulfilling regional to global soil evaluation and monitoring requirements [10]. The majority of research has concentrated on soil spectral studies in controlled indoor environments. There are concerns with test parameters, including light-source power, light-source distance, and irradiation angle during the test. The majority of soil samples used in the tests are stabilized soils. As a result, the spectral data is more unclear, making it harder to share the results of known inverse models of soil characteristics. The advantages of remote sensing technology include a vast coverage area for ground object information, as well as periodicity, currency, precision, and reliability [11]. The quantitative description of soil organic matter by remote sensing technology has always been a research hotspot of many scholars. According to research, the spectral characteristics of SOM are primarily reflected in the absorption of incident light energy by organic matter, and soil reflectance decreases as organic matter content increases [12,13].

There are still image factors, such as mixed pixels, water content, and spectral resolution in remote-sensing data, that must be taken into consideration for soil organic matter analysis. Therefore, the theoretical underpinning of using existing remote-sensing data for SOM mapping involves extracting adequate information from soil spectral data and generating a soil spectral index [14]. Researchers have made some progress in related fields. The application of traditional spectral index inversion theory was used to improve the estimation model's accuracy. Spectral indices could quantify the interrelationships between the SOM's characteristic bands utilizing spectral indices, enhancing weak band connections and reducing model complexity. Preprocessing transformations used to remove image-specific reflectance include soil moisture and particle size to transform soil spectral data, remove signal noise, and highlight features for quantitative model estimation [15]. The derivative algorithm is one of the common preprocessing transformations that reduces spectrum interference by eliminating baseline drift and improving spectral resolution, resulting in increased separation of overlapping peaks and less spectral interference [16]. Although a large number of studies have been carried out in related fields using differential spectral technology [17], relatively few studies have explored the predictive ability in monitoring soil nutrient content.

Previous studies have demonstrated that spectral preprocessing is an important component of multivariate modeling analysis and would improve the predictive performance of models [18–21]. A prediction model based on soil spectral information can effectively and rapidly estimate soil physical and chemical parameters. MLSR (multiple linear stepwise regression) has been developed on the basis of multiple linear regression. Considering the advantage of avoiding collinearity, MLSR has been used to develop models that estimate soil properties [22,23]. The regression equation was introduced using stepwise regression based on the effect, significance or contribution rate of global independent variables on the dependent variable. A linear regression model generates predictions about the dependent variable by eliminating independent variables that are not important to the dependent variable. PLS (partial least squares) is a widely used linear multivariate regression method in the field of soil spectroscopy [18,24,25]. PLS was more accurate than principal component regression or multiple linear regression in predictions of soil salinization using soil conductance in the semi-arid region of Brazil [26]. In PLS, the correlation between principal components is relatively insignificant, while the correlation with the dependent variable is the largest. At the same time, PLS can overcome the strong interpretation of independent variables by principal component analysis. It can effectively extract the comprehensive variables with substantial explanatory power to the system and improve the estimation ability of the model. Therefore, mathematical models are conducive to relate reflectance spectra to SOM content to predict soil nutrients [27,28].

Sentinel-2A is a high-resolution multispectral imaging satellite that covers 13 spectral bands. The bands vary in wavelength from 433 to 2280 nm, including ten bands in the visible near-infrared spectrum and three in the short infrared band. Sentinel-2A has an imaging bandwidth of 290 km, a spectral resolution of 15–180 nm and spatial resolutions including 10, 20 and 60 m. Compared with Landsat TM and other remote-sensing images, Sentinel-2A remote-sensing imaging has higher spectral and spatial resolutions and a shorter revisit cycle (the cycle is five days); it is primarily utilized in global ecological environment monitoring [29]. Morteza Sadeghi investigated soil moisture approaches using Sentinel-2 and Landsat-8 satellites and discovered that Sentinel-2 was more appropriate for the task [30]. Sentinel-2 is suitable for monitoring and mapping soil organic matter, but not soil texture (clay, silt, and sand content) [31]. Qi Gao presented two methodologies for the retrieval of soil moisture from remotely sensed SAR images, with a spatial resolution of 100 m [32]. However, few studies have used Sentinel-2A imagery to monitor soil nutrients.

Given the enormous range of remote-sensing imagery available due to periodic updates, a comprehensive grasp of image characteristics is critical when selecting which image to employ. Different researchers come to different conclusions in terms of the reflectance band and estimation model used to calculate organic matter content [33,34]. In order to comprehensively understand the prediction ability and feasibility of differential spectroscopy in soil nutrient content, the trial used differential processing of the high-resolution Sentinel-2A spectral data based on mathematical transformations to develop models to predict soil organic matter content in the study area.

On the basis of the above, the research aims to construct a SOM evaluation model based on spectral indices and compare the prediction accuracy of different methods in the study region, using Sentinel-2A remote-sensing images as the data source and measured test data of SOM content as analysis data. The objective of this research is: (1) to investigate the use of Sentinel-2A remote-sensing images as a reference for estimating soil organic matter; (2) to analyze the correlation of mathematical transformation (reciprocal, reciprocal logarithm, square root, and square and cubic transformation) with the first- and second-order differential of reflectance and SOM; (3) to construct single-band and multi-band MLSR and PLSR inversion models and evaluate the spectral indices and model performance in SOM estimation.

2. Materials and Methods

2.1. Experimental Site

The study area, Daqing, is located in the southwest of Heilongjiang Province in northeast China (45°46′–46°55′ N and 124°19′–125°12′ E). The region is located in the middle of the Songnen Plain, a Mesozoic subsidence area with a flat, slightly undulating, small ground slope. The landform in the region gradually declines from north to south and is generally plain, with a relative height difference of 10 to 35 m. It connects with the Suihua area in the east, faces Jilin Province (Songhua River) in the south, and borders the city of Qiqihar in the west and north. Winters are cold and snowy, whereas spring and autumn monsoons are more humid. The frost-free season lasts only a few weeks each year. With mean annual precipitation of 427.5 mm, a mean annual temperature of 4.2 °C, and a mean evaporation amount of 1635 mm; rain and heat are in the same season, which is beneficial for crop and forage grass growth. The cultivated land area of Daqing accounts for roughly 20% of the whole area and consists of an established farming industry (Figure 1).

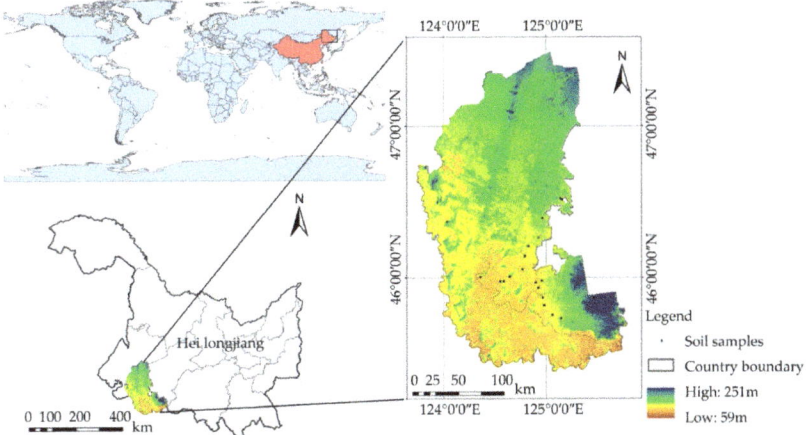

Figure 1. Distribution area diagram of the soil sampling area.

2.2. Soil Sampling

In this study, soil sampling was conducted in Daqing in July 2021, and 19 soil samples were randomly obtained. Five surface soil (soil depth of 20 cm) samples were collected and mixed within a 1 m radius of a specific sampling location, and approximately 500 g of soil per sampling site from the mixed models was used for chemical analysis. The actual longitudes and latitudes of the samples were recorded using a global positioning system (GPS) at the time of field sampling in order to obtain the reflectivity of the sampling point in the remote-sensing image. In the laboratory, all samples were air-dried and ground to pass through a 2-mm sieve to remove impurities such as gravel and animal and plant residues.

Chemical analysis was used to determine the SOM content. The concentrations of all soil samples from each sample point were measured through the potassium dichromate method [35]. The SOM content varied from 13.42 to 22.04 g kg^{-1}. The coefficient of variation (CV) was 0.15, indicating that SOM showed medium variability across all samples (i.e., 0.1 CV 1.0). To ensure the rationality of model establishment and validation, the data were randomly divided into 14 prediction sites and 5 validation sites. (Table 1).

Table 1. Descriptive statistics of SOM (g/kg).

Sample Set	Max.	Min.	Range	Mean	SD	CV
Calibration	22.02	13.42	8.60	17.59	2.70	15.4%
Validation	17.33	14.48	2.85	15.49	0.75	4.9%
Total	22.02	13.42	8.60	17.04	2.57	15.1%

Notes: SD, standard deviation; CV, coefficient of variation.

2.3. Remote-Sensing Image Processing

Based on region size, we selected six Sentinel-2A images in July 2021 for the experiment. Ten visible bands (B2, B3, B4, B5, B6, B7, B8, B8A, B11, B12), which are near-infrared and short-wave infrared bands with a resolution of 10 m, were selected from the images (Table 2). The preprocessing of remote-sensing images mainly included radiometric correction, atmospheric correction, geometric correction, image mosaic, and image clipping. Sentinel-2A remote-sensing images were processed with Sen2cor software for radiometric and atmospheric correction, an ESA plug-in dedicated to the creation of L2A level data that is used to reduce radiometric inaccuracies caused by atmospheric influence and to invert the true surface reflectance of objects. Compared with the typical atmospheric correction software (SMAC and 6S), Sen2cor operation is more straightforward, without human input parameters. The geometric correction of Sentinel-2A remote-sensing imaging was completed with ENVI software, and the geometric correction error was less than 1 pixel. The boundary of the research area was classified in ArcGIS software, and remote-sensing images were clipped and arranged as a mosaic. Remote-sensing images of the original research region were obtained by clipping remote-sensing image data of six scenes with vector boundary data from the research area.

To properly manipulate the data from the sample points, a vector map of the administrative divisions of Daqing was obtained using the BIGEMAP map loader. Furthermore, ArcMap and ArcGIS software were used to complete the longitude and latitude distribution map of sampling points by clicking add data and the directory window. Then ArcMap and ENVI were used in combination to extract Sentinel-2A images for each sampling point corresponding to a DN (digital number) value of each band.

Table 2. The relevant parameters of Sentinel-2A.

Sentinel-2A				
Band	Band Name	Central Wavelength/nm	Spectral Width/nm	Spatial Resolution/m
1	Coastal Aerosol	433	20	60
2	Blue	490	65	10
3	Green	560	35	10
4	Red	665	30	10
5	Vegetation Red edge	705	15	20
6	Vegetation Red edge	740	15	20
7	Vegetation Red edge	783	30	20
8	NIR	842	115	10
8A	Narrow NIR	865	20	20
9	Water Vapour	945	20	60
10	SWIR-Cirrus	1375	30	60
11	SWIR	1610	90	20
12	SWIR	2190	180	20

2.4. Statistical Modeling

2.4.1. Differential Algorithm

Differential spectral technology, which is a common spectrum processing approach, can effectively dig spectral effective information and provide better resolution than the original spectral reflectance. It also improves the correlation between spectral data and soil parameters, allowing for better monitoring of progress in soil nutrient content research and improved prediction accuracy. The reference formula is as follows [36].

The first derivative can be described as:

$$FDR_{(\lambda)} = \left[R_{(\lambda_{i+1})} - R_{(\lambda_i)}\right] / [\lambda_{i+1} - \lambda_i] \tag{1}$$

The second derivative (SDR) can be described as:

$$SDR_{(\lambda)} = \left[R'_{(\lambda_{i+1})} - R'_{(\lambda_i)}\right] / [\lambda_{i+1} - \lambda_i] \tag{2}$$

where λ_i is the wavelength of the i-th band, $R_{(\lambda_{i+1})}$, $R_{(\lambda_i)}$ are the reflectance at bands λ_{i+1}, λ_i, and $R'_{(\lambda_{i+1})}$, $R'_{(\lambda_i)}$ are the first derivative at bands λ_{i+1}, λ_i, respectively.

2.4.2. Multiple Linear Stepwise Regression

MLSR is mainly a comparative analysis of all independent variables according to influence or contribution size to all dependent variables through the F test [23]. Variables significant by the sum of squares are selected for the regression equation. Only one variable is introduced for each step. When a variable is introduced, the partial regression sum of squares of each variable is then tested. If the introduced variable is found to be insignificant, it is removed from the partial regression equation. If more than two variables are introduced in successive steps, it is determined whether or not any existing variables can be removed. Further, when no independent variables can be eliminated, a new independent variable with significant influence is selected for evaluation. This process is repeated until none of the introduced variables can be removed. The original independent variable is also tested, and the gradual regression equation ends.

The formula of the gradual regression equation is:

$$SOM = a_0 + \sum_{i=1}^{n} a_i R_{\lambda_i} \tag{3}$$

where a_0, $a_1 = 1$, n is the regression coefficient, i is the number of bands used for modeling, λ_i is the wavelength of the ith modeling band, and R_{λ_i} is the reflectance value at wavelength λ_i.

2.4.3. Partial Least Squares Regression

PLSR adopts the idea of extracting principal components from principal component analysis, which can simplify the data structure [25]. There are p dependent variables and m independent variables considered. The basic practice is to extract the first component x_i in the independent variable set and the first component u_i in the dependent variable set, requiring maximum correlation between x_i and u_i. The regression of the dependent variable with x_i is then established, and the algorithm is terminated until the equation reaches satisfactory accuracy. Otherwise, the extraction of the second pair component continues to achieve satisfactory accuracy. If the n components are finally extracted from the independent variable set from the independent variable set, the partial least squares regression will establish the regression equation between the dependent variable and $x_1, x_2, \ldots, x_n$. This represents the regression equation between the dependent variable and the original independent variable: the partial least squares regression equation. In PLS calibration, significant wavelengths can be assessed on the basis of variable important in projection (VIP), If the VIP score of a specific wavelength exceeds 1, then the wavelength is considered important [37,38].

2.5. Construction of Spectral Indexes

SOM exhibits unique spectral response properties in visible and near-infrared bands, and the soil spectral reflectivity and SOM content are generally significantly negatively correlated [39,40]. The increase and decrease of SOM content can be reflected from the soil reflection spectrum to a certain extent. The determination of soil spectral reflectance becomes a novel approach to assessing SOM content due to the particular response relationship. Furthermore, the soil spectrum and SOM content show a nonlinear variation caused by the interaction of soil structure and the spectral measurement environment, making the absorption belt and reflection belt of the spectral curve not visible. On the other hand, low-order (first-order, second-order) differential transformation of the spectrum is less sensitive to noise, eliminating some of the background and noise influence and improving the correlation between spectral data and organic matter content.

Therefore, the spectral data are processed by conventional mathematical transformation and differential processing to increase sensitivity to the SOM content of the spectral index. The original spectrum was subjected to six different types of traditional mathematical transformations and respective first and second derivatives. Spectral characteristic indicators were screened using the Pearson correlation analysis method. SOM content measured in the laboratory is the dependent variable of the function, with the characteristic spectral index as the independent variable. The model was constructed between SOM content and the transformed spectral data of the reflection spectrum. The correlation between SOM content and the reflectance of remotely sensed images was analyzed in SPSS. The correlation coefficient was calculated by Formula (4):

$$r = \sum_{i=1}^{n}(x_i - \overline{x})(y_i - \overline{y}) / \sqrt{\sum_{i=1}^{n}(x_i - \overline{x})^2} \sqrt{\sum_{i=1}^{n}(y_i - \overline{y})^2} \qquad (4)$$

where r is the correlation coefficient of SOM and reflectence, and x_i and $\overline{x}$ are the measured value and mean value of reflectivity, respectively; y_i and $\overline{y}$ are the measured value and mean value of SOM content, respectively. When $r > 0$, reflectivity is positively correlated with SOM, and when $r < 0$, reflectivity is negatively correlated with SOM. The closer r is to 1, the more stable the model is and the better the fit is [41].

The prediction of SOM model stability is measured by the determination coefficient R^2; the larger the R^2, the more stable the model; the accuracy is tested by RMSE. The smaller the RMSE, the higher the model accuracy [42,43].

The calculation formula is shown in (5) and (6):

$$R^2 = \sum_{i=1}^{n}(y_i - \hat{y}_i)^2 / \sum_{i=1}^{n}(y_i - \overline{y})^2 \qquad (5)$$

$$RMSE = 1/n \sum_{i=1}^{n} (y_i - \hat{y}_i)^2 \qquad (6)$$

where $\hat{y}_i$ indicates the values estimated by the model; y_i indicates the measured values; $\bar{y}$ indicates the average of the measured values; and n is the number of observations of the variable to be modelled.

2.6. Flow Chart

Figure 2 shows the flowchart of the research to estimate the model between SOM content and spectral index with differential transformations.

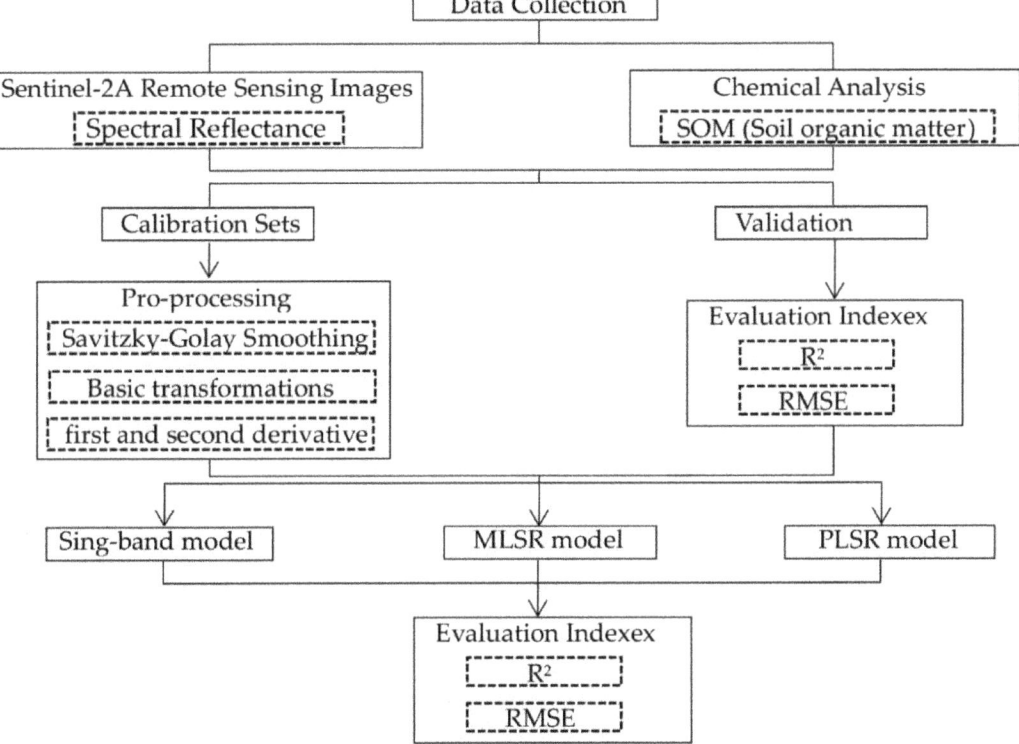

Figure 2. Conducting SOM prediction models.

3. Results and Analysis

3.1. Differential Analysis of the Multispectral Data

The first-order differential and second-order differential are processed by IDL software, with a remote-sensing third band image as an example (Figure 3). The image can better express the real situation of the object, and the first-order differential image better distinguishes the water body from the soil. Raw remote-sensing images contain much information, including noise, which can be excluded by differential image processing of remote-sensing images. However, the meaning of the information in the differential processing image cannot be seen directly, so it needs to be further analyzed based on actual data.

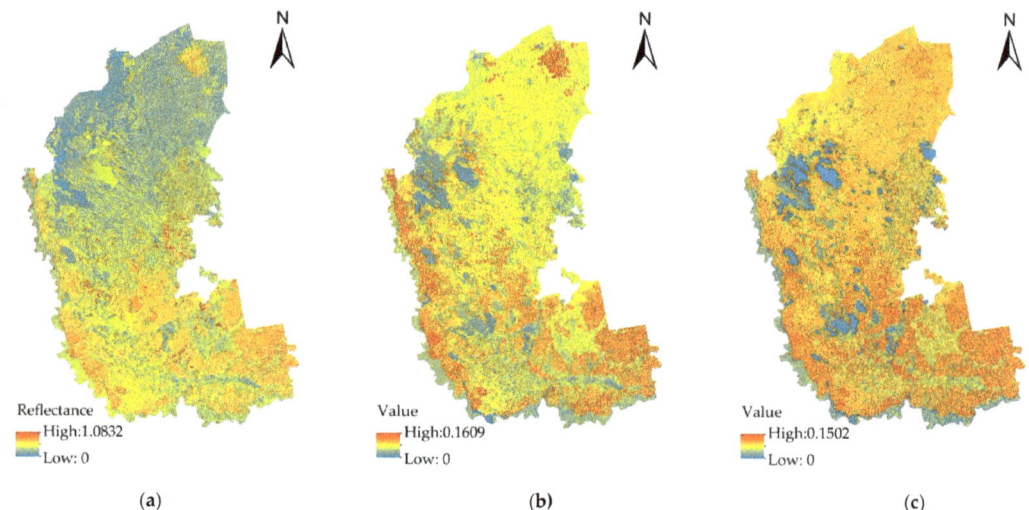

Figure 3. Derivative processing of remote image. (**a**) Original remote-sensing image; (**b**) First derivative processing of remote-sensing image; (**c**) Second derivative processing of remote-sensing image.

3.2. Correlation between SOM Content and Spectral Metrics

The remote-sensing estimation and inversion of site parameters are based on the relationship between remote-sensing data and site parameters. The correlation between multispectral reflectance data and measured SOM data was analyzed in SPSS to clarify the relationship and to find the spectral information sensitive to SOM content. All original remote-sensing bands exhibited a high degree of correlation (Table 3). According to the first-derivative image data, B3′ and B4′ have a higher correlation to SOM (Table 4), whereas the overall correlation was relatively low in the second-derivative image data (Table 5). In general, many bands have a high correlation with the original remote-sensing data, and the correlation value of the band is relatively large.

Table 3. The correlation between original image data and SOM.

	B3	B4	B5	B6	B7	B8	B8A	B11	SOM
B3	1	0.993 **	0.934 **	0.708 **	0.670 **	0.636 *	0.485 *	0.459	−0.738 **
B4		1	0.960 **	0.822 **	0.705 **	0.674 **	0.531	0.514	−0.779 **
B5			1	0.947 **	0.866 **	0.843 **	0.730 **	0.706 **	−0.852 **
B6				1	0.972 **	0.964 **	0.895 **	0.868 **	−0.854 **
B7					1	0.995 **	0.953 **	0.905 **	−0.763 **
B8						1	0.960 **	0.919 **	−0.762 **
B8A							1	0.982 **	−0.635 *
11								1	−0.640 **

Notes: * and ** represent the confidence level at 0.05 and 0.001, respectively.

Through the correlation table of each dataset and SOM content, the correlation of each band of original image data is more significant than 0.6, with B5 and B6 reaching 0.8. However, the first-derivative image data was less-associated with SOM content, with only 0.7 in the B3 band. No sensitive band exists with the SOM after the second-derivative image data. The correlation was significantly reduced compared to the original data in the differential image. According to the aforementioned relationship, the differential processing single-band model results in significant spectral information loss, and the relationship between multispectral data and SOM analysis is not ideal.

Table 4. The correlation between first-derivative image data and SOM.

	B3′	B4′	B5′	B6′	B7′	B8′	B8A′	B11′	SOM
B3′	1	0.428	−0.364	−0.016	0.144	0.184	−0.108	0.193	−0.770 **
B4′		1	0.286	0.210	−0.313	0.364	0.420	0.252	−0.595 **
B5′			1	0.847 **	0.142	0.633 *	0.687 **	0.066	−0.058
B6′				1	0.400	0.753 **	0.589 *	0.113	−0.354
B7′					1	0.721	0.880	0.685	−0.177
B8′						1	0.570 *	−0.038	−0.209
B8A′							1	0.643 *	−0.095
B11′								1	−0.263

Notes: * and ** represent the confidence level at 0.05 and 0.001, respectively.

Table 5. The correlation between second-derivative image data and SOM.

	B3″	B4″	B5″	B6″	B7″	B8″	B8A″	B11″	SOM
B3″	1	−0.847 **	−0.336	0.695 **	−0.276	0.156	−0.627 **	0.758 **	−0.485
B4″		1	0.077	−0.617 *	0.132	0.053	0.432	−0.626 *	0.504
B5″			1	−0.742 **	0.087	−0.135	0.355	−0.517	−0.292
B6″				1	−0.335	0.153	−0.530	0.749 **	−0.088
B7″					1	−0.911 **	0.490	−0.237	0.394
B8″						1	−0.494	0.139	−0.202
B8A″							1	−0.873 **	0.089
B11″								1	−0.109

Notes: * and ** represent the confidence level at 0.05 and 0.001, respectively.

3.3. Single-Band Inversion Model

The spectral reflectance of the different variation processing was correlated with SOM content to determine the sensitive bands according to the magnitude of the correlation coefficient using SPSS software (Figure 4). SOM exhibited significant spectral response properties in the visible and near-infrared wavelengths and was negatively correlated with spectral reflectance in Sentinel-2A remote-sensing images. The correlation coefficient of the original spectral reflectance peaked at around 740 nm ($r = -0.854$, $p < 0.001$). The fifth and sixth bands had remarkable correlation coefficients for the transformed converted spectral index. The correlation between the spectral index of the first-order differential (R') and SOM content was significant in the third waveband, with the weakest correlation coefficient ($r = -0.770$, $p < 0.05$). Differential processing significantly reduces the correlation compared to the other forms of the spectral index, which were linked considerably with SOM ($|r| > 0.800$, $p < 0.001$). Square root processing ($R^{1/2}$) showed the most-significant correlation occurring at 705 nm ($r = -0.858$, $p < 0.001$).

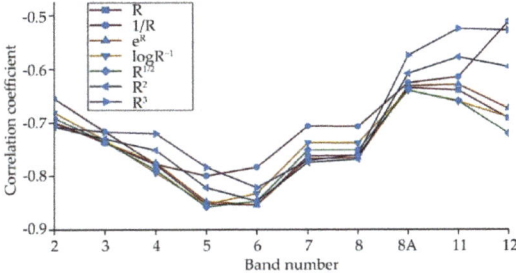

Figure 4. Correlation statistics of organic matter content and reflectivity (R represents reflectance spectra, $1/R$ represents inverted transformation, e^R is exponential transformation, $\log R^{-1}$ is logarithm the reciprocal of reflectance, $R^{1/2}$ is square-root transformation, R^2 is square transformation, R^3 is cubic transformation).

The bands with a correlation $r > 0.5$ with SOM were employed as independent variables to develop separate SOM prediction models, while the measured SOM contents were used as the dependent variable (Table 6). The correlation coefficient of calibration determination (R_c^2) and root mean square error of calibration (RMSEC) were used as an assessment indicator. The prediction coefficient of determination (R_p^2) and the root mean square error of the validation (RMSEP) set were used to assess accuracy of the final model. The single-band model based on the original spectrum's sensitive band reflectance and SOM produced satisfactory accuracy. In addition, the model with square root processing ($R^{1/2}$) had the best modeling efficiency ($R_c^2 = 0.74$, RMSEC = 1.50), but in validation sets performed poorly ($R_p^2 = 0.69$, RMSEP = 1.31). The single-band model based on the R' and $1/R$ processed spectra with SOM had inferior modeling ($R_c^2 = 0.60$ and 0.61, RMSEP = 1.86 and 1.75, respectively). The other five models ($1/R$, $\log R^{-1}$, $R^{1/2}$, R^2, R^3) all showed stronger modeling ($R_c^2 > 0.68$, RMSEP < 1.66). Compared to the original reflectivity, the R' transformation showed the best prediction, with an increase in R^2 of 0.03 ($R_p^2 = 0.82$), but showed extreme uncertainty (RMSECP = 3.72). The modeling of the organic matter single-band model was enhanced, but not dramatically, by spectral processing. Preprocessing of the original spectrum was used in the best suitable model utilizing the single-band ($R_p^2 = 0.79$, RMSEP = 2.18).

Table 6. Single-band inversion model of soil organic matter content based on spectral index.

Spectral Index	Sensitive Band	Correlation Coefficient	Inversion Model	Calibration		Validation	
				R_c^2	RMSEC	R_p^2	RMSEP
R	6	−0.854 **	Y = −34.206R + 25.651	0.71 **	1.52	0.79 **	2.18
$1/R$	5	−0.800 **	Y = 0.832/R + 12.289	0.61 **	1.75	0.74 *	1.66
$\log R^{-1}$	5	−0.853 **	Y = 14.507 logR^{-1} + 6.375	0.70 *	1.53	0.71	1.35
$R^{1/2}$	5	−0.858 **	Y = 30.729 − 31.491R$^{1/2}$	0.74 **	1.50	0.69 *	1.31
R^2	6	−0.847 **	Y = −67.546R + 79.856R^2 + 26.838	0.72 **	1.55	0.77 *	2.40
R^3	6	−0.822 **	Y = −12.333R − 261.662R^2 + 642.675R^3 + 24.151	0.68 *	1.66	0.75 *	2.66
R'	3	−0.770 **	Y = −149.981R + 18.636	0.60 **	1.86	0.82	3.72

Notes: * and ** represent the confidence level at 0.05 and 0.001, respectively.

3.4. Performance of Multiple Linear Stepwise Regression

MLSR models were utilized to evaluate the correlation between SOM content and spectral index, referring to the results of the single-band correlation analysis. Table 7 showed the sensitive band combinations used in the regression analysis. A variable variance significance level of 0.05 was set as the criterion for variable selection and exclusion. The maximum variance invasion factor (VIF) of each spectral band was less than 10, indicating no multicollinearity between bands. By comparing model accuracy, six better models were selected.

The band reflectance in the basic mathematical transformation was excluded as an opt-in variable to conduct the MLSR model, except for the raw spectra. However, the modeling is well-based on the differential treatment of mathematical transformations (Figure 5). The multi-band model demonstrated better predictive results than the single-band model in terms of accuracy and stability. Predictions of raw spectral data under MLSR models outperformed all single-band models ($R_c^2 = 0.78$, RMSEC = 1.32) in the calibration set. The raw and first-order differential processing (R') of the spectral indices resulted in a significantly improved model, but the effect of validation was poor and unstable ($R_p^2 = 0.16$ and 0.55, respectively). The MLSR models based on $(1/R)'$ and $(1/R)''$ showed better performance than the original spectrum. The inverse first-order differential $(1/R)'$ verification set, on the other hand, was poor at predicting and hence was not taken into account ($R_p^2 = 0.37$). The MLSR model constructed by $(\log R^{-1})'$ and $(\log R^{-1})''$ were slightly less effective than the original spectrum modeling in terms of modeling performance

(R_c^2 = 0.71 and 0.69, RMSEC = 1.51 and 1.71, respectively), but the accuracy and stability of validation sets were significantly improved (R_p^2 = 0.93 and 0.76). Overall, the second-order differential $(1/R)''$ model performed the best (R_c^2 = 0.91, RMSEPC = 1.54). However, the validation model performed a little worse (R_p^2 = 0.84, RMSEP = 1.23), but was considered to be unsuccessful at prediction. The validation of the model based on $(\log R^{-1})'$ performed incredibly well, with the R^2 improving by 0.09 and the RMSE reducing by 0.43 compared to the $(1/R)''$ transformation, even though it did not get the best match and model stability (R_c^2 = 0.71, RMSEC = 1.51, R_p^2 = 0.93, RMSEP = 1.11). The pre-processing method of the reciprocal logarithm first-order differential spectrum was used in the best suitable model utilizing the MLSR.

Table 7. Multi-band inversion model of soil organic matter content based on spectral index.

Sensitive Band	Spectral Index	Inversion Model	Calibration		Validation	
			R_c^2	RMSEC	R_p^2	RMSEP
R_6, R_{8A}	R	Y = −34.206a − 57.592b + 25.651	0.78 *	1.32	0.16	2.55
R_3, R_6, R_8, R_4	R'	Y = −146.835a − 75.734b + 184.192c − 110.819d + 24	0.89 **	0.92	0.55 *	2.04
R_6, R_{11}	$(1/R)'$	Y = −3.179a + 0.159b + 13.273	0.80 **	1.27	0.37	1.29
R_{8A}, R_3, R_4	$(1/R)''$	Y = 1.53a − 1.138b − 1.157c + 14.502	0.91 **	0.87	0.84 **	1.23
R_{8A}, R_7	$(\log R^{-1})'$	Y = 46.497a − 69.465b + 11.418	0.71 *	1.51	0.93 **	1.11
R_{8A}, R_4	$(\log R^{-1})''$	Y = 47.412a − 43.495b + 13.005	0.69 *	1.71	0.76 *	1.86

Notes: * and ** represent the confidence level at 0.05 and 0.001, respectively.

3.5. Performance of Partial Least Square Regression

PLSR was established for the eighteen spectral transformations using the Unscrambler software. The results showed that the number of factors obtained by PLSR analysis varies considerably. We performed a full cross-validation before establishing a predictive model. Cross-validation is a method of predicting how well a model will fit the hypothesis validation set. The number of PLSR factors based on the original reflectance was three, increasing to nine after reciprocal transformation. The correlation coefficient of cross-validation determination (R_{cv}^2) and root mean square error of cross-validation (RMSECV) were used as assessment metrics to optimize various spectrum post approaches. R_p^2 and RMSEP were used to assess the final effect [22,44]. PLSR regression models based on mathematical transformation with differentiation gave disappointing outcomes.

Among the basic processing, the R^2 model showed the most prediction accuracy in cross-validation (R_{cv}^2 = 69, RMSECV = 1.62) and in independent validation (RMSEP = 11.93), in which the prediction stability was not reliable (Table 8). Similarly, the PLSR model based on 1/R method also performed well in cross-validation (R_{cv}^2 = 0.68, RMSECV = 1.65) but had poor accuracy in independent validation (R_p^2 = 0.51). Prediction was worst using the R^3 method (R_{cv}^2 = 0.50, RMSECV = 2.06). A PLSR model based on $\log R^{-1}$ yielded the best prediction results (R_{cv}^2 = 0.66, RMSECV = 1.70, R_p^2 = 0.79, RMSEP = 1.55).

For the first-order derivative processing method, the accuracy and stability were reduced to varying degrees compared to the basic processing (Table 9). The prediction when using the $(1/R)'$ method was best in cross-validation (R_{cv}^2 = 0.65, RMSECV = 1.8) and in independent validation (though with poor prediction stability) (R_p^2 = 0.67, RMSEP = 0.84). In terms of the prediction results of the six calibration models, the $(\log R^{-1})'$ model validations performed the best (R_{cv}^2 = 0.62, RMSECV = 1.80, R_p^2 = 0.90, RMSEP = 0.51). Furthermore, PLSR regression models based on second-order derivative processing against SOM content had no practical significance (R_{cv}^2 < 0.60). The model based on $(\log R^{-1})''$ performed best among the processed sets (R_{cv}^2 = 0.57, RMSECV = 1.92, R_p^2 = 0.83, RMSEP = 1.28), suggesting that the PLSR model was not suitable for estimating the organic matter content of the region (Table 10).

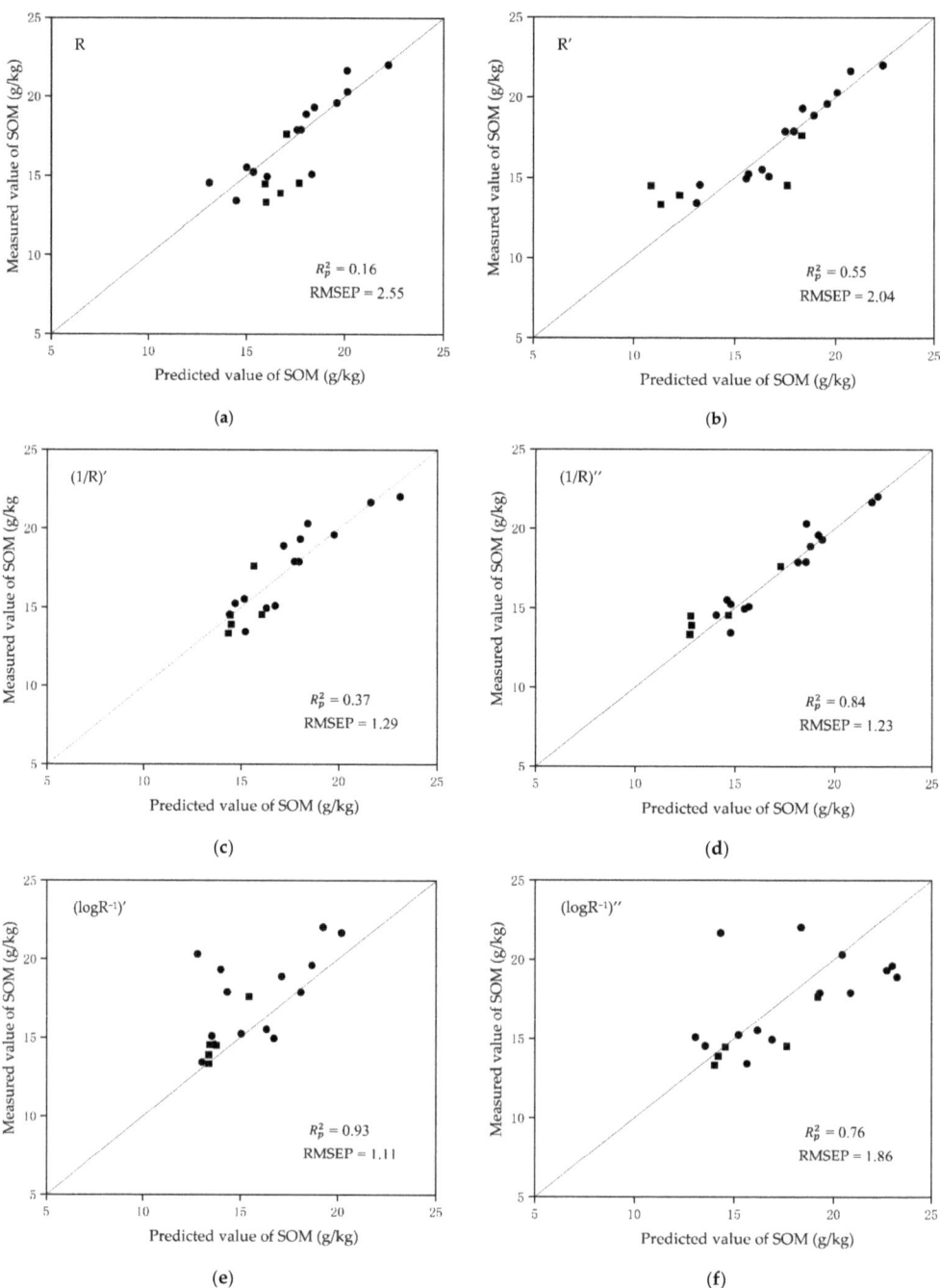

Figure 5. Model validation. (**a**) Original multi-band measured value; (**b**) First derivative of the multi-band measured value; (**c**) First derivative of reciprocal multi-band measured value; (**d**) Second derivative of reciprocal multi-band measured value; (**e**) First derivative of the reciprocal logarithm multi-band measured value; (**f**) Second derivative of the reciprocal logarithm multi-band measured value.

Table 8. Results of partial least squares regression analysis of the original spectral data and soil organic matter content.

Spectral Index	PLR Factors	Cross-Validation		Validation	
		R_{cv}^2	RMSECV	R_p^2	RMSEP
R		0.63 *	1.78	0.78 *	1.81
$1/R$		0.68 **	1.65	0.51	1.65
$\log R^{-1}$		0.66	1.70	0.79 *	1.55
$R^{1/2}$		0.63	1.76	0.78 **	1.63
R^2		0.69 *	1.62	0.87 *	11.93
R^3		0.50	2.06	0.61	2.98

Notes: * and ** represent the confidence level at 0.05 and 0.001, respectively.

Table 9. Results of partial least squares regression analysis of first-derivate transformation and soil organic matter content.

Spectral Index	PLR Factors	Cross-Validation		Validation	
		R_{cv}^2	RMSECV	R_p^2	RMSEP
R'		0.56	1.93	0.91 *	2.57
$(1/R)'$		0.65 **	1.72	0.67 *	0.84
$(\log R^{-1})'$		0.62 *	1.79	0.90 **	0.51
$(R^{1/2})'$		0.33	2.38	0.89 *	2.58
$(R^2)'$		0.64 *	1.75	0.22	3.14
$(R^3)'$		0.54 *	1.98	0.05	3.66

Notes: * and ** represent the confidence level at 0.05 and 0.001, respectively.

Table 10. Results of partial least squares regression analysis of second-derivate transformation and soil organic matter content.

Spectral Index	PLR Factors	Calibration		Validation	
		R_{cv}^2	RMSE	R_p^2	RMSEP
R''		—			2.26
$(1/R)''$		0.25	2.52	1.07	0.90
$(\log R^{-1})''$		0.57 *	1.92	1.41	0.83	1.08
$(R^{1/2})''$		0.26	2.50	1.08	0.01	0.35
$(R^2)''$		—			
$(R^3)''$		—			

Notes: * represents the confidence level at 0.05.

The PLSR model under the reciprocal logarithm first-order differential spectrum transformation showed the best correlation with SOM among the three methods. The information above indicates that $1/R$ transformation performed satisfactorily in data representing spectral features and quantitative inversion models. The results of multi spectral model conducted by PLS show that the prediction capability of PLSR modeling is high using different spectral pretreatment methods.

3.6. Spatial Pattern Analysis of Soil Organic Matter Content

The accuracy of the above single-band, MLSR and PLSR inversion models was investigated. The models created by MLSR and PLSR both had satisfactory prediction performance. The accuracy using PLSR is higher than that of MLSR, with excellent prediction. According to the validation sample detection model, the accuracy and stability of the MLSR model are relatively stable. This shows that MLSR regression is more stable and meets the application needs in predicting the SOM content in Daqing. Reciprocal logarithm first-order differential by MLSR regression model is the optimal model. The SOM content-inversion model based on the Sentinel-2A image spectral index was selected to invert and map the SOM content in the study area to obtain the SOM spatial distribution in Daqing (Figure 6).

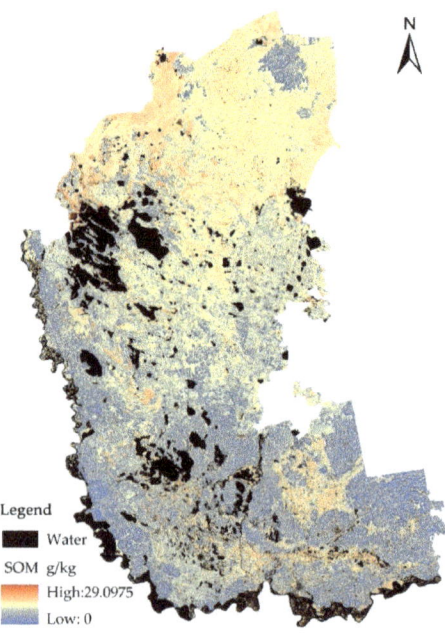

Figure 6. Inversion result of model.

4. Discussion

The inversion results are in line with the actual spatial distribution of SOM. The SOM content in the study area was generally low and uneven, with large spatial differences. The content was gradually distributed from northeast to southwest, and the SOM content in the northeast was generally higher than that in the southwest. According to the soil field survey, surface runoff accumulates in low-lying areas to form intermittent and permanent puddles due to concentrated precipitation. Poor drainage accelerates the process of salt accumulation, and the complex micro-topography causes uneven evaporation of soil water. The salt above the micro-topography is aggravated by strong evaporation and rainfall and irrigation water containing a certain amount of salt flow from high places to places in areas with poor soil permeability. Soil salinization occurs in low-lying areas after fraction evaporation. In addition, the Daqing area is located in a seasonally frozen soil area, and the freezing and thawing of the soil promotes the accumulation of salt. These aspects combined can lead to serious soil salinization. The high salt content of the soil is not conducive to the survival of vegetation, and the small amount of vegetation means that the content of humus is low and is not conducive to the survival of general decomposers [45]. Higher salinity in soil masks the spectral signature of SOM, resulting in a low inversion value of SOM content in the southwest of the study area [46,47].

The above results showed that Sentinel-2A remote-sensing images had a good correlation with SOM content in visible and near-infrared bands. The effect of multi-band modeling is better than that of single-band modeling. It is found that there is a sensitive band in the correlation between the first-order differential data and SOM content. The original remote-sensing data had the highest correlation in Band 6 ($r = -0.730$). In single-band modeling, the correlation between square root transformation and SOM is best, but the modeling is not as good as the original spectrum. The results of differential processing in the PLSR model are not satisfactory, probably because the differentially processed image eliminates some of the information of the original image. The correlation between reflectance and SOM is improved after basic transformation and differential processing in MLSR models. The single-band model only uses a very small amount of information, while for remote sensing, the data-rich multi-spectral band can only express extremely limited SOM information in a single band, which can easily cause the loss of some key information. The MLSR and PLSR models achieve convincing results, and the MLSR model has better predictive ability (based on fitting accuracy). The PLSR model, based on second-order differential processing, is ideal and relevant in practice. Reciprocal transformation will improve model prediction in MLSR and PLSR regression. Based on the reciprocal logarithm first-order derivative MLSR regression model, the inversion of SOM in the study area was carried out. The inversion result is in accordance with the actual situation, which is suitable for spectral inversion of soil organic matter content in a certain geographical area.

Compared to direct contact, remote sensing has advantages in the estimation of SOM content, such as predicting soil fertility without direct contact with the object of study, forecasting crop yields from visible and near-infrared bands, accessing the information on the surface of the earth more efficiently and affordably, and updating soil databases in many fields. The previous study found that for most places, the forecast accuracy based on high-resolution satellite was satisfactory for SOM content prediction, and mapped the soil organic matter more precisely than the airborne sensors [32]. The univariate model only considers a single variable to participate in the modeling [48], and the current research is mainly aimed at soil and crops. In this study, the single-band model built using the original spectra worked best—the mathematical transformed form reduced the model prediction—indicating the applicability of Sentinel-2A for predictions. The accuracy of SOM estimation using the MLSR model established by simple mathematical transformation and derivative transformation is higher than that of univariate model. Differential transformation is more beneficial to extract sensitive features in the soil spectrum [49]. The multivariable model integrates the features of multiple sensitive bands, alleviates the "multicollinearity" to a large extent, and improves the applicability and stability of the model. In previous studies, there have been inconsistencies between PLSR models and MLSR predictions [22,23,50,51]. The MLSR model possesses better prediction results than the PLSR model in this study, and high soil salinity in the study area may be a significant factor. In future work, we intend to develop the method further, for example, by expanding the number of soil samples, diversifying the soil types, and taking into account soil moisture and microorganisms. The multispectral examination of SOM by Sentinel-2A has received little attention and has not been thoroughly investigated. Future spectral modeling of the SOM could be integrated with different spectrum indices (e.g., salinity indices) to screen high-precision spectral parameters. Furthermore, indoor light sources can be used to generate reflectance spectra and provide a complete set of measurement data to eliminate the weather impact of spectral acquisition.

5. Conclusions

To maximize the correlation between spectral metrics and soil organic matter content, MLSR and PLSR modeling were applied to establish the SOM content model based on Sentinel-2A remote-sensing images. The effective and predictive capacities of different models, which combined basic transformation with differential processing, were validated. Sentinel-2A remote-sensing images had a good correlation with SOM content in visible and near-infrared bands. MLSR and PLSR models of SOM in the study area were established based on different processing and measured values, respectively. The correlation between SOM content and spectral data was improved by multi-spectral modeling after differential processing. However, the correlation between SOM content and reflectance was reduced after first-order differential, indicating that spectral information was partially lost due to differential treatment, and the relationship between spectral data and SOM was not ideal. Multi-band modeling made superior predictions compared to single-band. SOM content could be well-estimated using MLSR models. The MLSR model is more accurate and stable than PLSR, verified by the calibration and validation samples. The accuracy of the modeling results is high and can meet research requirements. These findings give a theoretical foundation and technological support for utilizing spectroscopy to estimate soil organic matter concentration, and indicates this method can substitute traditional experimental methods for measuring organic matter, thus enabling a larger scale of long-term monitoring of changes in soil organic matter content. In this study, Sentinel-2A images made it possible to retrieve surface soil organic matter with a high spatial and temporal resolution. For soil ecosystem observations, these prediction models will need to be assessed, optimized, and used more broadly.

Author Contributions: Conceptualization, C.D.; methodology, C.D. and S.T.; software, S.T.; validation, T.N.; writing—original draft, S.T.; writing—review and editing, T.N. and S.T.; funding acquisition, C.D. and T.N. All authors have read and agreed to the published version of the manuscript.

Funding: This research was supported by the basic scientific research business expenses of provincial colleges and universities in Heilongjiang Province of China (grant number 2018-KYYWF-1570).

Institutional Review Board Statement: Not applicable.

Informed Consent Statement: Not applicable.

Data Availability Statement: The data in this study is available on request from the corresponding author. As this data is the result of a project related to the remote sensing interpretation of soil organic matter, it is not allowed to be made public unless specifically requested.

Conflicts of Interest: The authors declare no conflict of interest.

References

1. Mcbratney, A.; Field, D.J.; Koch, A.J.G. The dimensions of soil security. *Geoderma* **2014**, *213*, 203–213. [CrossRef]
2. Kerdsueb, P.; Teartisup, P. The Use of Geoinformatics for Estimating Soil Organic Matter in Central Plain of Thailand. *Int. J. Environ. Sci. Dev.* **2014**, *5*, 282–285. [CrossRef]
3. Seleiman, M.F.; Al-Suhaibani, N.; Ali, N.; Akmal, M.; Alotaibi, M.; Refay, Y.; Dindaroglu, T.; Abdul-Wajid, H.H.; Battaglia, M.L. Drought Stress Impacts on Plants and Different Approaches to Alleviate Its Adverse Effects. *Plants* **2021**, *10*, 259. [CrossRef] [PubMed]
4. Lorenz, K.; Lal, R.; Preston, C.M.; Nierop, K.J.G. Strengthening the soil organic carbon pool by increasing contributions from recalcitrant aliphatic bio(macro)molecules. *Geoderma* **2007**, *142*, 1–10. [CrossRef]
5. Scharlemann, J.P.; Tanner, E.V.; Hiederer, R.; Kapos, V.J.C.M. Global soil carbon: Understanding and managing the largest terrestrial carbon pool. *Carbon Manag.* **2014**, *5*, 81–91. [CrossRef]
6. Gaius, R.S.; Billings, W.D.; Chapin, F.S. Global Change and the Carbon Balance of Arctic Ecosystems: Carbon/Nutrient Interactions should as as large constraints in global terrestrial carbon cycle. *BioScience* **1992**, *42*, 433–441.
7. Dotto, A.C.; Dalmolin, R.S.D.; Caten, T.; Geoderma, G.J. A systematic study on the application of scatter-corrective and spectral-derivative preprocessing for multivariate prediction of soil organic carbon by Vis-NIR spectra. *Geoderma* **2018**, *314*, 262–274. [CrossRef]
8. Bao, Y.; Meng, X.; Ustin, S.; Wang, X.; Tang, H.J.C. Vis-SWIR spectral prediction model for soil organic matter with different grouping strategies. *Catena* **2020**, *195*, 104703. [CrossRef]
9. Mingliang, L.I.; Xican, L.I.; Zhang, S. Grey Relation Estimating Pattern of Soil Water Content Based on Hyper-Spectral Data. *J. Geomat. Sci. Technol.* **2016**, *28*, 27–39.
10. Ben-Dor, E.; Banin, A. Near-Infrared Analysis as a Rapid Method to Simultaneously Evaluate Several Soil Properties. *Soil Sci. Soc. Am. J.* **1995**, *59*, 364–372. [CrossRef]
11. Ji, W.J.; Li, X.; Li, C.X.; Zhou, Y.; Shi, Z.J.S.; Analysis, S. Using Different Data Mining Algorithms to Predict Soil Organic Matter Based on Visible-Near Infrared Spectroscopy. *Spectrosc. Spectr. Anal.* **2012**, *32*, 2393.
12. Hummel, J. Soil moisture and organic matter prediction of b-horizon soils using an nirsoil sensor. *Am. J. Manag. Care* **2002**, *8*, 963–974.
13. Rinnan, R. Application of near infrared reflectance (NIR) and fluorescence spectroscopy to analysis of microbiological and chemical properties of arctic soil. *Soil Biol. Biochem.* **2007**, *39*, 1664–1673. [CrossRef]
14. Kasim, N.; Sawut, R.; Shi, Q.; Balati, M.; Kuerban, M.; Julaiti, S. Estimation of Soil Organic Matter Content Based on Optimized Spectral Index. *Trans. Chin. Soc. Agric. Mach.* **2018**, *49*, 155–163.
15. Hansen, P.M.; Schjoerring, J.K. Reflectance measurement of canopy biomass and nitrogen status in wheat crops using normalized difference vegetation indices and partial least squares regression. *Remote Sens. Environ.* **2003**, *86*, 542–553. [CrossRef]

16. Priori, S.; Fantappiè, M.; Lorenzetti, R.; Pellegrini, S.; Costantini, E.A. Field Scale Mapping of Soil Carbon Stock with Limited Sampling by Coupling Gamma Ray and Vis NIR Spectroscopy. *Soil Sci. Soc. Am. J.* **2016**, *80*, 954–964. [CrossRef]
17. Smith, K.L.; Steven, M.D.; Colls, J.J. Use of hyperspectral derivative ratios in the red-edge region to identify plant stress responses to gas leaks. *Remote Sens. Environ.* **2004**, *92*, 207–217. [CrossRef]
18. Mahesh, S.; Jayas, D.S.; Paliwal, J.; White, N. Comparison of Partial Least Squares Regression (PLSR) and Principal Components Regression (PCR) Methods for Protein and Hardness Predictions using the Near-Infrared (NIR) Hyperspectral Images of Bulk Samples of Canadian Wheat. *Food Bioprocess Technol.* **2015**, *8*, 31–40. [CrossRef]
19. Fernandes, M.; Coelho, A.P.; Fernandes, C.; Silva, M.; Marta, C. Estimation of soil organic matter content by modeling with artificial neural networks. *Geoderma* **2019**, *350*, 46–51. [CrossRef]
20. Dunn, B.W.; Batten, G.D.; Beecher, H.G.; Ciavarella, S. The potential of near-infrared reflectance spectroscopy for soil analysis—A case study from the Riverine Plain of south-eastern Australia. *Aust. J. Exp. Agric.* **2002**, *42*, 607–614. [CrossRef]
21. Kooistra, L.; Wanders, J.; Epema, G.F.; Leuven, R.; Buydens, L. The potential of field spectroscopy for the assessment of sediment properties in river floodplains. *Anal. Chim. Acta* **2003**, *484*, 189–200. [CrossRef]
22. Yu, X.; Liu, Q.; Wang, Y.; Liu, X.; Liu, X. Evaluation of MLSR and PLSR for estimating soil element contents using visible/near-infrared spectroscopy in apple orchards on the Jiaodong peninsula. *Catena* **2016**, *137*, 340–349. [CrossRef]
23. Sullivan, D.G.; Shaw, J.N.; Rickman, D. IKONOS Imagery to Estimate Surface Soil Property Variability in Two Alabama Physiographies. *Soil Sci. Soc. Am. J.* **2005**, *69*, 1789–1798. [CrossRef]
24. Goldshleger, N.; Chudnovsky, A.; Ben-Binyamin, R. Predicting salinity in tomato using soil reflectance spectra. *Int. J. Remote Sens.* **2013**, *34*, 6079–6093. [CrossRef]
25. Wold, S.; Sjöström, M.; Eriksson, L. PLS-regression: A basic tool of chemometrics. *Chemom. Intell. Lab. Syst.* **2001**, *58*, 109–130. [CrossRef]
26. Rocha Neto, O.C.D.; Teixeira, A.D.S.; de Oliveira Leão, R.A.; Moreira, L.C.J.; Galvão, L. Hyperspectral Remote Sensing for Detecting Soil Salinization Using ProSpecTIR-VS Aerial Imagery and Sensor Simulation. *Remote Sens.* **2017**, *9*, 42. [CrossRef]
27. Xu, S.; Shi, X.; Wang, M.; Zhao, Y. Determination of rice root density at the field level using visible and near-infrared reflectance spectroscopy. *Geoderma* **2016**, *267*, 174–184. [CrossRef]
28. Martens, H.; Næs, T. Multivariate Calibration. In *Chemometrics, Mathematics and Statistics in Chemistry*; Kowalski, B.R., Ed.; Springer: Dordrecht, The Netherlands, 1984; pp. 147–156.
29. Vajsová, B.; Fasbender, D.; Wirnhardt, C.; Lemajic, S.; Devos, W.J.R.S. Assessing Spatial Limits of Sentinel-2 Data on Arable Crops in the Context of Checks by Monitoring. *Remote Sens.* **2020**, *12*, 2195. [CrossRef]
30. Sadeghi, M.; Babaeian, E.; Tuller, M.; Jones, S.B. The optical trapezoid model: A novel approach to remote sensing of soil moisture applied to Sentinel-2 and Landsat-8 observations. *Remote Sens. Environ.* **2017**, *198*, 52–68. [CrossRef]
31. Gholizadeh, A.; Žižala, D.; Saberioon, M.; Borůvka, L. Soil organic carbon and texture retrieving and mapping using proximal, airborne and Sentinel-2 spectral imaging. *Remote Sens. Environ.* **2018**, *218*, 89–103. [CrossRef]
32. Gao, Q.; Zribi, M.; Escorihuela, M.J.; Baghdadi, N. Synergetic Use of Sentinel-1 and Sentinel-2 Data for Soil Moisture Mapping at 100 m Resolution. *Sensors* **2017**, *17*, 1966. [CrossRef] [PubMed]
33. Steinberg, A.; Chabrillat, S.; Stevens, A.; Segl, K.; Foerster, S. Premote sensing prediction of common surface soil properties based on vis-nir airborne and simulated enmap imaging spectroscopy data: Prediction accuracy and influence of spatial resolution. *Remote Sens.* **2016**, *8*, 613. [CrossRef]
34. Nawar, S.; Buddenbaum, H.; Hill, J.; Kozak, J.; Mouazen, A.M. Estimating the soil clay content and organic matter by means of different calibration methods of vis-NIR diffuse reflectance spectroscopy. *Soil Tillage Res.* **2016**, *155*, 510–522. [CrossRef]
35. Nelson, D.W.; Sommers, L.E. A rapid and accurate method for estimating organic carbon in soil. *Proc. Indiana Acad. Sci.* **1975**, *84*, 456–462.
36. Dawson, T.P.; Curran, P.J. Technical note A new technique for interpolating the reflectance red edge position. *Int. J. Remote Sens.* **1998**, *19*, 2133–2139. [CrossRef]
37. Chong, I.G.; Jun, C.H. Performance of some variable selection methods when multicollinearity is present. *Chemom. Intell. Lab. Syst.* **2005**, *78*, 103–112. [CrossRef]
38. Hong, Y. Application of fractional-order derivative in the quantitative estimation of soil organic matter content through visible and near-infrared spectroscopy. *Geoderma* **2018**, *337*, 758–769. [CrossRef]
39. Wijewardane, N.K.; Ge, Y.; Morgan, C. Prediction of soil organic and inorganic carbon at different moisture contents with dry ground VNIR: A comparative study of different approaches. *Eur. J. Soil Sci.* **2016**, *67*, 605–615. [CrossRef]
40. Al-Abbas, A.H.; Swain, P.H.; Baumgardner, M.F. Relating Organic Matter and Clay Content to the Multispectral Radiance of Soils. *Soil Sci.* **1972**, *114*, 65–82. [CrossRef]
41. Zhao, H.; Liu, P.; Qiao, B.; Wu, K. The Spatial Distribution and Prediction of Soil Heavy Metals Based on Measured Samples and Multi-Spectral Images in Tai Lake of China. *Land* **2021**, *10*, 1227. [CrossRef]
42. Wei, L.; Yuan, Z.; Zhong, Y.; Yang, L.; Zhang, Y. An Improved Gradient Boosting Regression Tree Estimation Model for Soil Heavy Metal (Arsenic) Pollution Monitoring Using Hyperspectral Remote Sensing. *Appl. Sci.* **2019**, *9*, 1943. [CrossRef]
43. Zhou, H.; Deng, Z.; Xia, Y.; Fu, M.; Neurocomputing, F.J. A new sampling method in particle filter based on Pearson correlation coefficient. *Neurocomputing* **2016**, *216*, 208–215. [CrossRef]

44. Liu, J.; Dong, Z.; Xia, J.; Wang, H.; Meng, T.; Zhang, R.; Han, J.; Wang, N.; Xie, J. Estimation of soil organic matter content based on CARS algorithm coupled with random forest. *Spectrochim. Acta Part A Mol. Biomol. Spectrosc.* **2021**, *258*, 119823. [CrossRef] [PubMed]
45. Zinck, G. Remote sensing of soil salinity: Potentials and constraints. *Remote Sens. Environ.* **2003**, *85*, 1–20.
46. Garcia, E.L.A. Detecting Soil Salinity in Alfalfa Fields using Spatial Modeling and Remote Sensing. *Soil Sci. Soc. Am. J.* **2008**, *72*, 201–211.
47. Kasim, N.; Tiyip, T.; Abliz, A.; Nurmemet, I.; Sawut, R.; Maihemuti, B. Mapping and Modeling of Soil Salinity Using WorldView-2 Data and EM38-KM2 in an Arid Region of the Keriya River, China. *Photogramm. Eng. Remote Sens.* **2018**, *84*, 43–52. [CrossRef]
48. Liesenfeld, R.; Richard, J.-F. Univariate and multivariate stochastic volatility models: Estimation and diagnostics. *J. Empir. Financ.* **2003**, *10*, 505–531. [CrossRef]
49. Xie, S.; Ding, F.; Chen, S.; Wang, X.; Li, Y.; Ma, K. Prediction of soil organic matter content based on characteristic band selection method. *Spectrochim. Acta Part A Mol. Biomol. Spectrosc.* **2022**, *273*, 120949. [CrossRef]
50. Escribano, P.; Palacios-Orueta, A.; Oyonarte, C.; Chabrillat, S. Spectral properties and sources of variability of ecosystem components in a Mediterranean semiarid environment. *J. Arid Environ.* **2010**, *74*, 1041–1051. [CrossRef]
51. Fontán, J.; Calvache, S.; López-Bellido, R. Soil carbon measurement in clods and sieved samples in a Mediterranean Vertisol by Visible and Near-Infrared Reflectance Spectroscopy. *Geoderma* **2010**, *156*, 93–98. [CrossRef]

Article

The Effect of Pelletized Lime Kiln Dust Combined with Biomass Combustion Ash on Soil Properties and Plant Yield in a Three-Year Field Study

Donata Drapanauskaitė [1,2], Kristina Bunevičienė [1], Regina Repšienė [1], Danutė Karčauskienė [1], Romas Mažeika [1] and Jonas Baltrusaitis [2,*]

1. Lithuanian Research Centre for Agriculture and Forestry, Instituto al. 1, Akademija, LT-58344 Kedainiai, Lithuania; donata.drapanauskaite@lammc.lt (D.D.); kristina.buneviciene@lammc.lt (K.B.); regina.repsiene@lammc.lt (R.R.); danute.karcauskiene@lammc.lt (D.K.); romas.mazeika@lammc.lt (R.M.)
2. Department of Chemical and Biomolecular Engineering, Lehigh University, B336 Iacocca Hall, 111 Research Drive, Bethlehem, PA 18015, USA
* Correspondence: job314@lehigh.edu; Tel.: +1-610-758-6836

Abstract: Extensive application of mineral fertilizers resulted in high soil acidity, which is one of the major problems for crop production and soil degradation. Industrial solid waste, such as lime kiln dust and wood ash, can be used as alternative liming materials to benefit sustainable agricultural development. In this work, pelletized lime kiln dust with and without wood ash was utilized as liming material and the results of the three-year field study were compared with conventional mineral-based liming materials. It was determined that pelletized lime kiln dust satisfies the requirements posed by the recent European Union regulations to qualify as liming materials. The application of 2000 kg/ha Ca equivalent pelletized lime kiln dust increased soil pH_{KCl} by ~0.55 pH units. Moreover, pelletized lime kiln dust significantly increased spring wheat grain yields ranging from 33.6% to 40.4%, depending on the pellet size. The usage of these liming materials not only increased crop yield but also decreased heavy metal concentration in soil. Due to high alkalinity, carbonate content, easy handling, and the transportation of pelletized lime kiln dust with and without wood ash, the materials have the potential to be used in agriculture as liming materials to reduce soil acidification and increase crop productivity or be used as soil amendments.

Keywords: lime kiln dust; pellets; soil chemical properties

Citation: Drapanauskaitė, D.; Bunevičienė, K.; Repšienė, R.; Karčauskienė, D.; Mažeika, R.; Baltrusaitis, J. The Effect of Pelletized Lime Kiln Dust Combined with Biomass Combustion Ash on Soil Properties and Plant Yield in a Three-Year Field Study. *Land* 2022, 11, 521. https://doi.org/10.3390/land11040521

Academic Editors: Chiara Piccini, Rosa Francaviglia and Richard Cruse

Received: 14 February 2022
Accepted: 1 April 2022
Published: 4 April 2022

Publisher's Note: MDPI stays neutral with regard to jurisdictional claims in published maps and institutional affiliations.

Copyright: © 2022 by the authors. Licensee MDPI, Basel, Switzerland. This article is an open access article distributed under the terms and conditions of the Creative Commons Attribution (CC BY) license (https://creativecommons.org/licenses/by/4.0/).

1. Introduction

High soil acidity is a significant problem impeding crop production and is associated with soil degradation globally. The total area of topsoils affected by soil acidity range from 3.78 to 3.95 billion ha [1]. While soil acidification is a natural process, it can be exacerbated by human factors, such as acid rain, leaching of nutrients, and human activities such as using acidic fertilizers or harvesting plant materials without returning them to the soil. The emerging cause of soil acidification due to nitrogen (N) fertilizers has been of increasing concern worldwide [2,3]. Ammonium salts strongly acidify soils through their nitrification [4]. In particular, acidification takes place when ammonia is converted to nitrites followed by nitrates that are then leached [5]. For example, in Chinese farmland areas, soil pH decreased by 0.3 pH units from 1981 to 2012 due to increased mineral fertilizer application, while a critical N fertilizer application amount of 200 kg/ha per year was also reached [6]. Soil acidification changes biodiversity and increases nutrient losses, such as potassium (K), sodium (Na), calcium (Ca), and magnesium (Mg), via leaching, thus reducing plant productivity and increasing greenhouse gas emissions [7]. Decreasing soil pH harms the productivity of many crops (barley, rapeseeds, clover, and sugar beet). For most plants, the optimum soil pH ranges from 5.5 to 7.0. Additionally, soil pH affects the availability of nutrients and enhances the solubility of toxic metals causing nutrition imbalance in plants. Soil acidification was behind increased cadmium (Cd), lead (Pb), and zinc (Zn) levels, while manganese (Mn) and aluminum (Al) solubility can often reach a toxic level with decreasing soil pH [4]. Soil acidification not only affects the availability of nutrients but also soil physical properties.

Various bioclimatic zones are associated with certain soil formation pathways resulting from the decomposition of primary minerals and secondary mineral formation, as well as the formation of secondary complex organomineral compounds followed by their accumulation and transport [8]. Water-soluble Ca and Mg cations in the upper layers of these minerals can leach out into the watershed, resulting in overall soil acidification. In this manner, Ca- and Mg-rich soils across the globe in localities associated with high precipitation amounts become more acidic. These soils with a higher propensity to acidify are chiefly Retisols and Luvisols comprising carbonates within the 2-meter depth; hence, it is intrinsically prone to acidification due to temporal and environmental factors. Liming has been shown as one of the most economical methods of decreasing soil acidity. Liming improves soil structure, oxygen infiltration, and aeration [9,10]. It also enhances biological N fixation and the mineralization of phosphorus (P) and sulfur (S) [11–13]. Several studies have shown that liming remarkably decreased Cd mobility in soil and accumulation in plants [14,15]. The application of liming materials not only reduces the solubility of heavy metals but also enhances the availability of phosphorus to the plants [16,17]. Finally, increasing soil pH has a direct effect on N-related greenhouse gas emissions. Khaliq et al. showed that the application of dolomite and lime can reduce N_2O emissions from the soil by 44% and 37%, respectively, in upland and 52% and 44%, respectively, in paddy soils [18]. Additionally, previous studies demonstrated that liming decreased N_2O production under some conditions in fluvial soils [19]. To mitigate the acidification processes in soil, limestone and dolomite ($CaMg(CO_3)_2$)—widely available natural minerals—are often applied and utilized for the dual purpose of acidity neutralization as well as soil fertilization [20]. Furthermore, other liming materials include those comprised of Ca- and Mg-containing oxides, hydroxides, carbonates, and even silicates [21,22]. Lime industry processing waste, such as lime kiln dust (LKD), has recently been proposed as a potential liming material to improve acid soil quality since it contains large amounts of calcium (Ca) and magnesium (Mg) [23]. Much less work has been carried out on comparing actual soil property enhancement when utilizing LKD in field experiments in comparison with different types of natural minerals on spring barley or spring wheat growth properties. This is particularly important since the granulation of powders is critical in economically and efficiently converting and transporting them into usable and recyclable raw materials [24–26]. Pelletization provides the benefits of slower nutrient release and the ease of handling these bulk materials. In particular, pelletization alleviates various problems associated with dust handling during transport and field applications. In particular, there are very few examples of compacted, granulated, and pelletized lime kiln dust [27] and biomass ash [28–30]. Sell and Fischbach described the pelletization of the cement kiln dust with the resulting pellets returned to the clinker-making process [31]. Yliniemi et al. granulated peat-wood ash using potassium silicate and sodium aluminate to produce lightweight aggregates suitable to use in civil engineering or lightweight concrete [32]. Other researchers granulated bioenergy production waste (fly ash and biochar) with organic-rich lake sediments for the sustainable reuse of waste materials and the possibility to use granules in agriculture [33].

Understanding whether the benefits of traditional soil liming materials, as well as those from industrial waste, originate due to the improvement in soil pH, increased Ca availability, or enhancement in soil structure is of great concern. Since approximately 51.0% of Eastern Lithuanian and 66.0% of Western Lithuanian agricultural land have soil pH values less than or equal to 5.5, fertilizer use in these acid soils is inefficient. The purpose of this study was to investigate the effects of recovered waste from lime processing plants as soil liming materials on soil properties and crop yield. In particular, pelletized LKD with and without biomass ash were utilized and their liming properties were compared to those of natural minerals, such as ground chalk and crushed dolomite.

2. Materials and Methods

2.1. Materials and Reagents

Liming materials after anthropogenic processing were obtained from industrial manufacturers in Lithuania. Specifically, ground chalk was obtained from *JSC Baltijos Klintis*, Lithuania while crushed dolomite from *SC Dolomitas*, Lithuania. LKD was obtained from *JSC Naujasis Kalcitas*, Naujoji Akmene, Lithuania, while the pelletization of LKD powders into several fractions (namely 0.1–2, 2–5, and 5–8 mm) was performed by *JSC Mortar Akmene*. PLKDWA was also obtained from *JSC Mortar Akmene* using lime kiln dust (LKD) and wood ash (WA), and pellet sizes vary from 2 to 5 mm.

These samples were stored in plastic containers. All chemicals for chemical analysis were obtained of reagent grade from Fischer Sci and used as received. Double distilled water was used in all experiments. All liming materials are summarized in Table 1.

Table 1. Sample description.

Abbreviation	Sample Preparation
GC	Ground chalk
CD	Crushed dolomite
PLKD 0.1–2	Pelletized LKD of 0.1–2 mm
PLKD 2–5	Pelletized LKD of 2–5 mm
PLKDWA 2–5	Pelletized LKD with wood ash of 2–5 mm

An evaluation of the chemical composition of the liming materials was performed. The samples were ground, and the elements were extracted using aqua regia and analyzed with Perkin Elmer Optima 2100 DV ICP-OES spectrometer. Atomic Absorption Spectroscopy (AAS) was used to find the concentration of Al, Fe, Ca, and Mg. The amount of Si was found using a gravimetric method.

The neutralizing value of liming materials was determined by treating a sample with 0.5 N HCl heated for 10 min and later potentiometrically titrated with 0.25 N NaOH u pH reached 7.0 for 1 min. The neutralizing value was estimated as follows:

$$N_V = \frac{0.014 \times (V_1 - 0.5 \times V_2) \times 100}{m} \quad (1)$$

where N_V—Neutralizing value (%);
V_1—the volume HCl (mL);
V_2—the volume NaOH (mL);
m—sample mass (g).

Reactivity was determined by treating the sample with water and quickly potentiometrically titrating with 5.0 N HCl until pH reached 2.0, and titration was finished after 10 min keeping pH 2.0. The reactivity was estimated as follows:

$$r_{ac} = \frac{c_{HCl} \times 14.02 \times 100}{m \times N_V} \quad (2)$$

where r_{ac}—reactivity (%);
C_{HCl}—the volume of 5.0 N HCl (mL);
N_V—neutralizing value of liming material (%);
m—sample mass (g).

Water content was determined gravimetrically according to the standard LST EN 12048:2003.

2.2. Field Experiments

Field experiments were conducted from 2017 to 2019 at the Lithuanian Research Centre for Agriculture and Forestry Vezaiciai Branch, West Lithuania, on naturally acidic moraine loamy soil (*Bathygleyic Distric Glossic Retisol*) [34]. The agrochemical characteristics of the upper soil layer were as follows: pH_{KCl}—5.06 ± 0.541; soluble Ca—899 ± 30.0 mg/kg; soluble Mg—127 ± 3.9 mg/kg; soluble K_2O—199 ± 4.6 mg/kg; soluble P_2O_5—164 ± 7.1 mg/kg; soluble Al—24.4 ± 14.07 mg/kg.

The field trial was set up in a randomized design with four replicates. The following experimental design was used. Namely, (1) control (without liming material), (2) 2000 kg Ca/ha ground chalk, (3) 2000 kg Ca/ha crushed dolomite, (4) 2000 kg Ca/ha pelletized lime kiln dust outer diameter (OD) 0.1–2 mm, (5) 2000 kg Ca/ha pelletized lime kiln dust OD 2–5 mm, and (6) 2000 kg Ca/ha pelletized lime kiln dust-wood ash OD 2–5 mm. The experimental plot size was 48 m^2 (12 × 4 m). The experimental site was limed (except for the control treatment) with liming materials in May 2017 before planting the seeds. The liming rate (2000 kg Ca/ha) was calculated using the amount of active element Ca in liming materials. Mineral fertilizers were added every year before sowing. The mineral fertilizer application rate for spring barley and spring wheat was 60 kg/ha N, 60 kg/ha P_2O_5, and 60 kg/ha K_2O before sowing and 60 kg/ha N at the bushing stage of spring barley and wheat. The mineral fertilizer application rate for pea was 20 kg/ha N, 40 kg/ha P_2O_5, and 60 kg/ha K_2O. On the day of sowing, the pea seeds were coated with bacterial product Rizogen and sown immediately. Liming materials and fertilizers were applied manually, e.g., spread by hand on the soil's surface and incorporated into the soil by the cultivator. Spring barley cv. 'Louke A' was grown in 2017, spring wheat cv. 'Granary' was grown in 2018, and peas cv. 'Respect' was grown in 2019.

2.3. Soil and Plant Sampling and Chemical Analyses

Soil samples for agrochemical analysis (pH_{KCl}, soluble P_2O_5, soluble K_2O, soluble Ca, soluble Mg, soluble Al, and total heavy metals) were collected from 0 to 20 cm depth of the topsoil layer. The sample for chemical analyses was collected from 10 to 15 spots via a "W" shaped pattern across the sampling area. Soil samples were taken from four replicates of each treatment every year during the spring and fall after harvest. Soil soluble K_2O, P_2O_5, Ca, and Mg were determined according to the Egner–Riehm–Domingo (A-L) method [35]. Soil soluble K_2O, P_2O_5, Ca, and Mg were extracted using a 1:20 (wt/vol) soil suspension of ammonium lactate–acetic acid extractant (pH = 3.7). The suspension was shaken for 4 h. Soluble P_2O_5 was determined in the extract using ammonium molybdate via the spectrometric method with a Shimadzu UV 1800 spectrophotometer, while soluble K_2O was determined using flame emission spectroscopy with a JENWAY PFP7 flame photometer; soluble Ca and soluble Mg were determined using atomic absorption spectrometer AAnalyst 200. Soil soluble Al was determined by the Sokolov method [36] via extraction from 1:2.5 (wt/vol) soil suspension in the 1 M KCl, shaken for 1 h, and later measured using the titrimetric method.

The determination of soil pH was performed using a 1:5 (vol/vol) soil suspension in the 1 M KCl. The mixture was shaken for 60 min and left to sit for 1 h. The pH of the suspension was measured at 20 ± 2 °C stirring with a pH meter.

The heavy metal (Cd, Cr, Ni, and Pb) content in soil was determined by extraction in aqua regia and analyzed with a Perkin Elmer Optima 2100 DV ICP-OES spectrometer and AAnalyst 200 AAS spectrometer. Heavy metals in soil were determined according to ISO 11466:1995, ISO 11047:1998, and ISO 22036:2008: Soil pH-ISO 10390:2005.

The crops were harvested at full maturity. Barley, wheat grain, and peas seeds properties were determined as follows: grain/seed yield (t/ha, calculated on a 14% grain/seeds moisture basis), grain/seeds number per spike/pod, and 1000 grain/seed weight (g). The thousand grain/seed weight was determined using an automatic seed counter.

2.4. Meteorological Conditions

The data presented in Figure 1 for the study period from 2017 to 2019 suggest differences in weather conditions relative to the standard climate norm (SCN). Averaged across the growing seasons, the air temperature was higher in the 2018 and 2019 years of the study compared to SCN (4.0 °C in May 2018 1.8 °C in June 2018, 2.1 °C in July 2018, 2.6 °C in August 2018 and 4.6 °C in June 2019, and 1.2 °C in August 2019). However, total precipitation during the plant growing seasons showed significant variations, especially during May 2017, when a 41.9 mm decrease in total precipitation compared to SCN was reported, and during the May 2019 growing season, when the total precipitation increased by 30.0 mm compared to the standard climate norm. In summary, for temperature and precipitation regimes, the weather conditions in 2017 were cool and wet; in 2018 and 2019, the conditions were warm and dry.

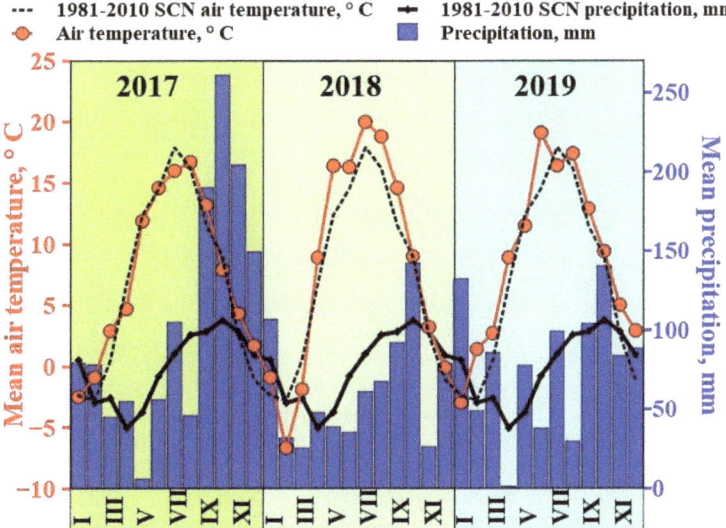

Figure 1. Mean monthly air temperature (°C) and precipitation (mm), according to the meteorological data from the Lithuanian Hydrometeorological Service under the Ministry of Environmental.

2.5. Statistical Analysis

A one-way analysis of variance was used to compare the soil characteristics and crop yield before and after soil liming. Means were compared using Fisher's least significant difference test at $p \leq 0.05$ and $p \leq 0.01$ and Duncan's multiple range test at $p \leq 0.01$. The statistical software package SAS [37] was used for analysis.

3. Results and Discussion

3.1. Chemical Composition of Liming Materials

Table 2 shows the physical properties of liming materials used in these field studies. Pelletized lime kiln dust (PLKD) alone and with wood ash was of highly alkaline pH (12.8–12.9) when compared to natural liming materials (GC and CD), which were less alkaline with a pH of 8.9 and 9.3, respectively. The water content in PLKD and PLKDWA varied from 11.9% to 6.36%, while in natural liming materials CD and GC, it varied from 9.3% to 8.9%. Water in the pelletization process was used as a binder, while pellets were dried afterward at 25 °C for 48 h. The neutralizing value measures the ability of the liming materials to reduce acidity, while reactivity indicates the rate of liming material to reduce the acidity of the soil. Hoşten and Gülsün showed that particle size and the dolomite content in the limestones were the most influential parameters in the reactivity of limestones [38]. According to

EU Regulation 2019/1009 for liming materials [39], the minimum reactivity (based on the hydrochloric acid test) cannot be less than 10% or neutralizing value (equivalent CaO) less than 15. As observed in Table 2, pelletized lime kiln dust (PLKD 0.1–2, PLKD 2–5, and PLKDWA 2–5) meets these requirements. The pellet strength may have a strong effect on the reactivity of granulated liming materials. PLKDWA 2–5 pellets were the hardest (51 ± 13.1 N/pellet), while PLKD 0.1–2 pellets were the weakest (18 ± 7.5 N/pellet). Effectively, pelletized liming materials can be hypothesized to contain different liming properties due to the more controlled release of nutrients, as opposed to CD, GC, LKD, and WA, which were applied as powders.

Table 2. Physical properties of liming materials and pelletized LKD.

	CD	GC	PLKD 0.1–2	PLKD 2–5	PLKDWA 2–5	LKD	WA
1-8 pH	9.3 ± 0.21	8.9 ± 0.07	12.8 ± 0.14	12.9 ± 0.07	12.8 ± 0.21	12.7 ± 0.07	12.9 ± 0.14
Water content, %	4.52 ± 0.141	0.10 ± 0.081	9.66 ± 0.162	11.9 ± 0.13	6.36 ± 0.219	1.25 ± 0.007	0.04 ± 0.021
Reactivity, %	10.0 ± 0.09	99.5 ± 0.49	24.8 ± 0.28	10.4 ± 0.42	19.3 ± 0.35	96.7 ± 0.35	42.2 ± 0.35
Neutralizing value, %	50.3 ± 0.21	52.2 ± 0.35	44.6 ± 0.42	41.5 ± 0.28	18.0 ± 0.42	45.4 ± 0.28	32.9 ± 0.14
Pellet strength, N/pellet	-	-	18 ± 7.5	37 ± 6.2	51 ± 13.1	-	-

Note: Colored columns for the table represent parent materials, which were used for pelletized liming materials.

LKD chemical composition typically varies and depends on the source and the process of lime being processed [40]. The typical composition of LKD varies from 31 to 55% CaO, 0–26% of free lime, 1.7–9.9% SiO_2, 0.7–4.1% Al_2O_3, 0.03–0.22% K_2O, and 0.5–25% MgO [41]. Figure 2a shows the chemical composition of the main nutrients present as well as alumosilicates measured in liming materials. The liming materials chiefly comprised Ca-containing compounds, with Ca accounting for 20% to 41% by weight of the measured nutrients. Other major plant nutrients, such as Mg and K, did not contribute significantly to the overall composition in CD reaching up to ~10%. Fe is an important micronutrient in small amounts needed to sustain plant growth and reproduction [42]. Alkaline soil, however, binds Fe and causes plant iron deficiency as it is immobilized and unavailable for plants [43]. The concentration of heavy metals in liming materials is shown in Figure 2b. Low levels of Cd were obtained in GC, CD, and PLKD without WA. In PLKDWA, the amount of Cd was 2.63 ± 0.120 mg/kg, and it slightly exceeded the allowable limit of 2 mg/kg according to EU Regulation 2019/1009. The higher Cd content in PLKDWA was due to the high concentration of 5.21 ± 0.134 mg/kg Cd in wood ash. The highest concentration of Pb (25.8 ± 0.21 mg/kg) compared to other liming materials was found in CD but did not exceed the 120 mg/kg allowable limit. Ni and Cr contents in liming materials also did not exceed allowable limits. Summarily, this work showed that most of the heavy metals measured were detected within concentrations lower than those defined in the regulatory documents describing fertilizers as well as liming materials for soil use. Hence PLKD and PLKDWA can be utilized as a source of liming material.

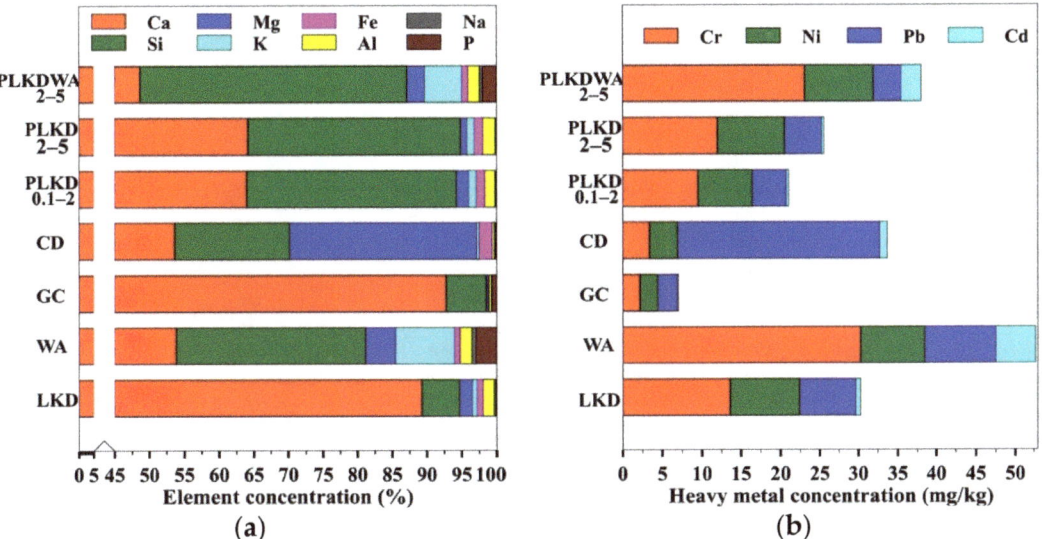

Figure 2. (**a**) Main nutrients and alumosilicates measured comprising liming materials. (**b**) Heavy metal concentration in liming materials.

3.2. Measured Soil Chemical Composition after the Liming Material Application

Liming has been widely recommended to manage soil acidification and improve plant yield and soil agrochemical parameters. The available reports suggest that crop yield can be increased using liming due to the improvement in the resulting soil's physical, chemical, and biological properties [44–46]. The soil pH_{KCl} in all experiments was ~5 before liming. The optimal soil pH_{KCl} range needed for plant growth is provided in Figure 3 and is between 5.5 and 7 [47,48]. Table 3 shows the pH_{KCl} values after liming.

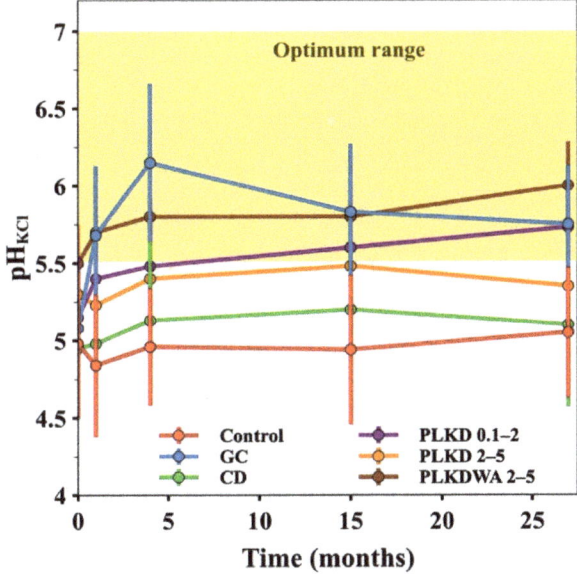

Figure 3. Measured soil pH values of different liming material treatments.

Table 3. pH_{KCl} values after 27 months of liming.

Treatment	pH_{KCl} Value ± SD
Control	5.1 ± 0.42 a
CD	5.1 ± 0.53 ab
GC	5.8 ± 0.38 cd
PLKD 0.1–2	5.7 ± 0.36 bcd
PLKD 2–5	5.4 ± 0.35 abcd
PLKDWA 2–5	6.0 ± 0.28 d

Note: different lowercase letters indicate a significant difference according to Duncan's multiple range test (DMRT $p \leq 0.05$).

Our results shown in Figure 3 suggest that different liming materials did not have the same effect on the neutralization of soil pH_{KCl}. The fastest and the highest increase in soil pH_{KCl} was with applied ground chalk (GC). GC increased soil pH_{KCl} from 5.08 to 6.15 after 4 months when applied at 2000 kg Ca/ha. However, after 27 months the pH_{KCl} decreased to 5.75, possibly due to the leaching of Ca^{2+} ions from the soil due to the high amount of rainfall in 2017, as shown in Figure 1. In agreement, long-term research in soil sorption complex of forest soils showed that soil pH_{KCl} decrease is related to H^+ increase and Ca^{2+} and Mg^{2+} ion decrease in soil sorption complex [49]. This is also supported by a strong positive correlation shown in Figure 5a (vide infra) between soluble Ca content in soil and soil pH_{KCl} (r = 0.875). In general, the fastest soil pH_{KCl} increase was observed when the milled GC was applied; the slowest was observed when the crushed CD was applied, while pelletized liming materials (PLKD 0.1–2; 2–5, PLKDWA 2–5) exhibited liming properties that were in between. Pelletized lime kiln dust (PLKD) 0.1–2 and pelletized lime kiln dust with wood ash (PLKDWA) 2–5 increased soil pH by ~0.5 after 3 years of liming. This corroborates the earlier studies on using ash as a liming material to increase soil pH [50,51]. Liming with 2000 kg Ca/ha of CD resulted in a very slightly statistically not significant increase in soil pH_{KCl}. This can be explained by the much faster removal of calcium compounds present in GC than those in dolomite due to the increased reaction kinetics and complex surface-limited reactions [52]. The hydrated calcium compounds in LKD can release calcium faster when in contact with the soil when compared to limestone and result in an efficient pH change of the soil [53]. de Vargas in a long-term field experimental study determined calcitic lime exhibits much more facile soil neutralization properties when compared to dolomite [54].

Soluble Ca concentration in soil was measured four times during the three years. Soluble Ca content in soil was ~900 mg/kg in all treatments before liming. The concentration of soluble Ca in soil depends on carbonating layer depression depth, which is rich in Ca- and Mg-carbonates. In Western Lithuania, this layer is at 1.5–3.0 m depth. Liming significantly ($p \leq 0.01$) increased soluble Ca content in the soil, as shown in Figure 4a. The application of PLKD 0.1–2 corresponding to 2000 kg Ca/ha increased soluble Ca concentration ~2.5 times more when compared to control after 4 months of liming. This is possible since the hydroxide amount in pelletized liming materials is higher, rendering it more reactive than calcium carbonate [55]. Moreover, PLKD 2–5 showed a statistically significant effect on soluble Ca content in the soil, but it was less than GC or PLKD 0.1–2. Liming with CD showed a statistically significant ($p \leq 0.01$) increase in soluble Ca in the soil, but the increase was the smallest of all tested materials. After 3 years of liming, the highest content of soluble Ca in the soil of 1500 mg/kg was measured after liming with PLKDWA 2–5. Importantly, when comparing liming performance among the materials after 3 years of liming, PLKD 0.1–2, 2–5 and PLKDWA 2–5 had a statistically significant ($p \leq 0.01$) effect on soluble Ca, which was higher than that of CD. Ultimately, soluble Ca level was assessed as low before liming and after three years, different liming materials increased this from low to average [56]. To this extent, the results presented here for Lithuanian Retisol are in agreement with previous work where various liming material powders of natural and industrial origin, including lime mud, carbide lime, wood ash, cement kiln dust, and natural calcitic and dolomitic lime, increased soil exchangeable Ca amount and enhanced microbial activity in soil [57,58]. Annually, soluble Ca from the soil is leached at about a rate of 200–300 kg/ha, which depends on the amount of rainfall, soil texture, the amount of CO_2 produced from plant roots, and other factors [59]. Due to the high precipitation from September 2017 to February 2018 shown in Figure 1, a large decrease in soluble Ca was observed after 15 months of liming compared to 4 months after liming (Figure 4a). This suggests that the addition of Ca ions in acid soils with liming was necessary.

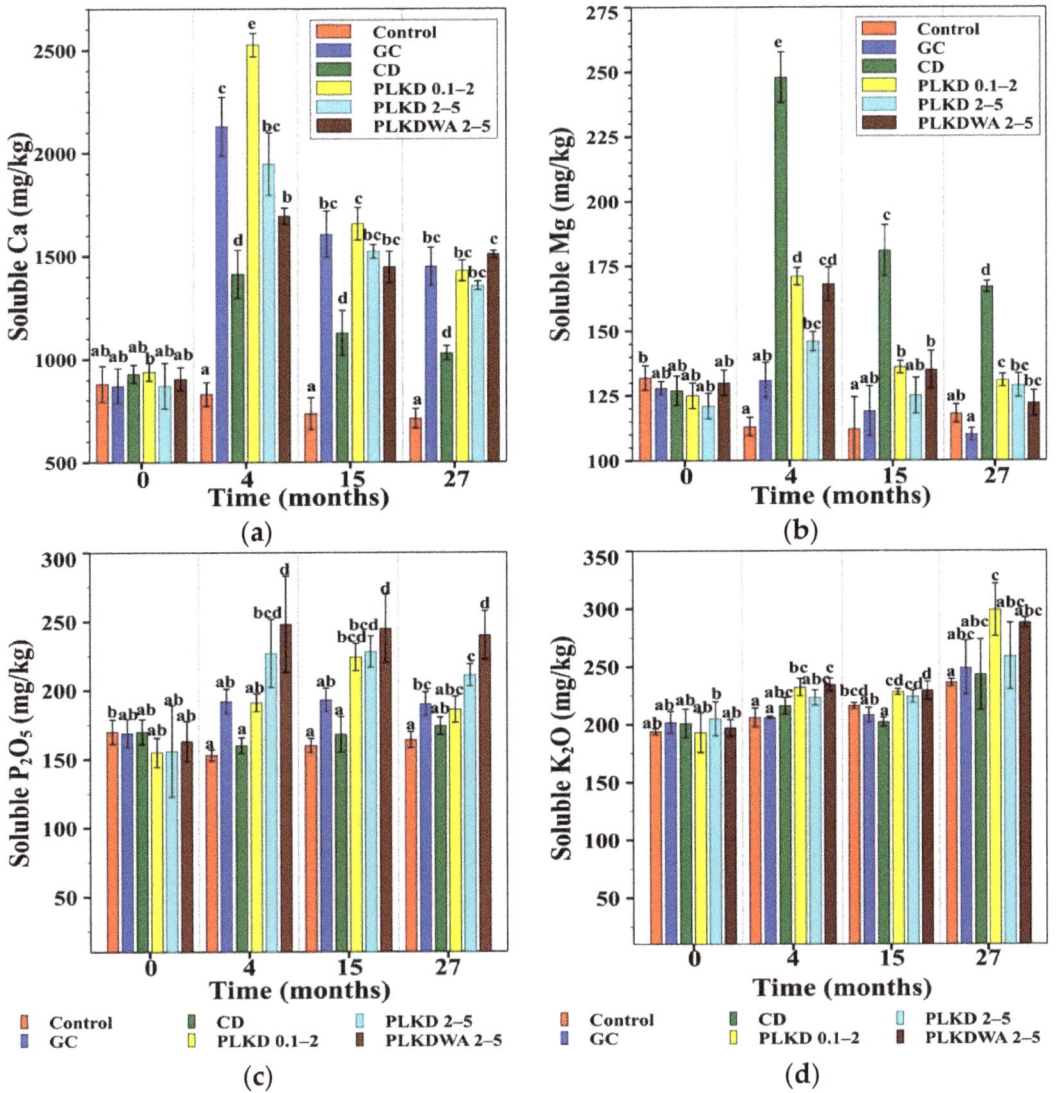

Figure 4. (a) Soluble Ca, (b) soluble Mg, (c) soluble P_2O_5, and (d) soluble K_2O in the soil during the 3-year liming experiment. *Soluble* is defined as being available for plants. Lowercase letters indicate a significant difference according to Duncan's multiple range test (DMRT $p \leq 0.01$).

Before liming, soluble Mg content in soil was low at around 130 mg/kg, as shown in Figure 4b. The largest increase in soluble Mg in the soil after liming was found with applied CD because it has the highest Mg content of 10.3%. However, liming with PLKD 0.1–2 and PLKDWA 2–5 also resulted in a statistically significant ($p \leq 0.01$) increase in soluble Mg in the soil after 4 months and significant ($p \leq 0.01$) after 15 months. Similar to Ca, after 27 months of liming with PLKD and PLKDWA 2–5, a statistically significant ($p \leq 0.01$) increase in soluble Mg was observed when compared to GC. Notably, soluble Mg is best absorbed by plants when the ratio of soluble Ca and soluble Mg is 1:5–8. When CD was applied, this ratio was 1:6 and was suitable for plants to absorb the soluble Mg. For other liming materials, the ratio was higher than 1:8 and the absorption of soluble Mg was blocked by Ca ions due to the antagonistic competition between Ca^{2+} and Mg^{2+} ions for cation exchange sites [60].

In acidic soil where there is an abundance of soluble Al and Fe, any P forms insoluble Al and Fe orthophosphates [61]. Before the field experiments, the soluble P_2O_5 concentration in soil was ~165 mg/kg. The further analysis of soluble P_2O_5 in soil showed that without liming, the amount of soluble P_2O_5 decreased throughout the experiment, as shown in Figure 4c. Only the application of 2000 kg Ca/ha with CD did not have a significant effect on soil soluble P_2O_5 concentration. Liming with PLKDWA 2–5 and PLKD 2–5 significantly ($p \leq 0.01$) increased soluble P_2O_5 content in soil compared to the control by 85 mg/kg and 71 mg/kg, respectively, after 4 months of application. Moreover, this soluble P_2O_5 increase was statistically significant ($p \leq 0.01$) when compared to GC and CD, not only to the control. Comparing PLKD with natural liming materials (GC, CD) after 3 years of application, PLKD 2–5 and PLKDWA 2–5 had a statistically significant ($p \leq 0.01$) effect on soluble P_2O_5 increase. An increase in soluble or available phosphorus amounts in acid soils after the application of liming materials was obtained in some other studies [44,62].

Measured soluble K_2O in the soil is shown in Figure 4d. Soil soluble K_2O varied from 194 mg/kg to 236 mg/kg before liming. After 4 months of liming with 2000 kg Ca/ha of PLKD 0.1–2 and PLKDWA 2–5, soil soluble K_2O content increased statistically and significantly ($p \leq 0.01$) by 26 mg/kg and 28 mg/kg, respectively. Additionally, application with PLKD 2–5, GC, and CD changed soil soluble K_2O statistically and insignificantly. After 27 months of liming, the overall highest soluble K_2O increase was observed in treatments with PLKD 0.1–2 and PLKDWA 2–5. For PLKDWA 2–5, these results may be related to the relatively high K_2O concentration of 2.9%. K in wood ash may be soluble and available to plants, and previous studies showed that the application of wood ash can increase K content in the soil [50,63].

Soluble Al concentration in the soil before the liming experiments was high or very high and varied in a wide range from 17.1 mg/kg to 52.14 mg/kg. The average soluble Al concentration before the application of liming materials was 24.4 ± 14.07 mg/kg. For the control treatment with no liming, the soluble Al content in the soil after 27 months increased from 24.4 ± 14.07 mg/kg to 37.2 ± 18.97 mg/kg. After liming in treatments where pH_{KCl} changed to 5.0 or higher, no soluble Al in soil was detected. Hence, liming reduced soluble Al content in the soil. Soluble Al concentrations were strongly affected by liming and exhibited a high negative nonlinear (polynomial) correlation with soil pH_{KCl}, as shown in Figure 5b. Soluble Al has a toxic effect on plant roots. The roots become poorly developed and weak. When there is an excess of soluble Al in the soil, plants hardly absorb P, Ca, S, and other elements. The relationship between soluble P_2O_5 and soluble Al was also found in this work to be described with a negative polynomial curve ($r = 0.836$) (Figure 5c). Similarly for other soils, Mrvić et al. observed a strong negative ($r = 0.952$) non-linear correlation between exchangeable Al and soil pH values in Stagnosols [64]. Moreover, Moir and Moot's research showed similar results to the relationship between exchangeable plant-available Al and soil pH in Brown soils [65].

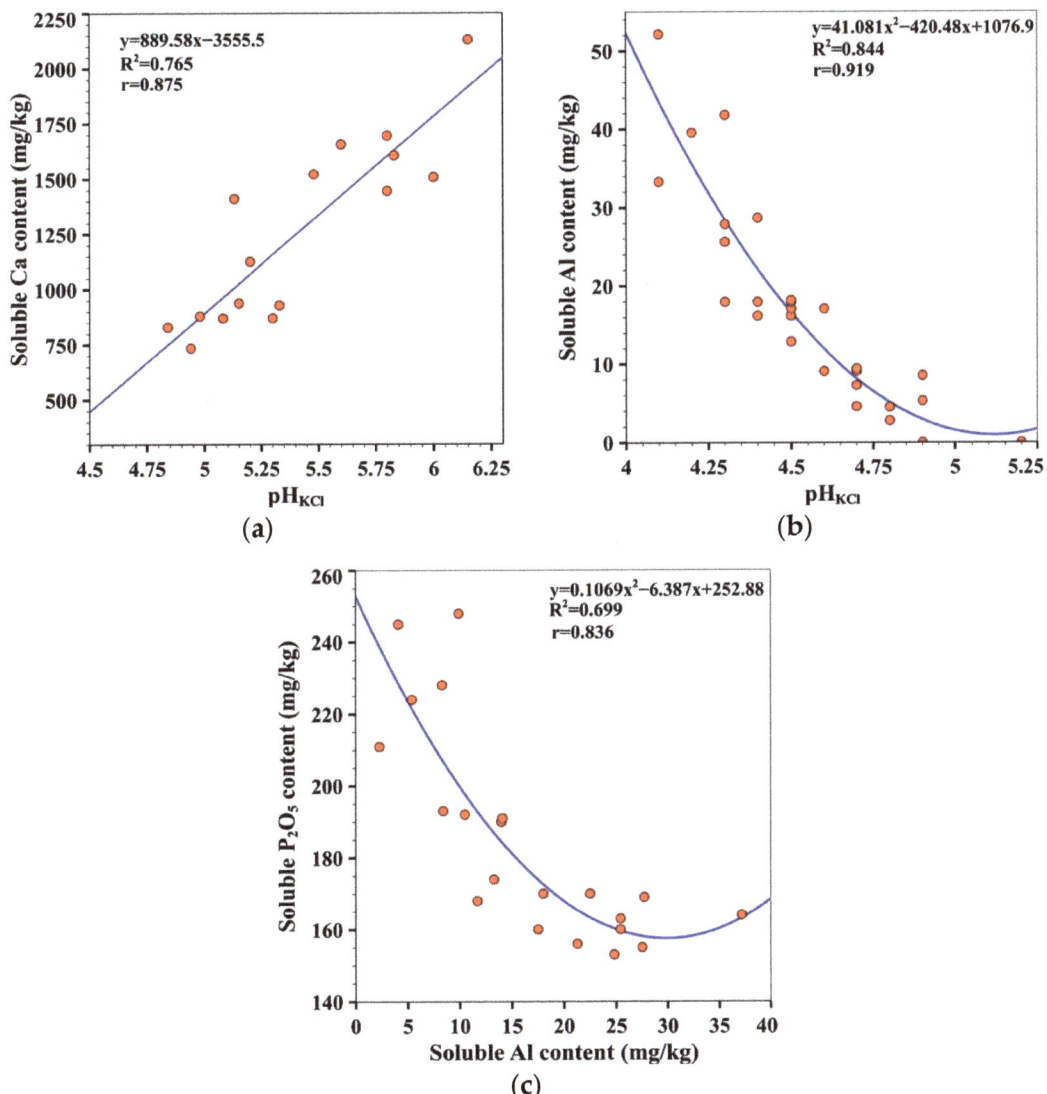

Figure 5. (**a**) Relationship between soluble Ca and pH_{KCl}, (**b**) between soluble Al and pH_{KCl}, and (**c**) between soluble P_2O_5 and soluble Al.

3.3. Grain Yield and Yield-Related Parameters

The effects of liming on grain yield improvement of crops in rotation are shown in Figure 6. In the first year, a statistically significant yield improvement was observed for spring barley in treatments with GC, PLKD 2–5, and PLKDWA 2–5. The biggest yield improvement of 10.4% and 9.9%, when compared to control, was obtained when liming with GC and PLKD 2–5, respectively. The highest statistically significant ($p < 0.01$) 1000th-grain mass of 54.4 ± 0.95 g was observed in the PLKD 2–5 treatment. However, liming with PLKD 0.1–2 did not improve spring barley yield. Grain yield was likely limited by the lower content of soil soluble P_2O_5 (Figure 4c) and pH_{KCl} in PLKD 0.1–2 treatment, although soil soluble P_2O_5 content was very similar to that obtained with GC.

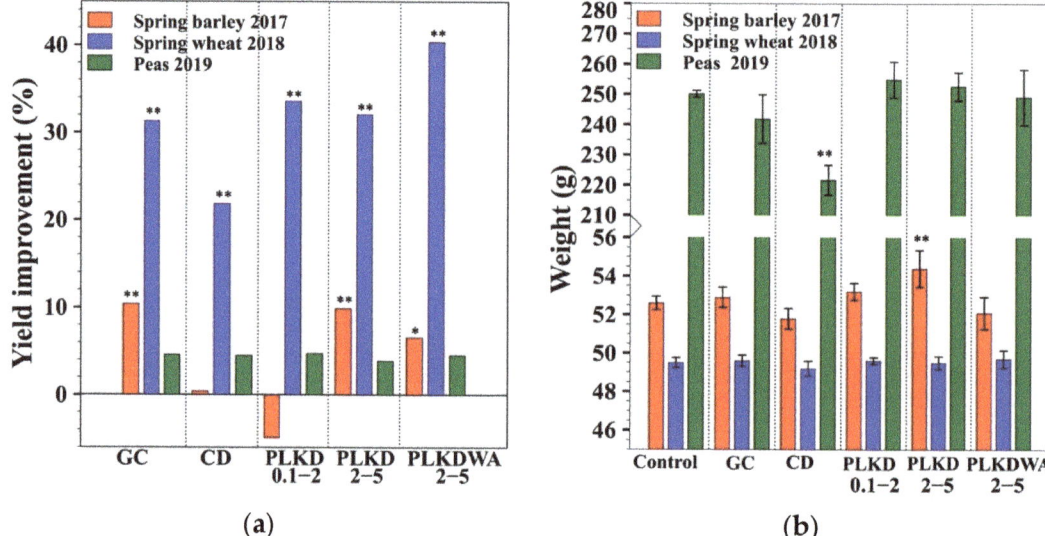

Figure 6. (a) The yield of spring barley, spring wheat and peas; (b) 1000-grain weight of spring barley, spring wheat and 1000-seed weight of peas. * and ** indicate significant differences according to Fisher's least significant difference test. Statistically significant at * $p < 0.05$ and ** $p < 0.01$ level.

The water deficit in 2017, shown in Figure 1, that occurred during barley booting could have influenced the number of grains per spike and spring barley grain yield. After 15 months of liming, a statistically significant ($p < 0.01$) spring wheat yield improvement, when compared to the control, was obtained in all treatments. However, the lowest statistically significant ($p < 0.01$) yield improvement of 21.9% of spring wheat was obtained in the treatment with CD while the smallest 1000th grain mass 42.9 ± 0.39 g was observed for the same treatment. The highest increase in spring wheat grain yield of 40.4% was obtained after liming with LKDWA. This result may be related to pH, which increases the availability of nutrients in the soil. Patterson et al. found that a 6 t/ha application of wood ash with nitrogen fertilizer increased barley and canola grain yield when compared to the control [66]. Moreover, other studies have shown an increase in oat biomass using pelletized wood ash [50] and an increase in oilseed productivity but a decrease in the quality of seed production [67]. In the third year of crop rotation, when the peas were grown, the liming increased yield for all treatments by about 4.5% compared to the control but the increase was not statistically significant. However, liming with CD was statistically significant ($p < 0.01$) for the reduced peas' 1000th seed mass.

Different crops exhibit significantly different tolerance to soil acidity and sensitivity to soil pH [20], as exhibited by various resulting properties such as plant height, plant density, germination, and reproductive performance [68,69]. The results of the number of plants per square meter are shown in Figure 7a. In 2017, when liming materials were applied, spring barley plants per square meter positively responded with CD (296 no. per m^2) and PLKD 0.1–2 (299 no. per m^2) compared to the control (283 no. per m^2) treatment. The second-year after liming spring wheat plants per square meter negatively responded to all treatments compared with the control. The lowest number of wheat plants per m^2 were obtained in treatments with CD and PLKD 2–5. It may be due to the higher than usual amount of precipitation in 2017 Fall and Winter, which caused high nutrient losses from soil. Additionally, a lower amount of rainfall and higher air temperature compared to SCN in 2018 may have an influence. After three years of liming, in 2019, liming had a positive effect on pea plants per square meter. It may be related to the pH increase and soluble

Al decrease in the soil after liming, which favors root proliferation. In agreement, other researchers reported a positive response of wheat plants per square meter with other liming material applications [70,71].

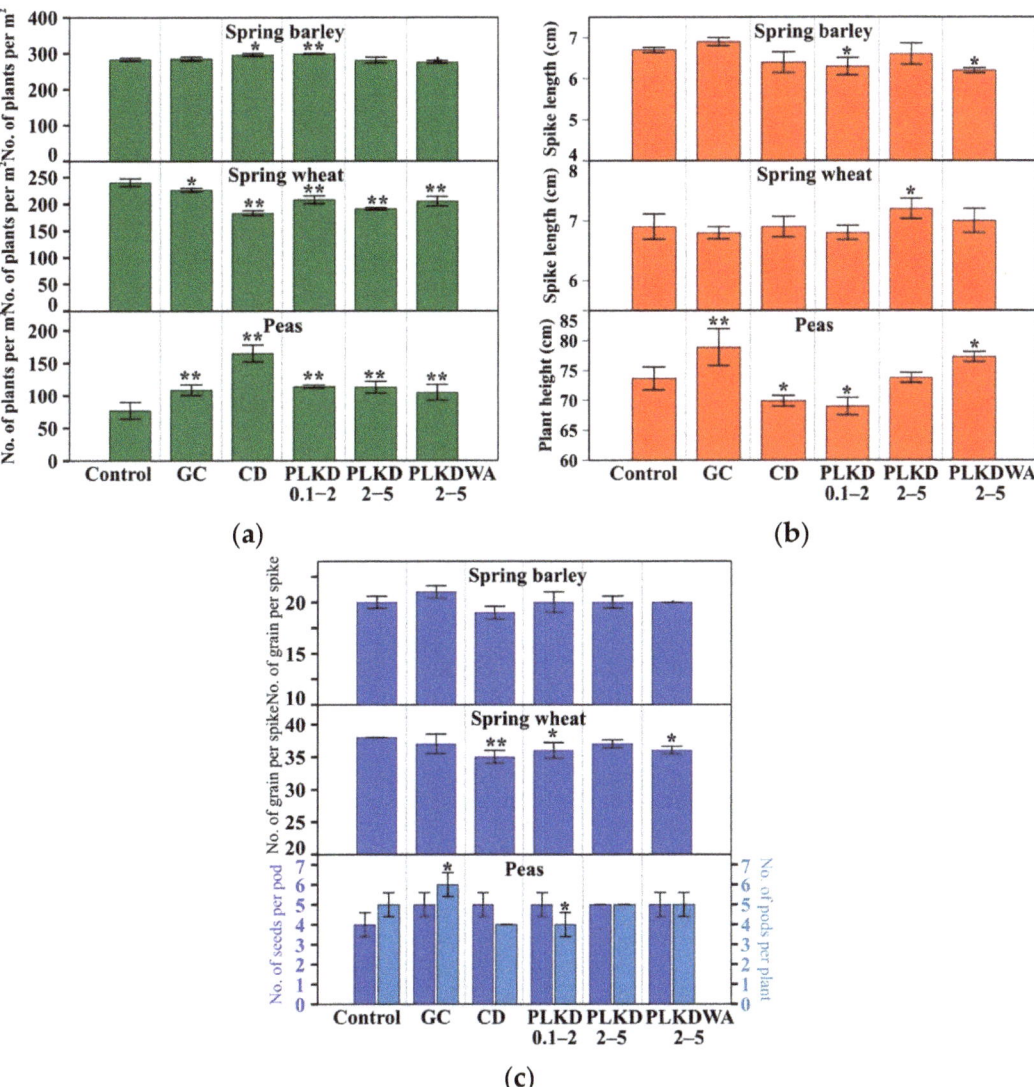

Figure 7. (**a**) The number of plants per square meter of spring barley, spring wheat, and peas; (**b**) the spike length of spring barley, spring wheat, and plant length of peas; (**c**) the number of grains per spike of spring barley, spring wheat, and the number of seeds per pod and number of pods per plant of peas. * and ** indicate significant differences according to Fisher's least significant difference test. Statistically significant at * $p < 0.05$ and ** $p < 0.01$ level.

The corresponding data for the pea plant height, spring barley, and wheat spike length are shown in Figure 7b. Plant height can potentially be improved due to the effect of liming. Liming increases soil pH, which affects root proliferation and increases nutrient availability, which can contribute to plant height. The application of 2000 kg Ca/ha of GC and PLKDWA 2–5 had a statistically significant effect on pea plant height when compared to control. In particular, pea plants limed with GC and PLKDWA 2–5 were 5.2 cm and 3.6 cm higher, when compared to the control. Moges et al. also showed that the application of 4 and 6 t/ha of lime significantly increased plant height [72]. However, the application of CD and PLKD 0.1–2 reduced plant height by 3.8 cm and 4.7 cm when compared to the control. Spring barley spike length and the number of grains per spike (Figure 7c) were greater with GC and PLKD 2–5 treatments compared to the control. The application of PLKD 2–5 also increases spike length and the number of grains per spike for spring wheat. Spike length and number of grains per spike were only affected by various environmental factors to a small degree since they strongly depend on the genotype [73].

3.4. Heavy Metals in the Soil

Changes in heavy metal (Cd, Cr, Ni, and Pb) concentrations in the soil during the liming period are shown in Figure 8. The total Cd content in the soil before liming was 1.95 ± 0.140 mg/kg (Figure 8a). According to The EU Commission Council Directive 86/278/EEC [74], the Cd content in the soil before liming does not reach the maximum allowable limit. After 15 months of application of liming materials, the total Cd content in soil was reduced 3 times from 0.45 ± 0.140 mg/kg to 0.14 ± 0.013 mg/kg. Cd concentration in the control treatment (unlimed) increased 1.3 times in 27 months, from 0.43 ± 0.035 mg/kg to 0.55 ± 0.040 mg/kg. Cd content after liming decreased because liming neutralized H^+ ions and reduced Cd bioavailability. Additionally, pH change can increase negative surface charge, which in turn could result in Cd adsorption and precipitation as Cd carbonates, reactions to $Cd(OH)_2$, and the reduction of Cd^{2+} to Cd^0. Cd content in unlimed soil–control increased due to the pH decrease, which may increase the enhanced solubility and mobility of cadmium in soil. Ramtahal et al. field and laboratory studies showed that liming reduced bioavailable Cd in soil and Cacoa beans [75]. Shaheen and Rinklebe used different low-cost alternative amendments to show that the application of cement kiln dust decreased soluble and exchangeable Cd content in the soil [76]. Total Cr content before the application of liming materials ranged from 9.77 mg/kg to 10.4 mg/kg and averaged at 10.1 ± 0.26 mg/kg (Figure 8b). The application of natural liming materials (GC and CD) slightly reduced Cr amounts in the soil while PLKD 0.1–2, PLKD 2–5, and PLKDWA 2–5 increased Cr content. In general, the amount of Cr in the soil can vary depending on the heterogeneity of soil and fertilization, and mineral fertilizers (especially phosphorus) can increase it. After 27 months of liming, the amount of Cr in soil increased 55% in the control treatment and 52% in the limed treatment with PLKDWA 2–5, while the smallest increase was obtained when limed with PLKD 0.1–2. This is due to H^+ competition for binding sites enhancing metal release from the soil matrix.

For GC and CD, the Cr content reduced after 4 months, while after 27 months, the amount increased by 31% and 36%, which is consistent with the change in soil pH values shown in Figure 3. Liming had a significant effect on reducing total lead (Pb) and total nickel (Ni) concentrations in soil (Figure 8c,d). The application of liming materials after 27 months reduced total Ni and total Pb amounts in soil by 1.5 as well as 1.3 times compared to untreated soil. Ni availability in the soil was reduced by increasing base-cation saturation, which consequently raises the soil's pH. Moreover, Ni solubility decreases when soil pH increases. Shaheen et al. showed that cement kiln dust and limestone decreased the water-soluble and soluble contents with exchangeable Ni concentration in the soil as a result of an increase in the sorbed and bound carbonate fraction [77]. Total heavy metal concentration in soil depends on the nature of the soil, its organic matter concentration, texture, and depth. As a consequence of adsorption of soil organic matter or atmospheric deposition, the highest concentration of some elements, such as Cd and Pb, are found on the soil's surface; for other elements (Ni, Fe, and V) that are associated with clays and hydrous oxides, they are concentrated in lower soil depths [78]. Liming changes not only the chemical properties of soil but also the morphological features, thus altering the size distribution of clay and silt with the soil's profile. Soil pH also has a direct effect on the availability of heavy metals by affecting their solubility and capacity to form chelates. An increase in soil pH after liming causes an increase in cation adsorption onto soil particles [79]. Tlustoš et al. pot and Rhizobox experimental results also showed that liming reduces 50% Cd, 20% Pb, and 80% of Zn and is effective for the immobilization of Cd, Pb, and Zn [80]. The efficiency of PLKD and PLKDWA in decreasing the mobilization of heavy metals may be explained by their high alkalinity and carbonate content, surface area, and oxide contents. The metals might decrease due to sorption and precipitation reactions.

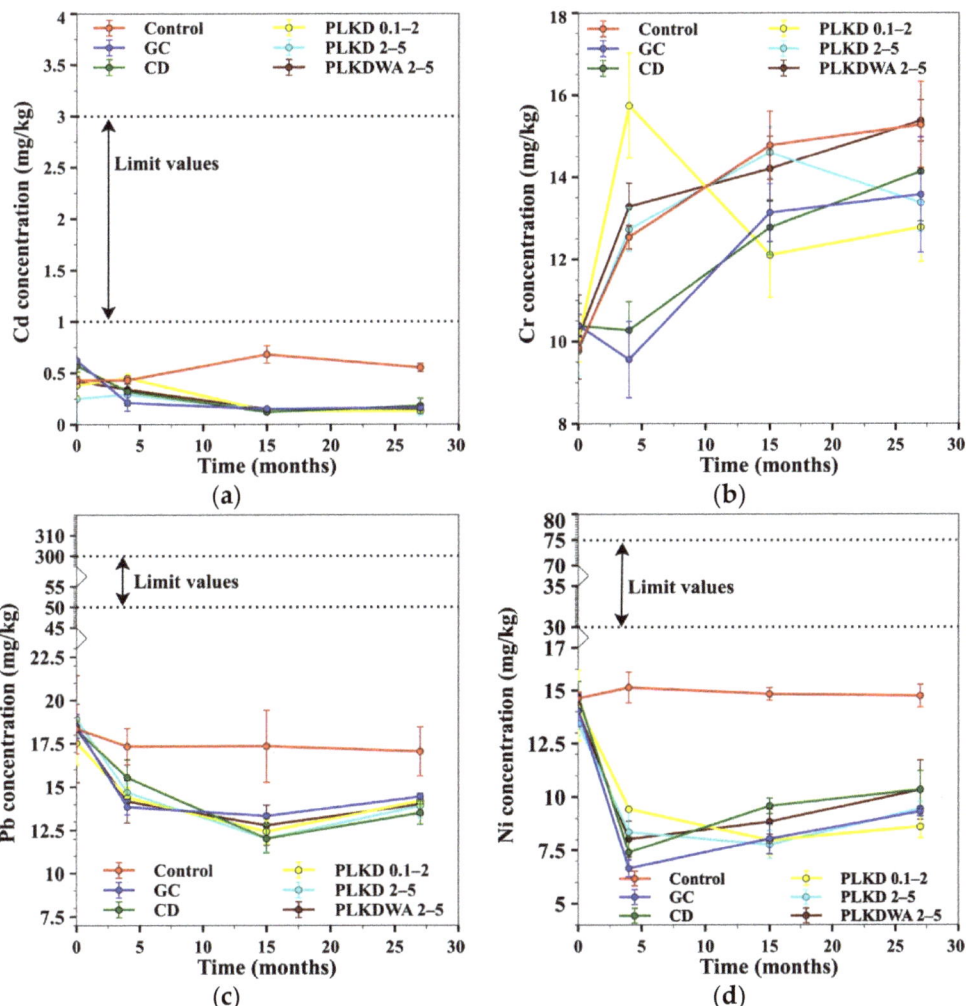

Figure 8. (**a**) Total Cd, (**b**) Cr, (**c**) Pb, and (**d**) Ni concentrations in the soil during the liming experiment. Regulatory limit values according to the EU Commission [74] standards are also shown.

4. Conclusions

- Application of 2000 kg/ha Ca of PLKD 0.1–2 and PLKDWA increased soil pH_{KCl} 0.58 and 0.50 pH units after three years of liming.
- Moreover, increased soil soluble Ca, Mg, P_2O_5, and K_2O contents and reduced soluble Al concentration in soil.

- After two years of application, PLKD (0.1–2; 2–5) and PLKDWA statistically significantly increased spring wheat grain yield by 33.6%, 32.1%, and 40.4%, respectively. After three years of liming, peas yield increased in all treatments ~4.5% compared to the control.
- Usage of these liming materials also decreased heavy metal concentration in soil. Liming reduced total Cd, Ni, and Pb contents in soil by 3, 1.5, and 1.3 times compared to unlimed treatment. However, liming did not reduce the total Cr content in the soil.
- Due to high alkalinity, carbonate content, easy handling, and the transportation of pelletized lime kiln dust and pelletized lime kiln dust with wood ash, the materials have the potential to be used in agriculture as liming materials.

Author Contributions: Conceptualization, K.B., R.R., D.K. and D.D.; methodology, R.M. and R.R.; investigation, D.D.; data Curation, R.M., D.D. and D.K.; writing—original draft preparation, D.D. and. J.B.; writing—review and editing, D.D. and J.B.; visualization, D.D. and K.B.; supervision, J.B. All authors have read and agreed to the published version of the manuscript.

Funding: This work is supported by Engineering for Agricultural Production Systems program, grant no. 2020-67022-31144 from the USDA National Institute of Food and Agriculture, and by the long-term research program 'Productivity and sustainability of agricultural and forest soils' supported by the Lithuanian Research Centre for Agriculture and Forestry.

Institutional Review Board Statement: Not applicable.

Informed Consent Statement: Not applicable.

Data Availability Statement: Data are contained within the article.

Conflicts of Interest: The authors declare no conflict of interest.

References

1. Sumner, M.E.; Noble, A.D. Soil Acidification: The World Story. In *Handbook of Soil Acidity*; Rengel, Z., Ed.; Mercel Dekker: New York, NY, USA, 2003; pp. 1–28.
2. Guo, J.H.; Liu, X.J.; Zhang, Y.; Shen, J.L.; Han, W.X.; Zhang, W.F.; Christie, P.; Goulding, K.W.T.; Vitousek, P.M.; Zhang, F.S. Significant Acidification in Major Chinese Croplands. *Science* **2010**, *327*, 1008–1010. [CrossRef] [PubMed]
3. Goulding, K.W.T. Soil Acidification and the Importance of Liming Agricultural Soils with Particular Reference to the United Kingdom. *Soil Use Manag.* **2016**, *32*, 390–399. [CrossRef] [PubMed]
4. Bolan, N.S.; Adriano, D.C.; Curtin, D. Soil Acidification and Liming Interactions with Nutrient and Heavy Metal Transformation and Bioavailability. *Adv. Agron.* **2003**, *78*, 215–272. [CrossRef]
5. Beeckman, F.; Motte, H.; Beeckman, T. Nitrification in Agricultural Soils: Impact, Actors and Mitigation. *Curr. Opin. Biotechnol.* **2018**, *50*, 166–173. [CrossRef]
6. Li, Q.; Li, S.; Xiao, Y.; Zhao, B.; Wang, C.; Li, B.; Gao, X.; Li, Y.; Bai, G.; Wang, Y.; et al. Soil Acidification and Its Influencing Factors in the Purple Hilly Area of Southwest China from 1981 to 2012. *Catena* **2019**, *175*, 278–285. [CrossRef]
7. Bouwman, A.F.; Boumans, L.J.M.; Batjes, N.H. Modeling Global Annual N$_2$O and NO Emissions from Fertilized Fields. *Glob. Biogeochem. Cycles* **2002**, *16*, 28-1–28-29. [CrossRef]
8. Eidukeviciene, M.; Vasiliauskiene, V.; Misevicius, J. *Lietuvos Dirvozemiai. Monografija*; Lietuvos mokslo redakcija: Kaunas, Lithuania, 2001; pp. 210–213.
9. Anikwe, M.A.N.; Eze, J.C.; Ibudialo, A.N. Influence of Lime and Gypsum Application on Soil Properties and Yield of Cassava (Manihot Esculenta Crantz.) in a Degraded Ultisol in Agbani, Enugu Southeastern Nigeria. *Soil Tillage Res.* **2016**, *158*, 32–38. [CrossRef]
10. Sheinberg, I.; Sumner, M.E.; Miller, W.P.; Farina, M.P.W.; Pavan, M.A.; Fey, M.V. Use of Gypsum on Soils: A Review. In *Advances in Soil Science*; Stewart, B.A., Ed.; Springer: New York, NY, USA, 1989; pp. 1–111.
11. Bolan, N.S.; Rowarth, J.; de la Luz Mora, M.; Adriano, D.; Curtin, D. Biological Transformation and Bioavailability of Nutrient Elements in Acid Soils as Affected by Liming. *Dev. Soil Sci.* **2008**, *32*, 413–446. [CrossRef]
12. Anderson, T.H. The Influence of Acid Irrigation and Liming on the Soil Microbial Biomass in a Norway Spruce (Picea Abies [L.] K.) Stand. *Plant Soil* **1998**, *199*, 117–122. [CrossRef]
13. Haynes, R.J.; Swift, R.S. Effects of Lime and Phosphate Additions on Changes in Enzyme Activities, Microbial Biomass and Levels of Extractable Nitrogen, Sulphur and Phosphorus in an Acid Soil. *Biol. Fertil. Soils* **1988**, *6*, 153–158. [CrossRef]
14. Tang, X.Y.; Katou, H.; Suzuki, K.; Ohtani, T. Air-Drying and Liming Effects on Exchangeable Cadmium Mobilization in Contaminated Soils: A Repeated Batch Extraction Study. *Geoderma* **2011**, *161*, 18–29. [CrossRef]

15. Yang, Y.; Chen, J.; Huang, Q.; Tang, S.; Wang, J.; Hu, P.; Shao, G. Can Liming Reduce Cadmium (Cd) Accumulation in Rice (*Oryza sativa*) in Slightly Acidic Soils? A Contradictory Dynamic Equilibrium between Cd Uptake Capacity of Roots and Cd Immobilisation in Soils. *Chemosphere* **2018**, *193*, 547–556. [CrossRef] [PubMed]
16. Trasar-Cepeda, M.C.; Carballas, T.; Gil-Sotres, F.; de Blas, E. Liming and the Phosphatase Activity and Mineralization of Phosphorus in an Andic Soil. *Soil Biol. Biochem.* **1991**, *23*, 209–215. [CrossRef]
17. Moreno-Jiménez, E.; Esteban, E.; Carpena-Ruiz, R.O.; Lobo, M.C.; Peñalosa, J.M. Phytostabilisation with Mediterranean Shrubs and Liming Improved Soil Quality in a Pot Experiment with a Pyrite Mine Soil. *J. Hazard. Mater.* **2012**, *201–202*, 52–59. [CrossRef]
18. Khaliq, M.A.; Khan Tarin, M.W.; Jingxia, G.; Yanhui, C.; Guo, W. Soil Liming Effects on CH4, N2O Emission and Cd, Pb Accumulation in Upland and Paddy Rice. *Environ. Pollut.* **2019**, *248*, 408–420. [CrossRef]
19. McMillan, A.M.S.; Pal, P.; Phillips, R.L.; Palmada, T.; Berben, P.H.; Jha, N.; Saggar, S.; Luo, J. Can PH Amendments in Grazed Pastures Help Reduce N2O Emissions from Denitrification?—The Effects of Liming and Urine Addition on the Completion of Denitrification in Fluvial and Volcanic Soils. *Soil Biol. Biochem.* **2016**, *93*, 90–104. [CrossRef]
20. Holland, J.E.; Bennett, A.E.; Newton, A.C.; White, P.J.; McKenzie, B.M.; George, T.S.; Pakeman, R.J.; Bailey, J.S.; Fornara, D.A.; Hayes, R.C. Liming Impacts on Soils, Crops and Biodiversity in the UK: A Review. *Sci. Total Environ.* **2018**, *610–611*, 316–332. [CrossRef]
21. Castro, G.S.A.; Crusciol, C.A.C.; da Costa, C.H.M.; Ferrari Neto, J.; Mancuso, M.A.C. Surface Application of Limestone and Calcium-Magnesium Silicate in a Tropical No-Tillage System. *J. Soil Sci. Plant Nutr.* **2016**, *16*, 362–379. [CrossRef]
22. Basak, B.B.; Biswas, D.R. Potentiality of Indian Rock Phosphate as Liming Material in Acid Soil. *Geoderma* **2016**, *263*, 104–109. [CrossRef]
23. Sreekrishnavilasam, A.; King, S.; Santagata, M. Characterization of Fresh and Landfilled Cement Kiln Dust for Reuse in Construction Applications. *Eng. Geol.* **2006**, *85*, 165–173. [CrossRef]
24. Tang, P.; Brouwers, H.J.H. Integral Recycling of Municipal Solid Waste Incineration (MSWI) Bottom Ash Fines (0–2 Mm) and Industrial Powder Wastes by Cold-Bonding Pelletization. *Waste Manag.* **2017**, *62*, 125–138. [CrossRef] [PubMed]
25. Gesoğlu, M.; Güneyisi, E.; Mahmood, S.F.; Öz, H.Ö.; Mermerdaş, K. Recycling Ground Granulated Blast Furnace Slag as Cold Bonded Artificial Aggregate Partially Used in Self-Compacting Concrete. *J. Hazard. Mater.* **2012**, *235–236*, 352–358. [CrossRef] [PubMed]
26. Li, J.; Xiao, F.; Zhang, L.; Amirkhanian, S.N. Life Cycle Assessment and Life Cycle Cost Analysis of Recycled Solid Waste Materials in Highway Pavement: A Review. *J. Clean. Prod.* **2019**, *233*, 1182–1206. [CrossRef]
27. Lanzerstorfer, C. Potential of Industrial De-Dusting Residues as a Source of Potassium for Fertilizer Production—A Mini Review. *Resour. Conserv. Recycl.* **2019**, *143*, 68–76. [CrossRef]
28. Holmberg, S.L.; Claesson, T. Mineralogy of Granulated Wood Ash from a Heating Plant in Kalmar, Sweden. *Environ. Geol.* **2001**, *40*, 820–828. [CrossRef]
29. Pesonen, J.; Kuokkanen, V.; Kuokkanen, T.; Illikainen, M. Co-Granulation of Bio-Ash with Sewage Sludge and Lime for Fertilizer Use. *J. Environ. Chem. Eng.* **2016**, *4*, 4817–4821. [CrossRef]
30. Holmberg, S.L.; Lind, B.B.; Claesson, T. Chemical Composition and Leaching Characteristics of Granules Made of Wood Ash and Dolomite. *Environ. Geol.* **2000**, *40*, 1–10. [CrossRef]
31. Sell, N.J.; Flschbach, F.A. Pelletizing Waste Cement Kiln Dust for More Efficient Recycling. *Ind. Eng. Chem. Process Des. Dev.* **1978**, *17*, 468–473. [CrossRef]
32. Yliniemi, J.; Nugteren, H.; Illikainen, M.; Tiainen, M.; Weststrate, R.; Niinimäki, J. Lightweight Aggregates Produced by Granulation of Peat-Wood Fly Ash with Alkali Activator. *Int. J. Miner. Process.* **2016**, *149*, 42–49. [CrossRef]
33. Vincevica-Gaile, Z.; Stankevica, K.; Irtiseva, K.; Shishkin, A.; Obuka, V.; Celma, S.; Ozolins, J.; Klavins, M. Granulation of Fly Ash and Biochar with Organic Lake Sediments—A Way to Sustainable Utilization of Waste from Bioenergy Production. *Biomass Bioenergy* **2019**, *125*, 23–33. [CrossRef]
34. WRB IUSS. *Working Group World Reference Base for Soil Resources 2014*; WEB IUSS: Rome, Italy, 2014.
35. Egnér, H.; Riehm, H.; Domingo, W.R. Untersuchungen Uber Die Chemische Bodenanalyse Als Grundlage Fur Die Beurteilung Des Nährstoffzustandes Der Böden. *K. Lantbr. Ann.* **1960**, *26*, 199–215.
36. Sokolov, A.V. (Ed.) *Agrochemical Methods of Soil Studies*; Nauka: Moscow, Russia, 1975.
37. SAS Institute. *The SAS System for Windows, Version 9.4*; SAS Institute: Cary, NC, USA, 2016.
38. Hoşten, Ç.; Gülsün, M. Reactivity of Limestones from Different Sources in Turkey. *Miner. Eng.* **2004**, *17*, 97–99. [CrossRef]
39. Regulation (EU) 2019/1009 of the European Parliament and of the Council of 5 June 2019, Laying down Rules on the Making Available on the Market of EU Fertilising Products and Amending Regulations (EC) No 1069/2009 and (EC) No 1107/2009 and Repealing Regul. 2019. Available online: https://www.legislation.gov.uk/eur/2019/1009/contents# (accessed on 1 April 2022).
40. Chesner, W.H.; Collina, R.J.; MacKay, M.H. *User Guidelines for Waste and By-Product Materials in Pavement Construction*; Federal Highway Administration: Washington, DC, USA, 1998.
41. Collins, R.J.; Emery, J.J. *Kiln Dust-Fly Ash Systems for Highway Bases and Subbases*; Federal Highway Administration: Washington, DC, USA, 1983.
42. Morrissey, J.; Guerinot, M. Lou Iron Uptake and Transport in Plants: The Good, the Bad, and the Ionome. *Chem. Rev.* **2009**, *109*, 4553–4567. [CrossRef] [PubMed]

43. Nikolic, M.; Kastori, R. Effect of Bicarbonate and Fe Supply on Fe Nutrition of Grapevine. *J. Plant Nutr.* **2000**, *23*, 1619–1627. [CrossRef]
44. Jaskulska, I.; Jaskulski, D.; Kobierski, M. Effect of Liming on the Change of Some Agrochemical Soil Properties in a Long-Term Fertilization Experiment. *Plant Soil Environ.* **2014**, *60*, 146–150. [CrossRef]
45. Laudelout, H. Chemical and Microbiological Effects of Soil Liming in a Broad-Leaved Forest Ecosystem. *For. Ecol. Manag.* **1993**, *61*, 247–261. [CrossRef]
46. Repsiene, R.; Karcauskiene, D. Changes in the Chemical Properties of Acid Soil and Aggregate Stability in the Whole Profile under Long-Term Management History. *Acta Agric. Scand. Sect. B Soil Plant Sci.* **2016**, *66*, 671–676. [CrossRef]
47. Puissant, J.; Jones, B.; Goodall, T.; Mang, D.; Blaud, A.; Gweon, H.S.; Malik, A.; Jones, D.L.; Clark, I.M.; Hirsch, P.R.; et al. The PH Optimum of Soil Exoenzymes Adapt to Long Term Changes in Soil PH. *Soil Biol. Biochem.* **2019**, *138*, 107601. [CrossRef]
48. Islam, A.K.M.S.; Edwards, D.G.; Asher, C.J. PH Optima for Crop Growth—Results of a Flowing Solution Culture Experiment with Six Species. *Plant Soil* **1980**, *54*, 339–357. [CrossRef]
49. Porebska, G.; Ostrowska, A.; Borzyszkowski, J. Changes in the Soil Sorption Complex of Forest Soils in Poland over the Past 27 Years. *Sci. Total Environ.* **2008**, *399*, 105–112. [CrossRef] [PubMed]
50. Park, N.D.; Michael Rutherford, P.; Thring, R.W.; Helle, S.S. Wood Pellet Fly Ash and Bottom Ash as an Effective Liming Agent and Nutrient Source for Rye Grass (*Lolium perenne* L.) and Oats (*Avena sativa*). *Chemosphere* **2012**, *86*, 427–432. [CrossRef]
51. Qin, J.; Hovmand, M.F.; Ekelund, F.; Rønn, R.; Christensen, S.; de Groot, G.A.; Mortensen, L.H.; Skov, S.; Krogh, P.H. Wood Ash Application Increases PH but Does Not Harm the Soil Mesofauna. *Environ. Pollut.* **2017**, *224*, 581–589. [CrossRef] [PubMed]
52. Kiani, D.; Silva, M.; Sheng, Y.; Baltrusaitis, J. Experimental Insights into the Genesis and Growth of Struvite Particles on Low-Solubility Dolomite Mineral Surfaces. *J. Phys. Chem. C* **2019**, *123*, 25135–25145. [CrossRef]
53. Tate, M. Lime Kiln Dust: An Overlooked Resource. In Proceedings of the Lime: Building on the 100-Year Legacy of The ASTM Committee C07, San Diego, CA, USA, 28 June 2012; Thomson, M., Brisch, J., Eds.; ASTM International: West Conshohocken, PA, USA, 2012; pp. 135–144.
54. de Vargas, J.P.R.; dos Santos, D.R.; Bastos, M.C.; Schaefer, G.; Parisi, P.B. Application Forms and Types of Soil Acidity Corrective: Changes in Depth Chemical Attributes in Long Term Period Experiment. *Soil Tillage Res.* **2019**, *185*, 47–60. [CrossRef]
55. Zhang, X.; Glasser, F.P.; Scrivener, K.L. Reaction Kinetics of Dolomite and Portlandite. *Cem. Concr. Res.* **2014**, *66*, 11–18. [CrossRef]
56. Marx, E.S.; Hart, J.; Stevens, R.G. *Soil Test Interpretation Guide*; Ministry of Agriculture: Corvallis, OR, USA, 1999.
57. Lalande, R.; Gagnon, B.; Royer, I. Impact of Natural or Industrial Liming Materials on Soil Properties and Microbial Activity. *Can. J. Soil Sci.* **2009**, *89*, 209–222. [CrossRef]
58. Ziadi, N.; Gagnon, B.; Nyiraneza, J. Crop Yield and Soil Fertility as Affected by Papermill Biosolids and Liming By-Products. *Can. J. Soil Sci.* **2013**, *93*, 319–328. [CrossRef]
59. Gasser, J.K.R. Processes Causing Loss of Calcium from Agricultural Soils. *Soil Use Manag.* **1985**, *1*, 14–16. [CrossRef]
60. Gunes, A.; Alpaslan, M.; Inal, A. Critical Nutrient Concentrations and Antagonistic and Synergistic Relationships among the Nutrients of NFT-Grown Young Tomato Plants. *J. Plant Nutr.* **1998**, *21*, 2035–2047. [CrossRef]
61. Barrow, N.J. The Effects of PH on Phosphate Uptake from the Soil. *Plant Soil* **2017**, *410*, 401–410. [CrossRef]
62. Özenç, N.; Özenç, D.B. Interaction between Available Phosphorus and Lime Treatments on Extremely Acid PH Soils of Hazelnut Orchards. *Acta Hortic.* **2009**, *845*, 379–386. [CrossRef]
63. Demeyer, A.; Voundi Nkana, J.C.; Verloo, M.G. Characteristics of Wood Ash and Influence on Soil Properties and Nutrient Uptake: An Overview. *Bioresour. Technol.* **2001**, *77*, 287–295. [CrossRef]
64. Mrvić, V.; Jakovljević, M.; Stevanović, D.; Cłakmak, D. The Forms of Aluminium in Stagnosols in Serbia. *Plant Soil Environ.* **2007**, *53*, 482–489. [CrossRef]
65. Moir, J.L.; Moot, D.J. Medium-Term Soil PH and Exchangeable Aluminium Response to Liming at Three High Country Locations. *Proc. N. Z. Grassl. Assoc.* **2014**, *76*, 41–45. [CrossRef]
66. Patterson, S.J.; Acharya, S.N.; Thomas, J.E.; Bertschi, A.B.; Rothwell, R.L. Barley Biomass and Grain Yield and Canola Seed Yield Response to Land Application of Wood Ash. *Agron. J.* **2004**, *96*, 971–977. [CrossRef]
67. Patterson, S.J.; Acharya, S.N.; Bertschi, A.B.; Thomas, J.E. Application of Wood Ash to Acidic Boralf Soils and Its Effect on Oilseed Quality of Canola. *Agron. J.* **2004**, *96*, 1344–1348. [CrossRef]
68. Gentili, R.; Ambrosini, R.; Montagnani, C.; Caronni, S.; Citterio, S. Effect of Soil PH on the Growth, Reproductive Investment and Pollen Allergenicity of *Ambrosia artemisiifolia* L. *Front. Plant Sci.* **2018**, *9*, 1–12. [CrossRef]
69. Schuster, B.; Diekmann, M. Changes in Species Density along the Soil PH Gradient—Evidence from German Plant Communities. *Folia Geobot.* **2003**, *38*, 367–379. [CrossRef]
70. Rahman, M.A.; Barma, N.; Sarker, M.; Sarker, M.; Nazrul, M. Adaptability of Wheat Varieties in Strongly Acidic Soils of Sylhet. *Bangladesh J. Agric. Res.* **2013**, *38*, 97–104. [CrossRef]
71. Shaheb, M.R.; Nazrul, M.I.; Ataur Rahman, M. Production Potential and Economics of Wheat as Influenced by Liming in North Eastern Region of Bangladesh. *Asian J. Agric. Biol.* **2014**, *2*, 152–160.
72. Moges, T.; Melese, A.; Tadesse, G. Effects of Lime and Phosphorus Fertilizer Levels on Growth and Yield Components of Malt Barley (*Hordeum distichum* L.) in Angolelana Tera District, North Shewa Zone, Ethiopia. *Adv. Plants Agric. Res.* **2018**, *8*, 582–589. [CrossRef]

73. Kirchev, H.; Delibaltova, V.; Yanchev, I.; Zheliazkov, I. Comparative Investigation of Rye Type Triticale Varieties, Grown in the Agroecological Conditions of Thrace Valley. *Bulg. J. Agric. Sci.* **2012**, *18*, 696–700.
74. European Commission Protection of the Environment, and in Particular of the Soil, When Sewage Sludge Is Used in Agriculture. *Off. J. Eur. Communities* **1986**, *4*, 6–12.
75. Ramtahal, G.; Chang Yen, I.; Hamid, A.; Bekele, I.; Bekele, F.; Maharaj, K.; Harrynanan, L. The Effect of Liming on the Availability of Cadmium in Soils and Its Uptake in Cacao (*Theobroma c acao* L.) In Trinidad & Tobago. *Commun. Soil Sci. Plant Anal.* **2018**, *49*, 2456–2464. [CrossRef]
76. Shaheen, S.M.; Rinklebe, J. Impact of Emerging and Low Cost Alternative Amendments on the (Im)Mobilization and Phytoavailability of Cd and Pb in a Contaminated Floodplain Soil. *Ecol. Eng.* **2015**, *74*, 319–326. [CrossRef]
77. Shaheen, S.M.; Rinklebe, J.; Selim, M.H. Impact of Various Amendments on Immobilization and Phytoavailability of Nickel and Zinc in a Contaminated Floodplain Soil. *Int. J. Environ. Sci. Technol.* **2015**, *12*, 2765–2776. [CrossRef]
78. Bañuelos, G.S.; Ajwa, H.A. Trace Elements in Soils and Plants: An Overview. *J. Environ. Sci. Health Part A* **1999**, *34*, 951–974. [CrossRef]
79. Carrillo-González, R.; Šimůnek, J.; Sauvé, S.; Adriano, D. Mechanisms and Pathways of Trace Element Mobility in Soils. *Adv. Agron.* **2006**, *91*, 111–178. [CrossRef]
80. Tlustoš, P.; Száková, J.; Kořínek, K.; Pavlíková, D.; Hanč, A.; Balík, J. The Effect of Liming on Cadmium, Lead, and Zinc Uptake Reduction by Spring Wheat Grown in Contaminated Soil. *Plant Soil Environ.* **2006**, *52*, 16–24. [CrossRef]

Article

Composted Sewage Sludge Sustains High Maize Productivity on an Infertile Oxisol in the Brazilian Cerrado

Adrielle Rodrigues Prates [1], Karen Cossi Kawakami [1], Aline Renée Coscione [2], Marcelo Carvalho Minhoto Teixeira Filho [1], Orivaldo Arf [3], Cassio Hamilton Abreu-Junior [4], Fernando Carvalho Oliveira [5], Adônis Moreira [6], Fernando Shintate Galindo [4], Zhenli He [7], Arun Dilipkumar Jani [8], Gian Franco Capra [9,10,*], Antonio Ganga [9] and Thiago Assis Rodrigues Nogueira [1,11]

[1] Department of Plant Protection, Rural Engineering, and Soils, São Paulo State University, Av. Brazil Sul no 56, Ilha Solteira 15385-000, SP, Brazil
[2] Center of Soils and Environmental Resources, Campinas Agronomic Institute, Av. Barão de Itapura no 1481, Campinas 13020-902, SP, Brazil
[3] Department of Plant Technology, Food Technology and Partner Economics, São Paulo State University, Av. Brazil Sul no 56, Ilha Solteira 15385-000, SP, Brazil
[4] Center for Nuclear Energy in Agriculture, Universidade de São Paulo, Av. Centenário no 303, Piracicaba 13416-000, SP, Brazil
[5] Tera Ambiental Ltda. Estrada Municipal do Varjão no 4.520, Jundiaí 13212-590, SP, Brazil
[6] Department of Soil Science, Embrapa Soja, Rodovia Carlos João Strass, Londrina 86001-970, PR, Brazil
[7] Indian River Research and Education Center, Institute of Food and Agricultural Sciences, University of Florida, Fort Pierce, FL 34945, USA
[8] Department of Biology and Chemistry, California State University, Monterey Bay, Seaside, CA 93955, USA
[9] Dipartimento di Architettura, Design e Urbanistica, Università degli Studi di Sassari, Polo Bionaturalistico, Via Piandanna no 4, 07100 Sassari, Italy
[10] Desertification Research Centre, Università degli Studi di Sassari, Viale Italia no 39, 07100 Sassari, Italy
[11] School of Agricultural and Veterinarian Sciences, São Paulo State University, Via de acesso Prof. Paulo Donato Castellane, s/n, Jaboticabal 14884-900, SP, Brazil
* Correspondence: pedolnu@uniss.it; Tel.: +39-079-228644

Abstract: Mato Grosso do Sul State in Brazil is characterized by the 'Cerrado' ecoregion, which is the most biologically rich Savannah globally. In agricultural terms, the region produces several commodities that are exported around the world. This level of productivity has been achieved through the large-scale use of synthetic fertilizers, which has created several economic and environmental concerns. New approaches in soil fertility management are required to avoid environmental degradation, pollution, and socio-environmental damages. A field experiment, lasting two years, was conducted to investigate the composted sewage sludge (CSS) effects on an infertile acidic soil (Oxisol) planted to maize (*Zea mays* L.). The following complete randomized complete block design with a 4 × 2 + 2 factorial scheme (four replications) was applied: four CSS increasing rates (from 5.0 to 12.5 Mg ha^{-1}, w.b.) following two application methods (whole area and between crop rows). A control, without CSS or synthetic fertilizers, and conventional synthetic fertilization without CSS were also investigated. Evaluated parameters were: (*i*) soil and leaf micronutrient concentrations; (*ii*) maize development, yield, and production. The CSS application increased: (*i*) the concentration of micronutrients in both soil and leaves; and (*ii*) the crop yield. Both were particularly true at the higher CSS applied rates. Such organic fertilizer can be safely used as a source of micronutrients for crops as an important low-cost and environmentally friendly alternative to mineral fertilizers, thus safeguarding soil health.

Keywords: circular economy; cleaner production; food security; micronutrients; urban by-products

1. Introduction

Savannahs cover approximately 20% of the global land surface and provide several ecosystem services, including storage of over 15% of terrestrial above-ground carbon [1] and support of the livelihoods of millions of people through agriculture, resource extraction, and tourism. Consequently, research aiming to improve soil use and management is critically important from an environmental and socio-economic perspective.

The Cerrado is a vast tropical Savannah ecoregion of Brazil, accounting for 23.3% of the country's land area. With approximately 10,000 plant species, it is classified as the most biologically rich Savannah globally [2]. Despite the presence of highly weathered soils, with low natural fertility in terms of both macro- and micronutrients [3], the efforts of researchers (from the beginning of the 90 s) to develop well-adapted cultivars for several tropical commodity crops have resulted in the Cerrado today providing more than 70% of beef in Brazil, in addition to large amounts of coffee (*Coffea* spp.), soybean (*Glycine max* (L.) Merr.), beans (*Phaseolus* sp.), and rice (*Oryza sativa* L.). In fact, the Cerrado is one of the most productive agroecosystems in the world [2].

Maize (*Zea mays* L.) yield in Brazil set a record of 87,000 thousand tons in 2020, making Brazil the third (after the US and China) and second (after the US) largest maize producing and exporting country in the world, respectively [4]. There are several reasons for this rise, including new maize varieties, increased demand for ethanol, and the expansion of production in Mato Grosso do Sul State [4], a region considered pivotal for the agricultural development of the entire world. The amounts of nutrients required for maize growth depend, in part, on soil and environmental conditions as well as yield expectations [5]. Consequently, all these factors must be considered when estimating nutrient needs for maize. For instance, Simão et al. [6] and Dias Borges et al. [7] claimed that B, Cu, Fe, Mn, and Zn are all indispensable for maize growth in the Cerrado ecoregion; however, the amounts and importance of these micronutrients at different growth stages depend on the aforementioned factors as well the maize variety.

One of the greatest challenges facing growers in the Cerrado is the low fertility status of soils in the region, thus strongly limiting crop productivity. Most of these soils are Oxisols and Ultisols low in SOM and plant available nutrients, limiting crop productivity [8]. Consequently, there is a strong reliance on synthetic fertilizers as well as in Cerrado cropping systems [9]. This management paradigm is responsible for several environmental and socio-economic concerns [10]. New approaches should be proposed that consider a circular economy perspective, i.e., the possible reuse of unconventional sources of fertilizers, such as by-products, leading to a change in the paradigm from a waste problem to a resource solution.

The inappropriate handling, storage, and disposal of human-produced waste generate severe environmental and human health concerns. Among these wastes, sewage sludge (SS) has garnered serious interest among scientists, policymakers, and the public because it supplies considerable amounts of organic matter [9,11] and both macro- [9,12] as well as micronutrients [13]. When used in agriculture, it has been shown to successfully replace commercial NPK mineral fertilizers by [14]: (*i*) maintaining soil fertility; (*ii*) enhancing microbial biomass and soil enzymatic activities; and (*iii*) preventing contamination and degradation of water resources [15,16].

Composting SS is a technique that significantly decreases pathogenic concentration, increases organic matter stabilization, and thus reduces the mobility of potentially toxic elements (PTE). Additionally, composted sewage sludge (CSS) is safer than SS in both agricultural and forestry applications [14,15] and is applied by many wastewater treatment plant (WTP) companies since it can reduce SS management costs [17]. The CSS can significantly improve the chemical quality of tropical soils [9,15]. In Brazil, CSS is considered an organic fertilizer if it meets the standards imposed by the national Normative [18].

We conducted a two-year field experiment on a low-fertility intensively cultivated Cerrado Oxisol with the aim of understanding how and to what extent CSS can influence soil properties and maize performance. The relationship and feedback between soil and plants were investigated as well through the application of multivariate statistics.

2. Materials and Methods

2.1. Study Area

The research (Figure 1a) was conducted for two consecutive crop seasons in 2017/18 and 2018/19 (Figure 1b,c). Investigated soil was a Rhodic Hapludox [19] with physical-chemical properties, as reported in Supplementary Material Table S1. Analyses were conducted on $\varnothing \leq 2.0$ mm soil samples collected in the Ap horizon (0.0–0.2 m); Brazilian official procedures were applied [20,21].

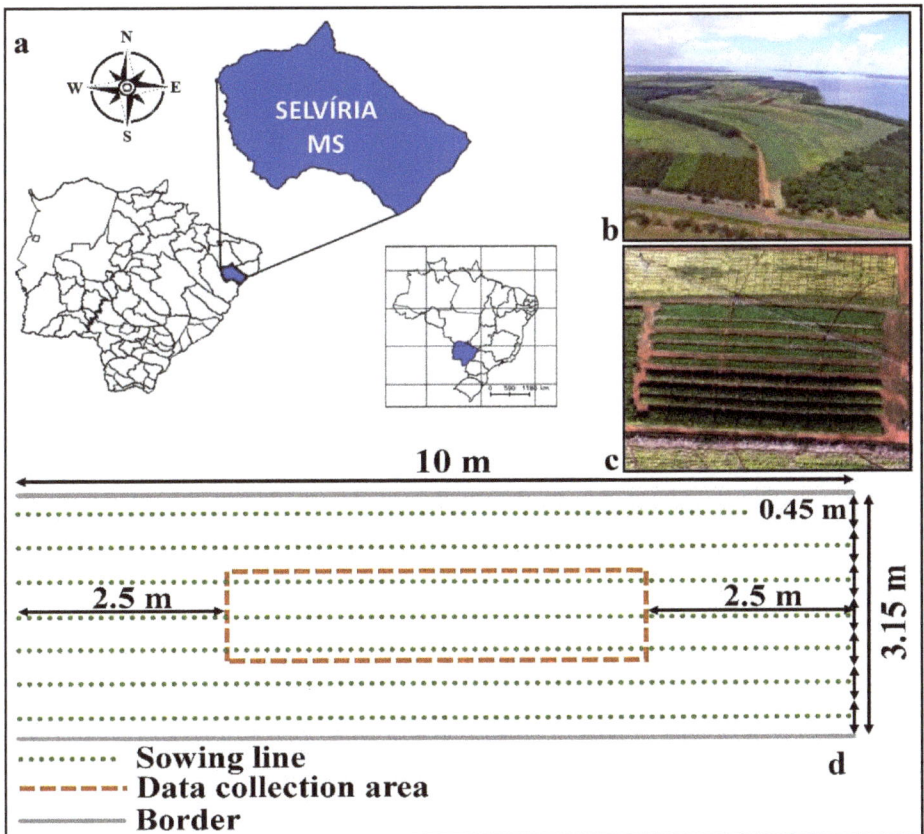

Figure 1. Experimental area at the Selvíria County (**a**: 20°20′35″ S, 51° 24′04″ W; 358 m asl; Mato Grosso do Sul State—MS, Brazil); (**b**) aerial view of the entire experimental area and (**c**) randomized plots; (**d**) schematic representation of a single plot with the individuation of the "useful area" for soil and plant data collection.

Experimental units were 3.15 × 10 m, with each maize row spaced at 0.45 m (Figure 1d). The three central rows were used to collect soil and plant data (Figure 1d). Before the experiment, maize was the only crop for ten consecutive years; during that period, conventional mineral fertilization and agronomic management were applied.

2.2. Field Experiment

The experimental design was set up according to a randomized complete block design following a 4 × 2 + 2 factorial arrangement: 1. CSS application rates: 5.0, 7.5, 10.0, and 12.5 Mg ha^{-1} on a wet basis; 2. application method: whole area (*WA*, hereafter) or between rows (*BR*); 3. two additional treatments: (a) a control where neither CSS nor mineral fertilizers were applied, (b) an area treated with conventional fertilization (CF) only (i.e., N, P, K, B, and Zn).

Soil was tilled to 0.30 m depth, and maize was planted in plots with four replications (Figure 1c).

2.3. Sewage Sludge Features

Sewage sludge was generated, during a process lasting approximately one year, in a common municipal wastewater treatment plant of the São Paulo State (Brazil). It was composted to reduce the pathogenic agent concentration and increase solid biomass by up to 25%. The whole process is made up of (*i*) periodic mixing and air drying, through a forced aeration system, for three consecutive months; (*ii*) a plaster and limestone addition to increase porosity and pH, respectively; (*iii*) a mixture cleaning to reach approximately 40% in moisture content. Finally, it was carefully sieved, and maturation was achieved during the final 15 days. Thus, it was fully characterized from a bio-physico-chemical viewpoint, as required by the Brazilian legislation [22]. For the sake of brevity, the following features are here reported (mean ± SE, n = 3): pH-CaCl$_2$: 7.0 ± 0.1; total moisture (%): 45.5 ± 0.2%; SOM: 309 ± 10 g kg^{-1}; total N: 139.0 ± 0.3 g kg^{-1}; C/N: 12.0 ± 0.8; CEC: 520 ± 20 mmol$_c$ kg^{-1}; total P: 12.3 ± 1.4 g kg^{-1}; total K: 6.0 ± 2.2 g kg^{-1}; B: 94.0 ± 4.5 mg kg^{-1}; Cu: 237.0 ± 16.5 mg kg^{-1}; Fe: 16400 ± 1300 mg kg^{-1}; Mn: 246 ± 37 mg kg^{-1}; Zn: 456 ± 8 mg kg^{-1}. Based on its chemical and biological properties, CSS was permitted for use as an amendment/fertilizer in agriculture, according to CONAMA [18].

2.4. Soil and Plant Preparation

Before the experiment began, 2.2 Mg ha^{-1} of lime (base saturation increased to 70%) and 1.8 Mg ha^{-1} of gypsum were applied [23].

Weed control was conducted by using 1.8 kg ha^{-1} (a.i.) of Glyphosate and 0.67 kg ha^{-1} (a.i.) of 2,4-Dichlorophenoxyacetic acid were applied; thus, CSS was applied seven days before and after sowing for *WA* and *BR* methods, respectively.

Maize seed (hybrid AG 7098, treated with insecticides and fungicides) was sown at approximately 73,333 plants per ha^{-1} (recommended rates; [24]).

Conventional fertilizer was applied at maize planning, with rates based on soil features, climatic conditions, maize hybrid, and research experience in the area. The following amounts were applied: 26 kg ha^{-1} of N (urea, 42% N), 90 kg ha^{-1} of P$_2$O$_5$ (triple superphosphate), 51 kg ha^{-1} of K$_2$O (KCl, 60% of K$_2$O), 1.0 kg ha^{-1} of B (boric acid), 2.0 kg ha^{-1} of Zn (zinc sulfate). An automatized irrigation system was designed for the whole investigated area to mitigate nutrient losses (volatilization processes) starting after the first CF application.

Weather conditions were recorded using a permanent daily-recording weather station installed in the field. The monthly rainfall, humidity, and temperature were recorded daily during all experimental periods (from November 2017 to October 2019). The highest rainfall (>300 mm per month) was observed in November and February, while lowest in June–July (<50 mm). Temperatures reached the highest values in September–October (>35 °C), while minimum values were observed in July–August (15 °C).

Before the experiment began, the soil was fully characterized for baseline conditions (Supplementary Material Table S1).

2.5. Soil and Plant Analysis

Five samples, collected at the end of the crop cycle, were randomly selected from the surface horizon (0–0.2 m) from each investigated plot. Micronutrient (bio)available concentrations were then assessed. In particular, Cu, Fe, Mn, and Zn were extracted with the DTPA-TEA method [20] and then analyzed by ICP–OES (inductively coupled plasma atomic emission spectroscopy). The barium chloride extraction method was used for B, with its concentration quantified by ultraviolet-visible (UV-Vis) spectroscopy. Analyses were performed in triplicates with blank samples to ensure accuracy. A standard reference material (SRM 2709a—San Joaquim) was used to test the precision of the applied analytical methods.

Ten different leaves were randomly collected from each investigated plot during the full bloom (R1) period [25]. Leaf micronutrient concentrations were determined according to Malavolta et al. [25]; HNO_3 and $HClO_4$ were used for dry material wet digestion. The azomethine-H colorimetric method for B determination. Atomic absorption spectrometry for Cu, Fe, Mn, and Zn.

2.6. Plant Development

The following plant parameters were assessed: plant height (PH), height from ear insertion (HEI), stem diameter (SD, all evaluated during the (R4) pasty grains phase), grain per ear (NGE), number of rows (NRE), and 1000 seed weight (SW, evaluated during the (R6) harvest period). Maize was harvested 143 days after seedling emergence and was reported at 13% moisture.

2.7. Statistical Analysis

The R statistical software [26] was used for univariate and multivariate statistics. The analysis of variance (ANOVA) for testing differences among mean values for both CSS rates and application method (*WA* or *BR*). In particular, in the case of F-test significance, a Tukey's test ($p \leq 0.05$) was applied. Significant differences ($p \leq 0.05$) among CSS vs. Control and CF were tested by the Dunnett test. Interactions or effects of CSS applied rates were evaluated through polynomial regression analysis. Bivariate and multivariate relationships were investigated by means of a Pearson's correlation matrix (CM) and factor analysis (FA), respectively. Before entering the multivariate statistic data, they were pretreated according to the method proposed by Capra et al. [27].

3. Results and Discussion

3.1. Soils

There was a CSS application rate by method interaction on soil micronutrient concentration, except for Cu and Mn after the first harvest year (Supplementary Material Table S2; Table 1). Specifically, at the end of the 2018 and 2019 harvests, CSS application in the *WA* showed a linear increase in Zn and B soil concentration, while the same held true only for B with CSS application under *BR* (Supplementary Material Table S2). For the 2018/2019 crop, a linear decrease was observed for Mn for CSS applied in the *WA*. Applications under *BR* promoted a B quadratic adjustment, while the peak (0.36 mg dm^{-3}) reached the highest (12.5 Mg ha^{-1}) CSS application rate (Table 1). This increase in B and Zn concentration with increasing CSS application rates was expected since this by-product was characterized by a higher concentration of both elements (*vide supra*).

Table 1. Boron, Cu, Fe, Mn, and Zn concentration (mg dm^{-3}) in soils at the end of the investigated crop seasons.

Treatment	B		Cu		Fe		Mn		Zn	
	WA	BR	WA	BR	WA	BR	WA	BR	WA	BR
	2018									
Control	0.27 ˣ		1.9 ˣ		20 ˣ		18.7 ˣ		0.7 ˣ	
CF	0.47 #		2.1 ˣ#		19 ˣ#		21.0 ˣ#		1.7 #	
5.0 CSS	0.33 abA ˣ	0.27 bB ˣ	2.3 aA #	2.4 aA	22 abB ˣ	25 aA	20.4 ˣ#	21.8 ˣ#	1.5 abA #	1.3 aA #
7.5 CSS	0.32 bA ˣ	0.35 aA	2.3 aA #	2.1 aA ˣ#	21 abA ˣ#	21 bA ˣ#	22.1 ˣ#	21.6 ˣ#	1.0 bA ˣ	1.2 aA ˣ#
10.0 CSS	0.34 abA	0.31 abA ˣ	2.4 aA	2.5 aA	20 bB ˣ#	25 aA	19.4 ˣ#	23.3 #	1.7 aA #	1.1 aA ˣ
12.5 CSS	0.38 aA	0.36 aA	2.3 aA	2.4 aA	25 aA	20 bB ˣ#	21.4 ˣ#	23.1 #	1.7 aA #	1.1 aA ˣ
F-test										
AM	3.82 NS		1.08 NS		0.76 NS		6.84 *		8.38 **	
CSS rates	8.61 **		4.11 *		2.32 NS		0.71 NS		2.73 NS	
(AM) × (CSS)	3.21 *		1.94 NS		13.55 **		2.22 NS		3.86 *	
CV (%)	8.5		6.0		7.4		8.1		21.3	
	2019									
Control	0.17 ˣ		1.1 ˣ		13.7 ˣ		12.0 ˣ		0.4 ˣ	
CF	0.40 #		1.2 #		12.0 ˣ#		12.7 ˣ#		1.1 #	
5.0 CSS	0.28 cA	0.28 bA	1.3 bB #	1.4 bA	17.0 aA	14.7 bB ˣ	13.7 aA ˣ#	13.7 aA ˣ#	1.2 cA ˣ#	1.2 cA #
7.5 CSS	0.24 cB	0.35 aA	1.7 aA	1.2 cB ˣ#	18.5 aA	11.0 cB ˣ#	12.4 abA ˣ#	10.6 bB ˣ	1.9 bA	1.1 cB #
10.0 CSS	0.39 bA #	0.37 aA #	1.4 bB	1.6 aA	14.3 bB ˣ	17.7 aA	11.1 bB ˣ#	15.1 aA	2.0 bA	2.0 aA
12.5 CSS	0.49 aA	0.34 aB	1.6 aA #	1.3 bcB #	18.0 aA	10.5 cB #	10.7 bA ˣ#	8.2 cB	3.1 aA	1.6 bB
F-test										
AM	2.30 NS		29.41 **		90.83 **		0.08 NS		74.93 **	
CSS rates	61.52 **		7.06 **		4.75 **		32.93 **		59.21 **	
(AM) × (CSS)	41.90 **		51.37 **		52.96 **		19.27 **		30.28 **	
CV (%)	7.0		4.8		6.9		7.8		11.7	
	Interpretation limits [1]									
Low	0–0.20		0–0.2		0–4		0–1.2		0–0.5	
Medium	0.21–0.60		0.3–0.8		5–12		1.3–5.0		0.6–1.2	
High	>0.60		>0.8		>12		>5.0		>1.2	

*, ** for $p \leq 0.05$, ≤ 0.01, respectively; NS = not significant; WA = whole area; BR = between rows. Different lowercase and uppercase letters indicate significant differences between CSS rates (see from 5.0 till to 12.5 Mg ha^{-1}, wet basis) or application methods (WA vs. BR), respectively. The absence of letters is for non-significant differences ($p < 0.05$). Different ˣ and # symbols along the same column show significant differences among treatments. wb = wet basis. [1] [21].

There was a significant increase in soil concentrations of Mn-Zn and Cu-Fe-Zn after the first and second crop years, respectively, when CSS was applied along WA (Table 1). In general, applying increasing CSS rates, particularly 12.5 Mg ha^{-1}, resulted in an increase in most soil micronutrients for both years (Table 1), regardless of the application method. By the end of 2018, soil B, Cu, and Fe concentrations were higher than in control plots. At an application rate of 12.5 Mg ha^{-1}, these elements reached their highest values. Most micronutrient concentrations were lower or statistically similar to CF plots. Additionally, soil in plots treated with CF received supplemental B and Zn and were thus already enriched with these elements. Additionally, CSS application resulted in an increase in B in these infertile Oxisols of the Cerrado ecoregion, which is noteworthy since these soils are usually poor in B due to their low SOM content [3,6].

In terms of "interpretation limits" (Table 1; [21]), we observed a significant increase in (bio)available concentration of all investigated micronutrients from the control (usually characterized by "low" and "medium" values) vs. CF and CSS treated soils ("high" values).

Our findings strongly support the application of CSS in maize cultivation since it increased soil micronutrient concentration, following both WA and BR methods. We also demonstrated that the WA method should be preferred, as it is more practical and cost-effective than BR, while resulting in strong crop performance.

3.2. Plant

3.2.1. Leaf

When the CSS application method was compared with the CSS applied rate, several interactions on leaf micronutrient concentrations at the end of both years of maize cultivation (Supplementary Material Table S3) were observed. Particularly, at the end of the first agronomic year, as CSS rates increased, a linear decrease of Cu concentrations in maize leaves ($R^2 = 0.95$) was observed if the *WA* method was applied. Iron showed a negative quadratic adjustment ($\hat{y} = 208.825 - 23.030x - 1.260x^2$; $R2 = 0.47$ **) at the end of the first year using the *WA* method. Conversely, B, Cu, and Mn concentrations showed a negative quadratic adjustment, in both agronomic years, by using the *BR* method. A linear increase in leaf Zn levels was only observed at the end of the first year. The reduction in B, Cu, and Mn concentrations with a concomitant increase in Zn concentration, may be attributed to well-known competitive/inhibition processes among these nutrients [28].

No significant differences were observed in micronutrient concentrations in maize leaves sampled in the plots treated with CSS, CF, or the control, regardless of application method (Table 2).

Table 2. Boron, Cu, Fe, Mn, and Zn concentration (mg dm^{-3}) in leaves [1].

Treatment	B		Cu		Fe		Mn		Zn	
	WA	BR	WA	BR	WA	BR	WA	BR	WA	BR
					2018					
Control	23 ˣ		19 ˣ		115 ˣ		66 ˣ		35 ˣ	
CF	18 #		16 ˣ#		118 ˣ		76 #		39 #	
5.0 CSS	14 aB	17 aA #	23 aA ˣ	21 aA ˣ#	121 aA ˣ	113 bcB ˣ	71 ˣ#	66 a ˣ	41 #	39 ab #
7.5 CSS	16 aA #	14 aA	21 abA ˣ#	14 bB ˣ#	119 aA ˣ	120 abA ˣ	73 ˣ#	66 a ˣ	40 #	36 b ˣ#
10.0 CSS	16 aA #	14 aA	16 bcA ˣ#	13 bA #	93 bB	108 cA ˣ	67 ˣ	57 b	39 #	39 ab #
12.5 CSS	17 aA #	16 aA #	15 cA ˣ#	18 abA ˣ#	122 aA ˣ	125 aA	71 ˣ#	68 a ˣ#	41 #	42 a #
F-test										
AM	0.24 NS		7.24 *		4.48 *		20.54 **		3.43 NS	
CSS rates	1.77 NS		9.84 **		57.59 **		7.25 **		4.52 *	
(AM) × (CSS)	4.34 *		4.95 **		12.97 **		1.38 NS		2.85 NS	
CV (%)	11.4		15.1		3.2		5.7		4.6	
					2019					
Control	5 ˣ		6 ˣ		87 ˣ		36 ˣ		28 ˣ	
CF	6 ˣ		7 ˣ		90 ˣ		37 ˣ#		24 ˣ#	
5.0 CSS	5 aB ˣ	13 aA	6 aA ˣ	6 bA ˣ	99 bA #	88 bB ˣ#	28 aA	22 cB	24 ˣ#	27 ˣ#
7.5 CSS	5 aA ˣ	5 cA ˣ	7 aA ˣ	7 bA ˣ	90 bB ˣ#	104 aA	29 aA	30 abA ˣ	27 ˣ#	26 ˣ#
10.0 CSS	4 aB ˣ	9 bA	6 aB ˣ	11 aA	117 aA	102 aB	27 aB	35 aA ˣ#	29 ˣ#	29 ˣ#
12.5 CSS	5 aA ˣ	5 cA ˣ	7 aA ˣ	6 bA ˣ	89 bB ˣ#	99 abA #	29 aA	25 bcA	28 ˣ#	27 ˣ
F-test										
AM	37.77 **		6.33 *		0.06 NS		0.03 NS		0.07 NS	
CSS rates	11.78 **		7.46 **		14.08 **		6.94 **		3.78 *	
(AM) × (CSS)	13.03 **		8.22 **		13.51 **		9.11 **		1.15 NS	
CV (%)	24.6		16.4		5.8		10.1		8.4	
OCR [2]	10–25		6–20		30–250		20–200		15–100	

For letters and symbols after mean values, see legend in Table 1; [1] Collection: full bloom (R1) period; OCR = optimal concentration range according to [2] Raij et al. [21].

When such values were compared with those (optimal concentration range, OCR; Table 2) suggested by Raij et al. [21], we observed that for all micronutrients, there were neither deficiencies nor toxicities, as all concentrations were within the proposed ranges. In fact, no visual symptoms of deficiencies or toxicities were observed in maize during the study.

Overall, the CSS application method by rate interactions on leaf micronutrient concentrations was micronutrient-dependent, suggesting that soil-plant feedbacks and micronutrient interactions through competitive/inhibition processes play a pivotal role in investigated micronutrient uptake. Most of the investigated micronutrients were within the adequate range [21], indicating that maize was not negatively influenced even at 12.5 Mg ha^{-1} CSS rate. Our findings suggest that CSS application promotes: (*i*) an adequate concentration of micronutrients, thus avoiding their deficiencies while (*ii*) avoiding toxicity problems. As for soils, few significant differences between application methods were observed in terms of maize micronutrient concentrations; thus, the *WA* application method must be recommended since it is the most cost-effective and less time-consuming to be implemented relative to *BR*.

3.2.2. Plant Parameters

A clear interaction between CSS rate and *WA* application method on the number of rows per ear (NRE) and crop yield (Supplementary Material Table S4) was observed at the end of the first year. A positive quadratic adjustment and a linear increase were respectively observed. Conversely, the *BR* method did not show significant residual effects.

We found that with CSS application, regardless of application method, mean values for all investigated parameters (PH, HEI, SD, NRE, NGE, SW, and Yield) were significantly higher than means under the control or similar means with CF (Table 3).

Table 3. Plant parameters and yield at the end of the investigated crop seasons.

Treatments	PH		HEI		SD		NRE		NGE		SW		Yield	
	WA	BR	WA	BR	WA	BR	WA	BR	WA	BR	WA	BR	WA	BR
	cm										g		kg ha^{-1}	
	2018													
Control	187 ⋈		114 ⋈		2.0 ⋈		17 ⋈		465 ⋈		233 ⋈		5304 ⋈	
CF	215 #		131 #		2.5 #		17 #		621 #		267 #		7767 #	
5.0 CSS	211 #	207 #	128 #	125 #	2.3	2.3	17 bB ⋈	18 aA #	581 #	608 #	266 #	264 #	6359 bB ⋈	7493 aA #
7.5 CSS	212 #	205 #	129 #	130 #	2.4 #	2.4 #	17 aA #	17 aA #	631 #	633 #	267 #	269 #	7466 aA #	7523 aA #
10.0 CSS	214 #	212 #	128 #	131 #	2.3	2.4 #	17 aA #	18 aA #	621 #	614 #	270 #	266 #	7706 aA #	7424 aA #
12.5 CSS	209 #	213 #	131 #	129 #	2.4 #	2.4 #	17 aB #	18.0 aA #	616 #	646 #	277 #	273 #	7921 aA #	7717 aA #
F-test														
AM	1.10 NS		0.00 NS		0.95 NS		18.51 **		1.30 NS		0.21 NS		0.86 NS	
CSS rates (wb)	0.98 NS		1.42 NS		2.27 NS		3.32 *		2.26 NS		1.14 NS		3.93 *	
(AM) × (CSS)	1.43 NS		0.85 NS		0.72 NS		4.86 **		0.62 NS		0.12 NS		2.97 *	
CV (%)	2.7		3.2		2.4		2.0		5.4		4.1		7.4	
	2019													
Control	214 ⋈		128 ⋈		1.8 ⋈		16 ⋈		523 ⋈		256 ⋈		8143 ⋈	
CF	232 #		143 ⋈		2.2 ⋈		16 ⋈		550 ⋈		290 #		9524 ⋈#	
5.0 CSS	221 ⋈#	224 ⋈#	135 ⋈#	138 #	2.1 ⋈	2.0 ⋈	16 ⋈	16 ⋈	524 ⋈	555 ⋈	287 ⋈#	284 ⋈#	9722 ⋈#	8707 ⋈#
7.5 CSS	222 ⋈#	229 ⋈#	138 #	138 #	2.2 ⋈	2.2 ⋈	17 ⋈	17 ⋈	540 ⋈	563 ⋈	282 ⋈#	294 #	9373 ⋈#	10176 #
10.0 CSS	228 ⋈#	227 ⋈#	136 ⋈#	137 #	2.1 ⋈	2.0 ⋈	17 ⋈	16 ⋈	553 ⋈	529 ⋈	302 #	290 #	9473 ⋈#	10224 #
12.5 CSS	228 ⋈#	230 ⋈#	139 #	139 #	2.1 ⋈	2.1 ⋈	16 ⋈	16 ⋈	556 ⋈	663 ⋈	294 #	299 #	9696 ⋈#	9963 #
F-test														
AM	1.16 NS		0.34 NS		1.04 NS		0.07 NS		0.51 NS		0.01 NS		0.42 NS	
CSS rates (wb)	1.72 NS		0.62 NS		1.06 NS		1.06 NS		0.57 NS		1.33 NS		0.94 NS	
(AM) × (CSS)	0.49 NS		0.18 NS		0.13 NS		1.06 NS		0.91 NS		1.18 NS		1.84 NS	
CV (%)	6.5		3.0		8.7		3.9		6.6		4.7		9.3	

For letters and symbols after mean values, see legend in Table 1; PH = plant height; HEI = height from ear insertion; SD = stem diameter; NRE = number of rows per ear; NGE = number of grains per ear; SW = 1000 seed weight; wb = wet basis.

These results demonstrated that infertile, dystrophic, acidic Tropical Oxisols could receive several benefits when CSS is applied. As a matter of fact, it can be a possible alternative to conventional fertilizers for maize production in the Cerrado ecoregion. While previous research has demonstrated higher maize productivity with SS [29,30], this research firstly demonstrated how maize productivity could be positively influenced after just two years.

It must be emphasized that even low CSS rates positively impacted maize performance. Our results indicate that CSS could be used in infertile Oxisols for maize, even at low rates, with several benefits. Barbosa et al. [31] observed a positive residual effect on maize yields when 36 Mg ha^{-1} of SS was applied; however, we observed more positive results at lower application rates.

3.3. Multivariate Statistics

The CM revealed the following results (Table 4): (*i*) only leaf B concentration (*vide infra*) seems to have a negative effect on most of the investigated plant parameters, confirming that with an increase, especially with the *BR* method (vide supra), a negative effect on plants occurred; (*ii*) most of the other investigated elements in leaves did not affect plant parameters, with the exception of Zn, in which an increase resulted in elevated values for most plant parameters; (*iii*) an increase in most soil parameters (Fe excluded) favored an increase in all of the investigated plant parameters, particularly for B, Cu, and Zn.

Table 4. Correlation matrix according to Pearson's correlations coefficients.

	B_L	Cu_L	Fe_L	Mn_L	Zn_L	B_S	Cu_S	Fe_S	Mn_S	Zn_S
PH	−0.53 ***	NS	NS	NS	0.58 ***	0.45 **	0.51 ***	NS	NS	0.53 ***
HEI	−0.54 ***	NS	NS	NS	0.36 *	0.36 *	0.33 *	NS	NS	0.40 *
SD	−0.46 **	NS	NS	NS	0.53 ***	0.55 ***	0.47 **	NS	NS	0.49 **
NRE	NS	NS	NS	NS	NS	NS	0.46 **	NS	NS	NS
NGE	−0.46 **	NS	NS	NS	0.45 **	0.38 *	0.44 **	NS	NS	0.38 *
SW	−0.55 ***	NS	NS	NS	0.41 **	0.39 *	0.49 **	NS	0.37 *	0.37 *
Yield	−0.38 *	−0.40 *	NS	NS	NS	0.52 ***	NS	NS	0.34 *	0.37 *

*, **, *** for $p \leq 0.05$, ≤ 0.01, ≤ 0.001, respectively; ^{NS} = not significant; _L, green = micronutrients concentration in leaf; _S, orange = micronutrients concentration in soils; Plant parameters are reported in grey: PH = plant height; HEI = height from ear insertion; SD = stem diameter; NRE = number of rows per ear; NGE = number of grains per ear; SW = 1000 seed weight.

The aforementioned results should be interpreted in the context of soil–plant relationships and feedbacks. Such mechanisms are not always easily explainable without additional analyses, such as scanning electron microscopy-based energy-dispersive X-ray spectroscopy, particle-induced X-ray emission, X-ray fluorescence microscopy, laser ablation inductively coupled plasma-mass spectrometry, nanoscale secondary ion mass spectroscopy, etc. [32]. Thus, further investigations will need to investigate such specific aspects, which are beyond the scope of the current research. However, some early outcomes can be outlined. First, looking at nutrient plant concentration and behavior, the reduction in B, Cu, and Mn concentrations with a concomitant increase in Zn, may be attributed, as previously reported, to well-known competitive/inhibition processes among these nutrients [33]. For example, Zn is extremely active in biochemical processes with other elements [34]. In plants, it can interfere with the control of ion absorption, causing a decrease in plant accumulation of other elements. This was particularly true, among other elements, for B, Cu, and Mn, whose uptake was especially depressed in the case of Zn presence in leaves. Such antagonistic interactions, in which the uptake of one element was competitively inhibited by the other, might indicate the same carrier sites in the absorption mechanisms of these metals. Thus, Zn presence in plants would be expected to reduce uptake of most nutrients, Fe and P included. Indeed, Zn vs. Fe antagonism is widely known, with its mechanism similar to the depressing effects of other trace metals [33]. There are two possible mechanisms for this interaction [35]: (*i*) the competition between Zn and Fe in uptake processes; (*ii*) the interference in chelation processes during the Fe-uptake and roots to shoots translocation. On the other hand, Zn–Fe can be featured by a synergistic interaction in the cases of adequate P supply; indeed, a relatively high accumulation of P and Zn in roots could promote the precipitation of $FePO_4$ in root tissue, thus accounting for Fe uptake [36]. In terms of micronutrient behavior in the soil, it is well-known that the addition of sewage sludge to soil modifies the distribution pattern of several nutrients (e.g., B, Cu, and Zn), with a significant increase in easily assimilable and exchangeable forms [33]. This seems to be a major reason why improvements in most plant parameters (PH, HEI, SD, NRE, NGE, and SW) were observed (Table 4).

Through the factor analyses (FA; Table 5), five significant (eigenvalues > 1) factors were produced. The obtained five-component model accounts for more than 65% of all data variation. The F1 (variance: 19%) showed that most soil micronutrients were positively related to each other, confirming that these elements increased in the soil at increasing rates of their main sources with particular reference to CSS. This factor clearly showed that an increase in soil micronutrient concentrations was mainly due to CSS sources. Factors from F2 up to F5, even of minor importance, reported important correlations that: (*i*) were already explained through CM (F2, F4, and F5); (ii) emphasized the soil Zn role in increasing maize yield (F3). Overall, these four factors underline the pivotal role of some soil/plant micronutrient concentrations on selected plant parameters (for example, F3 represents the key role of soil Zn in increasing maize yield).

Table 5. Factor analysis (FA) was extracted through the principal factor analysis (PFA) and rotation method (bold loadings > 0.5).

Parameters	Factors				
	F1	F2	F3	F4	F5
PH	−0.029	−0.010	−0.080	0.132	**−0.872**
HFP	0.140	−0.261	0.035	**0.840**	−0.063
NPP	−0.106	**−0.849**	0.050	0.128	−0.081
NGP	−0.014	**−0.917**	0.070	−0.067	−0.054
SW	0.214	0.187	0.376	0.103	−0.225
FPP	−0.184	0.374	0.316	**0.695**	0.072
Yield	0.126	−0.108	**0.889**	−0.093	0.077
B_L	−0.129	0.135	0.134	0.307	**0.634**
Cu_L	0.132	0.035	−0.299	**0.706**	0.047
Fe_L	0.118	0.156	−0.093	0.090	0.126
Mn_L	0.131	0.340	−0.268	0.129	**0.510**
Zn_L	0.126	0.005	0.225	0.104	−0.078
B_S	**0.878**	−0.039	0.122	0.098	0.016
Cu_S	**0.884**	0.107	0.141	0.031	0.055
Fe_S	**0.695**	0.125	−0.344	0.018	−0.269
Mn_S	0.313	0.159	−0.482	−0.032	0.014
Zn_S	**0.613**	−0.036	**0.636**	0.038	0.127
Variance (%)	19	14	14	10	8
Cumulative variance (%)	19	33	46	57	65
Eigenvalues	3.180	2.387	2.306	1.778	1.405

Plant parameters are reported in grey (see Legend in Table 4); leaf micronutrient concentration in green; and soil micronutrient concentration in orange.

Overall, the factor analysis confirmed the pivotal role played by some specific micronutrients in the observed variability. These elements, which were more concentrated in CSS treatments and more effective with the *WA* method, exert a strong influence on plant parameters; in particular, they improved plant performance with specific reference to crop yield. Our results showed that low fertility Tropical Oxisols, cultivated with maize and treated with CSS following the *WA* application method, had higher B, Cu, and Zn concentrations than commercial treatments with mineral fertilizers.

Our outcomes warrant further investigation from a soil health perspective, too. Soil health, i.e., the continued capacity of soil to function as a vital living ecosystem that sustains plants, animals, and humans [36], is dramatically under threat, in part due to the misuse of synthetic fertilizers. As recently well summarized by Pahalvi et al. [37], managing soil health is a pivotal way of improving sustainable agricultural productions, thus safeguarding the overall biodiversity and ecosystem quality. Soil health is regulated and can be monitored by investigating physical-chemical properties. Thus, comparing soil physical-chemical properties in areas cultivated for maize production and treated with synthetic fertilizers vs. the same areas treated with increasing application of CSS provides important information on how soil health can be affected. Modern agriculture as practices in Brazil is largely dependent on the massive use of synthetic fertilizers [38]. Indeed, even if increasing soil crop productivity, their continuous application can bring (*i*) a decline in SOM; (*ii*) crust formation and pH alteration with particular emphasis for acidification; (*iii*) increase in pests, microbial activity, and diversity decrease; (*iv*) soil, air, and water pollution as well as greenhouse gas emissions. Thus, as demonstrated in the present research, a net decrease in the use of synthetic fertilizer application can be achieved by applying low-cost, environmentally friendly by-products that align with a circular-economy perspective [39–41]. By using CSS instead of CF, we obtained statistically comparable maize yields (Table 3) of c.a. 7400–7700 kg ha^{-1}; however, and extremely important, this is true also at the lowest CSS applied rate (5.0 Mg ha^{-1}; Table 3), meaning that even if a low amount of CSS is applied the same crop yield observed in soils treated with the common amount of CF used for the Cerrado unfertile soils, can be achieved. Additionally, using CSS instead of CF, we reached maize yields ranging from 6300 (by applying 5.0 Mg ha^{-1} of CSS) to 10,200 kg ha^{-1} (10.0 Mg ha^{-1}), meaning an increase ranging from +18% to +88% in maize yield when compared with the mean productivity for Brazilian agricultural areas treated with conventional commercial fertilizers (~5400 kg ha^{-1}; 2019–2020 harvest period [4]). Overall, this has been achieved without neither negatively affecting soil physical-chemical properties or creating soil pollution issues (*vide supra*). Substituting CF with CSS can enable economic benefits, while also enhancing soil health in the Cerrado.

4. Conclusions

Considering the global agricultural importance of the Cerrado, improving soil fertility while safeguarding its health represent a key socio-economic strategy for the entire world. This is particularly true for an area suffering from natural unfertile soils that have been intensively cultivated in part due to the availability of synthetic fertilizers, thus creating important environmental concerns. We demonstrated that CSS could efficiently replace conventional mineral fertilizers for the growth of a pivotal crop, such as maize. Our results demonstrated that CSS application led to excellent plant and agronomic performance without creating soil pollution issues, thus additionally safeguarding soil health.

Supplementary Materials: The following supporting information can be downloaded at: https://www.mdpi.com/article/10.3390/land11081246/s1, Supplementary Material Table S1. Soil surface (Ap; 0.0–0.25 m) main physical and chemical properties before the experiment started (mean ± SE, n = 3); Supplementary Material Table S2. Equation for micronutrient behavior in soils; function was obtained considering soil elements concentration after 2 years of maize cultivation vs. CSS applied rates. Supplementary Material Table S3. Equation for investigated micronutrient behavior in leaves; function was obtained considering leaf element concentration after 2 years of maize cultivation vs. CSS applied rates; Supplementary Material Table S4. Equation for investigated plant parameters; function was obtained considering investigated plant parameters after 2 years of maize cultivation vs. CSS applied rates.

Author Contributions: Conceptualization, A.R.P. and T.A.R.N.; data curation, A.M., F.S.G., G.F.C. and T.A.R.N.; formal analysis, A.R.P. and K.C.K.; funding acquisition, T.A.R.N.; investigation, A.R.P. and K.C.K.; methodology, A.R.P., A.R.C., M.C.M.T.F., O.A. and T.A.R.N.; project administration, T.A.R.N.; resources, T.A.R.N.; software, G.F.C. and A.G.; supervision, T.A.R.N.; validation, C.H.A.-J. and F.C.O.; visualization, G.F.C.; writing—original draft, A.R.P., Z.H., A.D.J., G.F.C. and T.A.R.N.; writing—review and editing, Z.H., A.D.J., G.F.C. and T.A.R.N. All authors have read and agreed to the published version of the manuscript.

Funding: This research was funded by the São Paulo Research Foundation (FAPESP), grant numbers 18/15152-4 and 19/02198-9.

Data Availability Statement: The information and database for this research are currently not on a platform or website. They can be provided by the corresponding author.

Acknowledgments: We thank the GENAFERT (Grupo de Estudo em Nutrição, Adubação e Fertilidade do Solo) for technical support. We also would like to thank the companies Biossolo Agricultura e Ambiente Ltda and Tera Ambiental Ltda, for compost supplying. The authors acknowledge the National Council for Scientific and Technological Development (CNPq) for the Research Grant to the last author (Grant Number 308374/2021-5).

Conflicts of Interest: The authors declare no conflict of interest.

References

1. Moncrieff, G.R.; Scheiter, S.; Langan, L.; Trabucco, A.; Higgins, S.I. The future distribution of the savannah biome: Model-based and biogeographic contingency. *Phil. Trans. R. Soc. B* **2016**, *371*, 20150311. [CrossRef]
2. Lima, F.F.; Lescano, C.H.; Oliveira, I.P. *Fruits of The Brazilian Cerrado. Composition and Functional Benefits*; Springer: Amsterdam, The Netherlands, 2021.
3. Demattê, J.A.M.; Giasson, E.; Couto, E.G.; Samuel-Rosa, A.; de Castro, S.S.; Dalmolin, R.S.D.; Brilha, J.; Botelho, R.G.M.; Azevedo, A.C.; Cerri, C.E.P.; et al. The Soils of Brazil. *Geoderma Reg.* **2022**, *29*, e00503. [CrossRef]
4. FAO. Food Outlook. In *Biannual Report on Global Food Markets*; FAO: Rome, Italy, 2021. Available online: http://www.fao.org/3/cb4479en/cb4479en.pdf (accessed on 24 May 2022).
5. Gebrehiwot, K. Soil Management for Food Security. In *Natural Resources Conservation and Advances for Sustainability*; Jhariya, M.K., Meena, R.S., Banerjee, A., Meena, S.N., Eds.; Elsevier: Amsterdam, The Netherlands, 2022; pp. 61–71.
6. Simão, E.P.; Resende, A.V.; Neto, M.M.G.; Silva, A.F.; Godinho, V.P.C.; Galvão, J.C.C.; Oliveira, A.C.; Giehl, J. Nitrogen fertilization in off-season corn crop in different Brazilian Cerrado environments. *Pesqui. Agropec. Bras.* **2020**, *55*, e01551. [CrossRef]
7. Borges, I.D.; von Pinho, R.G.; de Andrade Pereira, J.L. Micronutrients accumulation at different maize development stages. *Ciênc. Agrotecnol.* **2009**, *33*, 1018–1025. [CrossRef]
8. Lopes, A.S.; Guimarães, G.L.R. A career perspective on soil management in the Cerrado region of Brazil. *Adv. Agron.* **2016**, 1–72. [CrossRef]
9. Prates, A.R.; Coscione, A.R.; Teixeira Filho, M.C.M.; Miranda, B.G.; Arf, O.; Abreu-Junior, C.H.; Oliveira, F.C.; Moreira, A.; Galindo, F.S.; Sartori, M.M.P.; et al. Composted sewage sludge enhances soybean production and agronomic performances in naturally infertile soils (*Cerrado* region, Brazil). *Agronomy* **2020**, *10*, 1677. [CrossRef]
10. Jankowski, K.; Neill, C.; Davidson, E.A.; Macedo, M.N.; Costa, C.; Galford, G.L.; Santos, L.M.; Lefebvre, P.; Nunes, D.; Cerri, C.E.P.; et al. Deep soils modify environmental consequences of increased nitrogen fertilizer use in intensifying Amazon agriculture. *Sci. Rep.* **2019**, *8*, 13478. [CrossRef]
11. Abreu-Junior, C.H.; Brossi, M.J.L.; Monteiro, R.T.; Cardoso, P.H.S.; Mandu, T.S.; Nogueira, T.A.R.; Ganga, A.; Filzmoser, P.; Oliveira, F.C.; Firme, L.P.; et al. Effects of sewage sludge application on unfertile tropical soils evaluated by multiple approaches: A field experiment in a commercial Eucalyptus plantation. *Sci. Total Environ.* **2018**, *655*, 1457–1467. [CrossRef]
12. Jakubus, M.; Graczyk, M. Microelement variability in plants as an effect of sewage sludge compost application assessed by different statistical methods. *Agronomy* **2020**, *10*, 642. [CrossRef]
13. Jakubus, M. *Sewage Sludge Origin and Administration*, 1st ed.; Poznan University of Life Sciences Press: Poznań, Poland, 2012; pp. 1–56. (In Polish)
14. Abreu-Junior, C.H.; Firme, L.P.; Maldonado, C.A.B.; Moraes Neto, S.P.; Alves, M.C.; Muraoka, T.; Boaretto, A.E.; Gava, J.L.; He, Z.; Nogueira, T.A.R.; et al. Fertilization using sewage sludge in unfertile tropical soils increased wood production in Eucalyptus plantations. *J. Environ. Manag.* **2017**, *203*, 51–58. [CrossRef]
15. Florentino, A.L.; Ferraz, A.V.; De Moraes Gonçalves, J.L.; Asensio, V.; Muraoka, T.; Dos Santos Dias, C.T.; Abreu-Junior, C.H.; Nogueira, T.H.R.; Capra, G.F. Long-term effects of residual sewage sludge application in tropical soils under *Eucalyptus* plantations. *J. Clean. Prod.* **2019**, *220*, 177–187. [CrossRef]
16. Bloem, E.; Albihn, A.; Elving, J.; Hermann, L.; Lehmann, L.; Sarvi, M.; Schaaf, T.; Schick, J.; Turtola, E.; Ylivainio, K. Contamination of organic nutrient sources with potentially toxic elements, antibiotics and pathogen microorganisms in relation to P fertilizer potential and treatment options for the production of sustainable fertilizers: A review. *Sci. Total Environ.* **2017**, *607–608*, 225–242. [CrossRef] [PubMed]
17. Hamdi, H.; Hechmi, S.; Khelil, M.N.; Zoghlami, I.R.; Benzarti, S.; Monkni-Tlili, S.; Hassen, A.; Jedidi, N. Repetitive land application of urban sewage sludge: Effect of amendment rates and soil texture on fertility and degradation parameters. *Catena* **2019**, *172*, 11–20. [CrossRef]
18. CONAMA–Conselho Nacional do Meio Ambiente. *Resolução n° 375, de 29 de Agosto de 2006*; Springer: Berlin/Heidelberg, Germany, 2020. Available online: http://www.mma.gov.br/port/conama/res/res06/res37506.pdf (accessed on 24 May 2022).
19. Soil Survey Staff. *Keys to Soil Taxonomy*, 12th ed.; USDA; Natural Resources Conservation Service: Washington, DC, USA, 2014.
20. Teixeira, P.C.; Donagemma, G.K.; Fontana, A.; Teixeira, W.G. *Manual de Métodos de Análise de Solo*, 3rd ed.; Embrapa Informação Tecnológica: Brasília, Brazil, 2017.
21. Van Raij, B.; Andrande, J.C.; Cantarella, H.; Guaggio, J.A. *Análise Química Para Avaliação da Fertilidade de Solos Tropicais*; Instituto Agronômico: Campinas, Brazil, 2001.

22. MAPA–Ministério da Agricultura, Pecuária e Abastecimento. Limites Máximos de Contaminantes Admitidos em Fertilizantes Orgânicos e Condicionadores de Solo. Instrução Normativa N° 7, de 12 de Abril de 2016. 2016. Available online: https://www.gov.br/agricultura/pt-br/assuntos/insumos-agropecuarios/insumos-agricolas/fertilizantes/legislacao/in-sda-27-de-05-06-2006-alterada-pela-in-sda-07-de-12-4-16-republicada-em-2-5-16.pdf (accessed on 24 May 2022).
23. Van Raij, B.; Cantarella, H.; Quaggio, J.A.; Furlani, A.M.C. *Recomendações de Adubação e Calagem Para o Estado de São Paulo*, 2nd ed.; Instituto Agronômico de Campinas: Campinas, Brazil, 1997.
24. Raij, B.V. *Fertilidade do Solo e Manejo de Nutrientes*; International Plant Nutrition Institute: Piracicaba, Brazil, 2011.
25. Malavolta, E.; Vitti, G.C.; Oliveira, S.A. *Avaliação do Estado Nutricional das Plantas: Princípios e Aplicações*, 2nd ed.; POTAFOS: Piracicaba, Brazil, 1997.
26. RStudio Team. *RStudio: Integrated Development for R. RStudio*; PBC: Boston, MA, USA, 2020. Available online: http://www.rstudio.com/ (accessed on 24 May 2022).
27. Capra, G.F.; Coppola, E.; Odierna, P.; Grilli, E.; Vacca, S.; Buondonno, A. Occurrence and distribution of key potentially toxic elements (PTEs) in agricultural soils: A paradigmatic case study in an area affected by illegal landfills. *J. Geochem. Explor.* **2014**, *145*, 169e180. [CrossRef]
28. He, M.M.; Tian, G.M.; Liang, X.Q. Phytotoxicity and speciation of copper, zinc and lead during the aerobic composting of sewage sludge. *J. Hazard. Mater.* **2009**, *163*, 671–677. [CrossRef] [PubMed]
29. Simonete, M.A.; Kiehl, J.C.; Andrade, C.A.; Teixeira, C.F.A. Efeito do lodo de esgoto em um Argissolo e no crescimento e nutrição de milho. *Pesq. Agropec. Bras.* **2003**, *38*, 1187–1195. [CrossRef]
30. Nascimento, C.W.A.; Barros, D.A.S.; Melo, E.E.C.; Oliveira, A.B. Alterações químicas em solos e crescimento de milho e feijoeiro após aplicação de lodo de esgoto. *Rev. Ciênc. Agron.* **2004**, *28*, 385–392. [CrossRef]
31. Barbosa, G.M.C.; Tavares Filho, J.; Brito, O.R.; Fonseca, I.C.B. Efeito residual do lodo de esgoto na produtividade do milho safrinha. *Rev. Bras. Ciênc. Solo* **2007**, *31*, 601–605. [CrossRef]
32. Kopittke, P.M.; Lombi, E.; van der Ent, A.; Wang, P.; Laird, J.S.; Moore, K.L.; Persson, D.P.; Husted, S. Methods to visualize elements in plants. *Plant Physiol.* **2020**, *182*, 1869–1882. [CrossRef]
33. Kabata-Pendias, A.; Pendias, H. *Trace Elements in Soils and Plants*, 4th ed.; CRC Press: Boca Raton, FL, USA, 2011.
34. Balafrej, H.; Bogusz, D.; Abidine Triqui, Z.E.; Guedira, A.; Bendaou, N.; Smouni, A.; Fahr, M. Zinc hyperaccumulation in plants: A review. *Plants* **2020**, *9*, 562. [CrossRef]
35. Mousavi, S.R.; Galavi, M.; Rezaei, M. The interaction of zinc with other elements in plants: A review. *Int. J. Agric. Crop Sci.* **2012**, *4*, 1881–1884.
36. Rudani, K.; Patel, V.; Kalavati, P. The importance of zinc in plant growth-A review. *Int. Res. J. Nat. App. Sci.* **2018**, *5*, 38–48.
37. Lehmann, J.; Bossio, D.A.; Kögel-Knabner, I.; Rillig, M.C. The concept and future prospects of soil health. *Nat. Rev. Earth Environ.* **2020**, *1*, 544–553. [CrossRef] [PubMed]
38. Pahalvi, H.N.; Rafiya, L.; Rashid, S.; Nisar, B.; Kamili, A.N. Chemical Fertilizers and Their Impact on Soil Health. In *Microbiota and Biofertilizers*; Dar, G.H., Bhat, R.A., Mehmood, M.A., Hakeem, K.R., Eds.; Springer: Cham, Switzerland, 2021; Volume 2. [CrossRef]
39. Colussi, J.; Schnitkey, G.; Zulauf, C.; Department of Agricultural and Consumer Economics, University of Illinois at Urbana-Champaign. Rising Fertilizer Prices to Affect Brazil's Largest Corn Crop. *Farmdoc Dly.* **2021**, *11*, 154. Available online: https://farmdocdaily.illinois.edu/2021/11/rising-fertilizer-prices-to-affect-brazils-largest-corn-crop.html (accessed on 8 July 2022).
40. Guerrini, I.A.; Croce, C.G.G.; Bueno, O.C.; Jacon, C.P.R.P.; Nogueira, T.A.R.; Fernandez, D.M.; Ganga, A.; Capra, G.F. Composted sewage sludge and steel mill slag as potential amendments for urban soils involved in afforestation programs. *Urban For. Urban Green* **2017**, *22*, 93–104. [CrossRef]
41. Silva, R.S.; Jalal, A.; Nascimento, R.E.N.; Elias, N.C.; Kawakami, K.C.; Abreu-Junior, C.H.; Oliveira, F.C.; Jani, A.D.; He, Z.; Zhao, F.; et al. Composted sewage sludge application reduces mineral fertilization requirements and improves soil fertility in sugarcane seedling nurseries. *Sustainability* **2022**, *14*, 4684. [CrossRef]

Article

Bacterial Microbiota and Soil Fertility of *Crocus sativus* L. Rhizosphere in the Presence and Absence of *Fusarium* spp.

Beatrice Farda, Rihab Djebaili, Matteo Bernardi, Loretta Pace, Maddalena Del Gallo and Marika Pellegrini *

Department of Life, Health and Environmental Sciences, University of L'Aquila, Via Vetoio, 67100 Coppito, Italy
* Correspondence: marika.pellegrini@univaq.it; Tel.: +39-0862-433-258

Abstract: Intensive agricultural practices have led to intense soil degradation and soil fertility losses. Many soil-borne diseases affect these intensive agricultural soils, worsening the physical-chemical and fertility imbalances. Among the numerous pathogens, the genus *Fusarium* includes members that destroy many crops, including *Crocus sativus* L., which also impairs the composition and functions of the microbial communities. This work aimed to investigate, for the first time, the bacterial communities of the rhizosphere of saffron in the presence and absence of fusariosis. The rhizosphere of the saffron fields in the territory of L'Aquila (Italy) with and without fusariosis was sampled and subjected to a microbiological analysis. Culture-dependent methods characterized the fusariosis. The dehydrogenase activity assay was estimated. The metabarcoding of the *16S rRNA* gene, a metagenome functioning prediction, and a network analysis were also carried out. The results showed that fusariosis, when it is linked to intensive agricultural practices, causes alterations in the microbial communities of the rhizosphere. The culture-dependent and independent approaches have shown changes in the bacterial community in the presence of fusariosis, with functional and enzymatic imbalances. The samples showed a prevalence of uncultured and unknown taxa. Most of the known Amplicon Sequence Variants (ASVs) were associated with the Pseudomonadoa (syn. Proteobacteria) lineage. The composition and richness of this phylum were significantly altered by the presence of *Fusarium*. Moreover, pathogenesis appeared to improve the ASVs interconnections. The metagenome functions were also modified in the presence of fusariosis.

Keywords: saffron; fusariosis; soil microbial diversity; DHA assay; 16S metabarcoding; PICRUSt 2; rhizosphere

Citation: Farda, B.; Djebaili, R.; Bernardi, M.; Pace, L.; Del Gallo, M.; Pellegrini, M. Bacterial Microbiota and Soil Fertility of *Crocus sativus* L. Rhizosphere in the Presence and Absence of *Fusarium* spp. *Land* **2022**, *11*, 2048. https://doi.org/10.3390/land11112048

Academic Editors: Chiara Piccini and Rosa Francaviglia

Received: 8 October 2022
Accepted: 13 November 2022
Published: 15 November 2022

Publisher's Note: MDPI stays neutral with regard to jurisdictional claims in published maps and institutional affiliations.

Copyright: © 2022 by the authors. Licensee MDPI, Basel, Switzerland. This article is an open access article distributed under the terms and conditions of the Creative Commons Attribution (CC BY) license (https://creativecommons.org/licenses/by/4.0/).

1. Introduction

The growing demand for healthy food from a growing human population requires intensive and efficient land management practices and crop control to reduce the disease losses [1]. However, intensive farming practices are leading to the degradation of agricultural soils and a gradual loss of their fertility [2]. Soil degradation leads, in turn, to the loss of its functions. Climate change also increases the uncertain and complex management of agricultural soil, jeopardizing its long-term viability, its biodiversity, and consequently, its functions. The use of chemical fertilizers is considered to be the fastest way to increase agricultural production. However, their cost and other constraints are increasingly discouraging farmers from using them [3]. These products also cause environmental pollution with negative consequences for human health [4].

A lack of knowledge about soil biodiversity has been identified as the main limitation to its management. The diversity of soil microbial communities can be critical for soil resilience to various abiotic and biotic stressors [5]. Microorganisms in agricultural soils have a significant impact on soil fertility, on the availability of nutrients for the plant and on the suppression of soil-borne plant diseases [6]. The conservation and sustainable use of soil microbial diversity are crucial for increasing agricultural productivity [7]. The loss of biodiversity has a detrimental impact of productivity, stability, and services [4]. According to a recent meta-analysis, fields that undergo organic management practices had between 32% and 84% higher soil microbial biomasses (carbon, nitrogen, total phospholipid fatty acids) and enzymatic activities (dehydrogenase, urease, protease) than the conventional systems do. Crop rotation, legume intercropping, and organic inputs have all been linked to an increased microbial richness in agricultural soils [8,9].

The loss of soil biodiversity is also linked to the increase in soil-borne diseases, especially in agricultural ecosystems, resulting in higher production costs [6]. Among the numerous pathogens, the genus *Fusarium* includes members that cause diseases in many plants. *Fusarium* diseases are mainly associated with vascular wilt, but several species can cause the seedling wilt, crown, lower stem, root and seed rot, and head and seed plague [10]. *Fusarium* spp. live saprophytically on the roots, stems, leaves, flowers, and seeds of diseased and dead plants [11]. The fungus can survive on seeds (internal and external) or as spores or mycelium in the dead or infected tissues [12]. Within the *Fusarium* genus, *Fusarium oxysporum* is responsible for wilting of plants in nurseries and field crops, causing significant losses [11].

Saffron (*Crocus sativus* L.) is one of the valuable crops that is affected by *F. oxysporum*. Several fungal species belonging to *Fusarium, Rhizoctonia, Penicillium, Aspergillus, Sclerotium, Phoma, Stromatinia, Cochliobolus,* and *Rhizopus* genera affect saffron [13]. *Fusarium* corm rot, which is caused by *F. oxysporum*, is the most destructive disease [14]. Infected plants die early, thus reducing the corm yield, quality, and flower and stigma production [15]. *F. oxysporum* causes vascular wilt, as shown by yellowing of the leaf, the loss of turgidity, necrosis, wilting, and the plant's death.

A *Fusarium* infection occurs when the mycelium or germinating spores penetrate the plant's roots, enter the xylem, and produce microconidia. Vascular vessels become clogged by the accumulation of mycelium, spores, and the oxidation of the degradation products of enzymatic lysis. Toxins can cause vein clearing (the loss of chlorophyll production along the veins), a reduction in the photosynthesis rate, and tissue damage that leads to excessive water loss through transpiration [16]. Fusariosis also harms microbial communities' composition and functions. The recent study by Wang and collaborators highlighted the increase in the carbon cycle, the Calvin cycle, and the expression of hemicellulose and chitin degradation genes in watermelon soil in the presence of *Fusarium* [17].

The literature lacks studies which investigate the effect of *Fusarium* on the quality of the saffron rhizosphere. We hypothesized that *Fusarium* is closely associated with microbial biodiversity loss and a loss of the soil enzymatic activity. This work is aimed at investigating the bacterial communities in the saffron rhizosphere in the presence and absence of fusariosis. The rhizospheres of four saffron fields in the L'Aquila area (Italy) with different extensions of fusariosis were sampled. We performed the metabarcoding of *16s rRNA* and the dehydrogenase activity assay. The same analyses were also carried out on the rhizosphere of six saffron fields without fusariosis.

2. Materials and Methods

2.1. Soil Sampling

Ten saffron fields in the L'Aquila territory (Abruzzo region) were subjected to rhizosphere sampling at 20 cm depth in March 2021. Four fields showed evident fusariosis (ZF1, ZF2, ZF3, and ZF4) and six fields showed no evident pathogenesis (ZB1, ZB2, ZB3, ZB5, ZB6, and ZB7). Field ZF3 presented a less evident presence of the pathogen. Figure 1 shows an example of an evident fusariosis. Five soil sub-samples were collected per field following a non-systematic pattern. The soil samples were sieved (<2 mm) to remove large particles and plant debris. Fresh homogeneous aliquots of each soil sample were immediately processed for culturable approaches and enzymatic activity estimations. Ten aliquots of each soil sample were stored at $-80°$ until they were processed for DNA extraction.

Figure 1. Geolocalization map of the sampling area and examples of a field and a corm with an evident *Fusarium* pathogenesis.

2.2. Fusariosis Pathogenesis Confirmation

The Fusarium pathogenesis was confirmed by the corms inspection and microbial culturable approaches. Three aliquots of each rhizosphere were processed in saline with 1% of Tween 20 (1:10 ratio) in a bag mixer for 30 min. After centrifugation at 4000 for 10 min, the supernatants were subjected to serial dilutions up to 1×10^{-7}. One hundred µL of each serial dilution were plated on Selective *Fusarium* Agar (SFA) [18] and incubated at 25 °C for five days. We confirmed the presence of Fusarium by macro- and microscopic observations of hyphae and spores and by spores sub-culturing on Potato Dextrose Agar PDA (Sigma-Aldrich, St. Louis, USA).

2.3. DNA Extraction and 16S rRNA Metabarcoding

The genomic DNA was extracted using 500 mg of homogenous samples according to the manufacturer's protocol (NucleoSpin®Soil, Macherey Nagel, Germany). The DNA content and purity were verified using a Nanodrop spectrophotometer (Thermo Scientific™, Waltham, MA, USA) and a Qubit fluorometer (Thermo Scientific™, Waltham, MA, USA). For each sample, the individual replicates were combined in an equimolar ratio. We performed paired-end *16S rRNA* community sequencing on the Mi-Seq Illumina technology (Bio-Fab Investigation, Rome, Italy), focusing on the V3 and V4 regions of the *16S rRNA* gene [19]. The filtering was performed, and the readings were evaluated for reliability, and they were counted. Using QIIME2 (qiime2-2020.2 version), the DADA2 plugin was used to build ASV (Amplicon Sequence Variant) [20]. The V3–V4 specific area was taken from the 16S file that was obtained from the SILVA database (https://www.arb-silva.de/ accessed on 14 October 2021) and used to train the classifier using the fit-classifier-naive-Bayes plugin.

2.4. Prediction of Metagenomic Functions

PICRUSt 2 (Phylogenetic Investigation of Communities by Reconstruction of Unobserved States) was used to predict the functional abundances based on *16S rRNA* gene sequencing data [21]. Pathways (PWYs), Enzyme on (EC) numbers and KEGG Orthologs (KOs) were predicted based on the Amplicon Sequence Variants (ASVs) sequence profiles/abundances (BIOM file format obtained from qiime2). PICRUSt 2 was run as a plugin of qiime2 with default parameters. We used the ALDEx2 (ANOVA-like differential expression) to perform the differential abundance testing between the two conditions with

1000 Monte Carlo samples and a One-way ANOVA test. An effect size that is greater than 1 was used as a significance cutoff with or without the BH correction of the raw p values.

2.5. Network Analysis

The network analyses were performed following Barberán et al. [22]. Briefly, the network was inferred by all of the possible Spearman rank correlation comparisons between the ASVs with more than 5 sequences (Spearman's correlation coefficient > 0.6 and statistically significant p value < 0.01). The networks were reconstructed with 90% identity ASVs as nodes and strong and significant correlations between the nodes as edges. The network topology was estimated by a metrics calculation (i.e., average node connectivity and path length, diameter, cumulative degree distribution, clustering coefficient, and modularity) [23]. All of the statistical analyses were performed in the R program using the Igraph [24] package. The networks were investigated and visualized using the interactive platform Cytoscape v 3.9.1 [25] and the Network analyzer v 4.4.8 tool [26].

2.6. Dehydrogenase Activity of Soil Samples

The soil dehydrogenase activity (DHA) was estimated using fresh soil samples [27]: Three aliquots of each soil sample (6 g) were placed in test tubes and mixed with 4 mL of distilled water. Each mixture was supplemented with 120 mg of $CaCO_3$ and 1 mL of 2,3,5-triphenyltetrazolium chloride (TTC 3% v/w) and incubated at 30 °C for 20 h. The samples were filtered, and triphenylformazane (TPF) was extracted using ethanol. The samples were then mixed and placed in the dark for 1 h. After incubation, the supernatant was recovered by centrifugation and analyzed at λ = 485 nm (Multiskan GO™—Thermo Scientific, Waltham, MA, USA). The results are expressed as µg TPF g^{TM1} min^{TM1} using a calibration curve (y = 0.0132x + 0.0083, R^2 = 0.999) [28].

2.7. Statistical Analysis

The data were analyzed by One-way Analysis of Variance (ANOVA) using the XLSTAT 2016 software (Addinsoft, Paris, France). Significant differences were calculated with Tukey's post hoc test at p < 0.05. The Primer 7 and PAST 4.03 software allowed the realization of the taxonomic bar plots of ASVs at the phylum (1%) and genus (1.5%) level and the calculation of alpha-diversity metrics (i.e., Simpson, Shannon, and Chao1 indices) of the different samples.

3. Results

3.1. Fusariosis Pathogenesis Confirmation

The presence of *Fusarium* spp. was confirmed by the microbiological approaches in all of the field where the pathogenesis was evident (ZF1-ZF4). Culturable fungal microflora that were developed on SFA showed a huge presence of *Fusarium*. Based on the morphology of the colonies that were observed, many species of *Fusarium* were present. Some of the isolates were allegedly identified as *Fusarium oxysporum* based on the shapes and sizes of the macro- and microconidia, the presence or absence of chlamydospores, the colony pigments, and the growth rates on PDA. No *Fusarium* isolates were observed from the fields where a pathogenesis was not evident.

3.2. DNA Extraction and 16S rRNA Metabarcoding

The *16S rRNA* gene metabarcoding results were used to investigate the diversity of the samples. As shown in Table 1, a high diversity was present both in the presence and absence of *Fusarium* (Shannon H values higher than 3.5). Sample ZB1 showed more taxa numbers (1454), individuals (36,299), and a high diversity index (Chao-1) when it was compared to the other field with fusariosis. Sample ZF3 presented the highest taxa values, individuals, and diversity indices.

Table 1. Diversity indices calculated on *16S rRNA* metabarcoding results using PAST 4.03 software. Soil samples were labelled as follows: ZB1–ZB7 labels refer to saffron soil samples without evident *Fusarium* pathogenesis; ZF1–ZF4 labels refer to saffron soil samples with *Fusarium* pathogenesis.

	ZB1	ZB2	ZB3	ZB5	ZB6	ZB7	ZF1	ZF2	ZF3	ZF4
Taxa_S	1454	1283	1141	1309	958	1345	1155	1270	2078	1440
Individuals (Richness ASVs level)	36,299	32,009	27,465	28,502	19,823	27,429	25,296	25,719	54,646	33,625
Shannon_H	6.657	6.538	6.41	6.594	6.281	6.657	6.501	6.587	6.994	6.761
Evenness_e^H/S	0.5354	0.5385	0.5329	0.5583	0.5577	0.5788	0.5765	0.5711	0.5248	0.5996
Chao-1	1455	1285	1142	1310	958.7	1347	1156	1271	2080	1444

In the Table: ASVs, Amplicon Sequence Variants.

The *16S rRNA* metabarcoding results were also investigated for their structure and abundance. Figure 2 depicts the ASVs composition and abundances at the phylum level. Most of the ASVs were associated with Pseudomonadota (syn. Proteobacteria), which was followed by Actinobateriota. Latescibaterota and Entotheonellaeota were only present in ZB1 and ZB3, respectively. Firmicutes was only present in ZB2, ZB3, ZB6, and ZF2. Except for ZF1, Nitrospirota was absent in all of the ZF samples. Patescibacteria was not found in ZB3 and all of the ZF samples (except for ZF3). Except for ZB2, Planctomycetota was always present. The other phyla were shared by all of the samples.

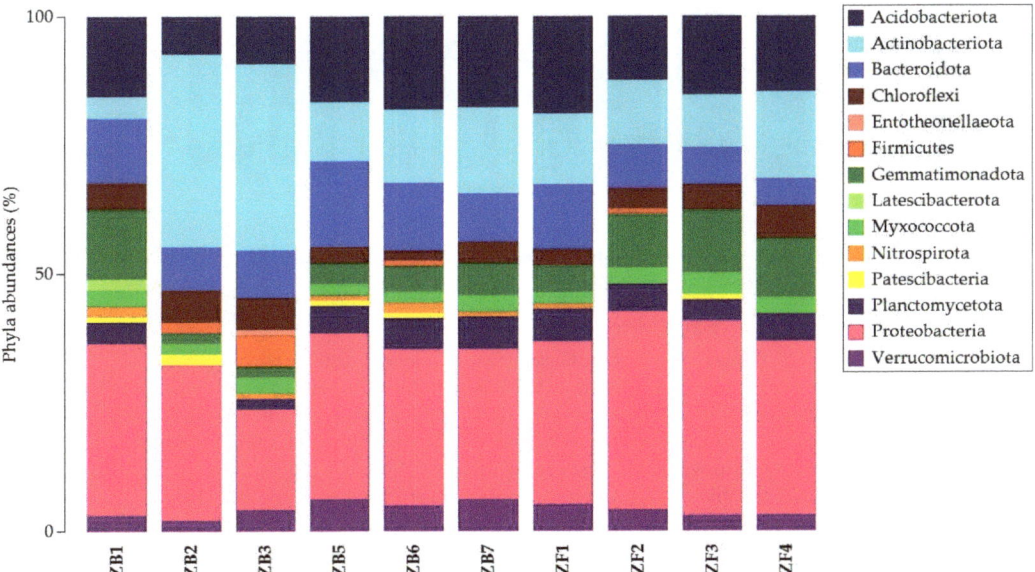

Figure 2. Taxonomic bar plot of the relative abundances of bacterial phyla associated with individual soil samples.

Given the relevance of the Pseudomonadota phylum within the bacterial communities in all of the fields, we carried out a comparison of the abundances and the composition of the ASVs based on the *Fusarium* presence/absence variable. Figure S1 shows the stacked boxplot of the comparison. In the presence of *Fusarium*, the abundances of the ASVs were lower than those that were observed in the absence of pathogenesis. This finding suggested a strong impact of the pathogenesis on richness of the ASVs associated with this phylum.

At the genus level, the common ASVs were those that were associated with uncultured and unknown taxa, which was followed by *Sphingomonas*. (Figure 3). *Vicinamibacteraceae*,

WD2101_soil_group, *RB41*, and *Rubrobacter* were also present in almost all of the samples. However, the occurrence of some genera was absent in the presence of *Fusarium* pathogenesis, i.e., *Streptomyces*, *Bacillus*, *Pseudomonas*, 67-14, *Nitrospira*, *Nocardioides*, *Adhaeribacter*, *Flavisolibacter*, *Flavobacterium*, *Gaiella*, KD4-96, MB-A2-108, *Stenotrophomonas*, *Terrimonas*, and UTCFX1. *Ellin6067* and *Massilia* were only present in the samples under the *Fusarium* pathogenesis condition.

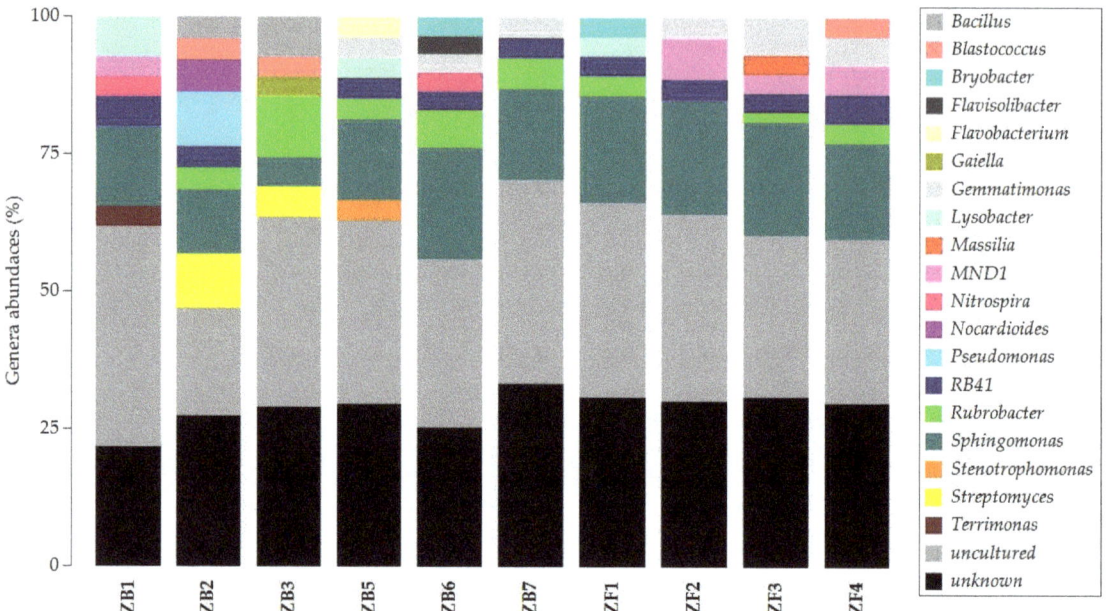

Figure 3. Taxonomic bar plot of the relative abundances of bacterial genera associated with individual soil samples.

3.3. Prediction of Metagenomic Functions

Some of the metabolic predictions showed differential abundances in the presence of fusariosis. Figures 4–6 show the Bland–Altman and Effect plots that shows the relationship between the effect size and the BH-adjusted p values (0.05 and 0.01) in the tests that were carried out for the ECs, KOs, and PWYs. Among the ECs (Figure S2), the most significant differences were observed for feature 1 (EC:1.1.1.21—aldose reductase) and 10 (EC 1.12.2.1—cytochrome-c3 hydrogenase), which were higher in the presence of *Fusarium*, and 11 (EC:1.3.1.87—3-(*cis*-5,6-dihydroxycyclohexa-1,3-dien-1-yl)propanoate dehydrogenase) and 61 (EC:4.3.1.29—D-glucosaminate-6-phosphate ammonia-lyase), which were higher in the absence of fusariosis.

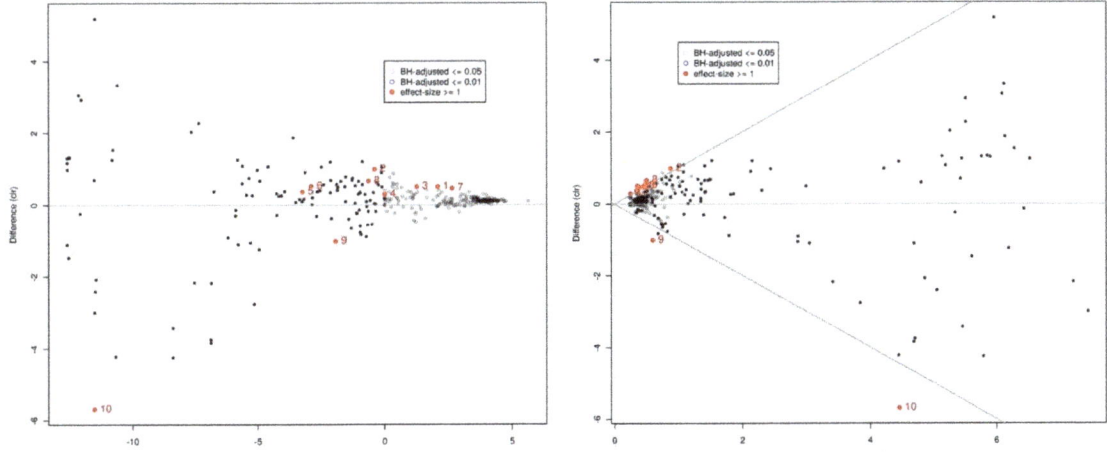

Figure 4. The panel on the left displays the Bland–Altman plot that shows the relationship between Abundance and Difference of the predicted pathways (PWYs) in the presence (lower part) and absence (upper part) of fusariosis. The panel on the right displays the Effect plot that shows the relationship between Difference and Dispersion of the PWYs between *Fusarium* and not *Fusarium* groups. In both of the plots, the 'not significant' features are shown in grey and black. Features that are statistically significant are in red.

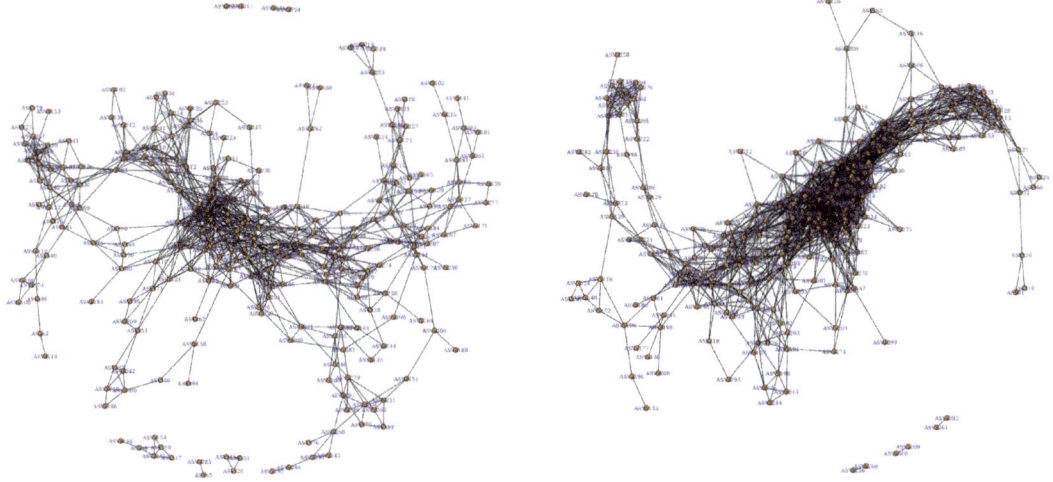

Figure 5. Network analyses carried out on saffron rhizosphere samples in the absence (on the **left**) and presence (on the **right**) of fusariosis. Jaccard similarity coefficient: 4.

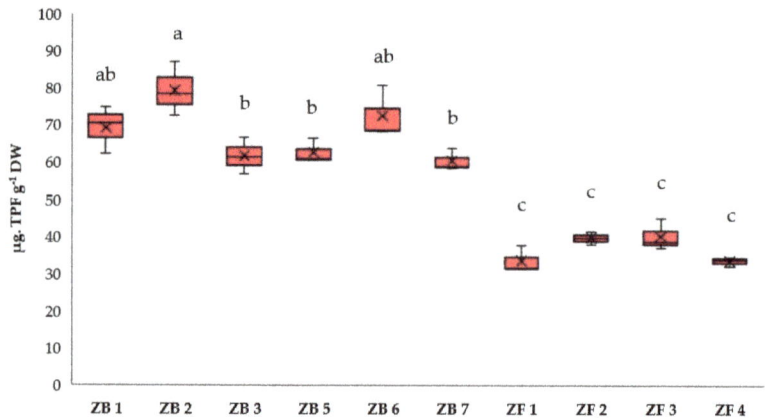

Figure 6. Dehydrogenase activity expressed as µg TPF g^{-1} DW. Results followed by the same case letter (a-c) are not significantly different according to Tukey's HSD post hoc test ($p > 0.05$).

Among the KOs (Figure S3), in the presence of *Fusarium*, higher counts were recorded for features 2 (K00011—aldehyde reductase), 24 (K02205—arginine/ornithine permease), 63 (K11601—manganese transport system substrate-binding protein), and 65 (K11638—two-component system, CitB family, response regulator CitT). In the absence of pathogenesis, the higher counts were recorded for the features 28 (K02791—maltose/glucose PTS system EIICB component), 32 (K03078—3-dehydro-L-gulonate-6-phosphate decarboxylase), and 35 (K03290—MFS transporter, SHS family, sialic acid transporter).

Among the PWYs (Figure 4), the features 9 (PWY-922—mevalonate pathway I) and 10 (THREOCAT-PWY—L-threonine metabolism) showed higher values in the presence of fusariosis. The pathogenesis altered the other PWYs, with low counts for features 1 (P124-PWY—fructose 6-phosphate pathway), 2 (P125-PWY—superpathway of (R,R)-butanediol biosynthesis), 3 (P161-PWY—acetylene degradation—anaerobic), 4 (PWY-5415—catechol degradation I), 5 (PWY-5529—superpathway of bacteriochlorophyll a biosynthesis), 6 (PWY-5531—3,8-divinyl-chlorophyllide a biosynthesis II—anaerobic), 7 (PWY-7254—TCA cycle VII—acetic acid-producers), and 8 (PWY-7315—dTDP-N-acetylthomosamine biosynthesis).

3.4. Network Analysis

The DNA sequencing results were also processed through a network analysis. Figure 5 shows the networks that were obtained for the soil samples with the presence and absence of Fusariosis. At a Jaccard similarity coefficient of four, the samples without *Fusarium* had a total number of 270 nodes and 989 edges, with an average number of neighbors of 7647. In the presence of pathogenesis, higher counts of all of the features were observed (295 nodes; 2750 edges; 19,010 average number of neighbors). A complete dataset of both groups was also processed, creating a network with the sample distribution base on the ASVs features. Figure S4 shows the interconnections among all of the samples based on shared ASVs occurrences, highlighting a close relationship among all of the samples.

3.5. Dehydrogenase (DHA) Activity

The results of the dehydrogenase activity analysis are presented in Figure 6. The samples without fusariosis showed the highest values of DHA ($p < 0.05$), with results of up to 79.43 µg TPF g^{-1} DW. Conversely, the samples with *Fusarium* pathogenesis recorded the lowest values ($p < 0.05$). No significant values among the fields with *Fusarium* pathogenesis were recorded ($p > 0.05$), with an average value of 36.84 µg TPF g^{-1} DW. Moreover, these samples presented the lowest values when they were compared to those from the field without the presence of *Fusarium*.

4. Discussion

The microbial diversity of the rhizosphere of numerous plants, including saffron, has been thoroughly studied using culture-dependent and -independent methodologies [29–32]. In this study, we investigated the changes that occur in the saffron rhizosphere in the presence of the *Fusarium* pathogenesis. The L'Aquila territory (Abruzzo, Italy) and the "*Zafferano dell'Aquila*" (a fine saffron variety with a protected designation of origin) were taken as a case study. Overall, the results suggest that pathogenesis does not affect the rhizosphere microbiota diversity and richness. However, the microbial communities' composition, structure, and functions were altered in the presence of the *Fusarium* pathogenesis.

A presence of uncultured and unknown taxa were found by the *16S rRNA* gene metabarcoding. Uncultured microorganisms are widespread in many environments. They play a crucial role in the biodegradation of various pollutants [33]. They constitute a buried group with a genetic resource encoding for unique valuable functions [34]. The uncultured microorganisms are detected in numerous degradation processes, allowing for efficient bioremediation by targeting specific eco-physiological niches [33]. The metagenomic analysis of chronically polluted coastal sediments revealed the presence of aromatic-ring-hydroxylating oxygenase, which is related to the biodegradation of polycyclic aromatic hydrocarbon as reported by Loviso et al. [35]. Likewise, the genus *Sphingomonas* is a part of the rhizospheric population, and it is linked with several biogeochemical cycles in soil and different metabolic processes [36].

In addition to the uncultured and unknown taxa, most of the ASVs were associated with Proteobacteria. In the presence of fusariosis, the abundances and taxa associated with this phylum were lower than they were in the healthy soils. Proteobacteria is one of the major phyla in soil ecosystems [37–40], with them having crucial roles in fixing the atmospheric nitrogen and mineralizing numerous soil nutrients [36]. This decrease in Proteobacteria is in line with the findings of Zhou et al., who described the same behavior for the banana rhizobacteria microbiota that were infected by *Fusarium* [41]. Proteobacteria have been closely associated with fungal pathogenesis in other plant species. Shen et al., for example, found that the prevalence of Proteobacteria is linked to the epidemic stage of wheat take-all disease [42]. In our case, this phylum is the most prevalent in the saffron rhizosphere, with it comprising up to 54% of the population [43].

At the genus level, the exclusive presence of *Bacillus*, *Nitrospira*, *Pseudomonas*, and *Streptomyces* in the healthy rhizospheres may indicate the presence of beneficial bacteria. These genera are usually associated with plant growth-promoting rhizobacteria (PGPR), with important biostimulant and biocontrol abilities [44–47]. Conversely, the exclusive presence of *Massilia* in the rhizospheres of samples with the pathogenesis indicates an unhealthy status. This lineage exploits the succession of communities within niches [48] and colonizes fungal hyphae with biocontrol effects [49]. A similar situation has been described by Bejarano-Bolívar et al., who described the presence of genera with biocontrol abilities (e.g., *Myxococcus* or *Lysobacter*) in the rhizosphere of an avocado that was affected by *Fusarium oxysporum* [50].

Metabolic predictions have highlighted interesting differences between the two groups. Among the most relevant, the increase in the mevalonate pathway I shows the increase in isoprenoids production. These compounds induce plant growth and development and improve the plant's response to environmental stresses [51]. The increase in the metabolic pathway of L-threonine indicates a high functionality of the community in the degradation of this amino acid [52]. These aspects suggest an attempt to counteract the pathogenesis by the microbial community of the rhizosphere.

Conversely, low counts of the other pathways related to the degradation of sugars, aromatic compounds, and hydrocarbons, the production of acetic acid and chlorophylls, and the production of sucrose metabolites were found. In line with previous reports, these decreases show less functionality in the presence of pathogenesis. The study by Wu et al., for example, described a higher carbohydrate and energy biosynthesis and secondary metabolites in the *Panax notoginseng* rhizosphere in the presence of root-rot fungal pathogens [53].

The network analyses also confirmed the attempt to counter the pathogenesis by the rhizosphere microbial community. Pathogenesis appeared to improve the ASVs interconnections. As reported by the recent review by Siles et al. [54]. Conversely, in the presence of pathogenesis, the organic matter increases due to the plant's degradation. This organic supply can increase the saprotrophic and symbiotrophic interactions, producing a more interconnected network [54].

Estimating the soil enzymatic activity is another approach to studying soil microbial community alterations [55–58]. Among the soil enzymes, dehydrogenase converts hydrogen from an organic material to inorganic acceptors, oxidizing the soil organic substances [59,60]. Soil DHA is an early indicator of alterations in the biological activities of the soil [55]. In the presence of Fusariosis, we found a significant decrease in DHA, which is in line with the results of the literature. Low DHA values have been described for the tomato rhizosphere in the presence of fusariosis by Dukare et al. [61]. A negative correlation between the DHA and pathogenesis was also found in the tomato rhizosphere in the presence of *Ralstonia solanacearum* pathogenesis [62]. This finding confirms the lower metabolic functions of the saffron rhizosphere in the presence of fusariosis which is underlined by the prediction of metagenome functions.

5. Conclusions

In this study, we investigated changes in the saffron rhizosphere in the presence of *Fusarium* pathogenesis. The territory of L'Aquila (Abruzzo, Italy) and *Zafferano dell'Aquila* were taken as a case study. We found alterations in the microbial communities' composition, structure, and functions in the presence of the *Fusarium* pathogenesis. Conversely, the diversity and richness of the rhizosphere microbiota were not affected. A predominance of uncultured and unknown taxa was reported using *16S rRNA* gene metabarcoding, and most of the ASVs were attributed to Proteobacteria. Additionally, the taxa that are associated with this phylum were less abundant in the presence of fusariosis when they were compared to those in the healthy soil. A noteworthy presence of beneficial bacteria in the healthy rhizospheres and genera with biocontrol activity in the samples with the pathogen was signaled. The microbial taxa interconnections have also improved to face the pathogen attack. To our knowledge, this is the first study on the saffron rhizosphere. Therefore, our findings help to enrich the knowledge on the subject. These results can be used as a starting point for future investigation on the microbial taxa of the rhizosphere that are involved in the suppression of *Fusarium* wilt disease to be used as sustainable disease control agents. Intensive agricultural practices are the most common reasons for fusariosis. Intensive managements, that are associated with agrochemical use and mechanizations, unbalance the soil microbiota and lead to outbreaks of fungal pathogenesis. For this reason, future studies should also investigate the agricultural practices that are used in fields to highlight the possible variables that induce Fusariosis and to develop strategies to avoid or control *Fusarium* outbreaks early.

Supplementary Materials: The following supporting information can be downloaded at: https://www.mdpi.com/article/10.3390/land11112048/s1, Figure S1: Stacked bar plot that shows the comparison of the abundances of ASVs associated with Pseudomonadota (syn. Proteobacteria) phylum of the group with *Fusarium* pathogenesis (on the left) and without (on the right). In the bar plot, the main classes and orders of the phylum are shown. Figure S2: Comparisons of the enzymes (ECs) predicted in the presence (lower part) and absence (upper part) of fusariosis. The panel on the left displays the Bland–Altman plot that shows the relationship between Abundance and Difference of the ECs. The panel on the right displays the Effect plot that shows the relationship between Difference and Dispersion of the ECs. In both plots, the 'not significant' features are shown in grey and black. Features that are statistically significant are in red. Figure S3: Comparisons of the gene copies (KOs) predicted in the presence (lower part) and absence (upper part) of fusariosis. The panel on the left displays the Bland–Altman plot that shows the relationship between Abundance and Difference of the KOs. The panel on the right displays the Effect plot that shows the relationship between Difference and Dispersion of the KOs. In both plots, the 'not significant' features are shown in grey and black. Features that are statistically significant are in red. Figure S4: Co-occurrence network analysis of sample carried out on the complete dataset of saffron rhizospheres with the absence and presence of fusariosis. Maximum Jaccard similarity coefficient.

Author Contributions: Conceptualization, M.P. and M.D.G.; methodology, L.P.; software, M.P.; validation, M.P. and M.D.G.; formal analysis, B.F. and M.B.; investigation, M.D.G. and M.P.; resources, L.P.; data curation, R.D. and B.F.; writing—original draft preparation, R.D., B.F., and M.B.; writing—review and editing, M.P. and M.D.G.; visualization, L.P.; supervision, R.D.; project administration, M.P.; funding acquisition, M.P. and M.D.G. All authors have read and agreed to the published version of the manuscript.

Funding: This research was funded by Regione Abruzzo, project reference: CUP: C19J21047230002 "Sviluppo di Protocolli Colturali per la Produzione di Zafferano Volti al Miglioramento dello Stato Fisiologico delle piante e alla Salvaguardia della Coltura Contro il Fungo Patogeno *Fusarium oxysporum*—PRO.ZAFF".

Institutional Review Board Statement: Not applicable.

Informed Consent Statement: Not applicable.

Data Availability Statement: The datasets generated and/or analyzed during the current study are available from the corresponding author on reasonable request.

Acknowledgments: We thank Luca Lepidi and Pio Feneziani, for the sampling's permissions and their availability, and Daniela Maria Spera for her professional support.

Conflicts of Interest: The authors declare no conflict of interest. The funders had no role in the design of the study, in the collection, analyses, or interpretation of data, in the writing of the manuscript, or in the decision to publish the results.

References

1. Hemathilake, D.M.K.S.; Gunathilake, D.M.C.C. Agricultural Productivity and Food Supply to Meet Increased Demands. In *Future Foods*; Elsevier: Amsterdam, The Netherlands, 2022; pp. 539–553.
2. Ambrosini, A.; de Souza, R.; Passaglia, L.M.P. Ecological Role of Bacterial Inoculants and Their Potential Impact on Soil Microbial Diversity. *Plant Soil* **2016**, *400*, 193–207. [CrossRef]
3. Aktar, W.; Sengupta, D.; Chowdhury, A. Impact of Pesticides Use in Agriculture: Their Benefits and Hazards. *Interdiscip. Toxicol.* **2009**, *2*, 1–12. [CrossRef] [PubMed]
4. Saleem, M.; Hu, J.; Jousset, A. More Than the Sum of Its Parts: Microbiome Biodiversity as a Driver of Plant Growth and Soil Health. *Annu. Rev. Ecol. Evol. Syst.* **2019**, *50*, 145–168. [CrossRef]
5. Dubey, A.; Malla, M.A.; Khan, F.; Chowdhary, K.; Yadav, S.; Kumar, A.; Sharma, S.; Khare, P.K.; Khan, M.L. Soil Microbiome: A Key Player for Conservation of Soil Health under Changing Climate. *Biodivers. Conserv.* **2019**, *28*, 2405–2429. [CrossRef]
6. Kennedy, A.C.; Smith, K.L. Soil Microbial Diversity and the Sustainability of Agricultural Soils. *Plant Soil* **1995**, *170*, 75–86. [CrossRef]
7. Rahobisoa, J.J.; Ratrimo, V.R.; Ranaivoarisoa, A. Mitigating Coastal Erosion in Fort Dauphin, Madagascar. In *Sustainable Living with Environmental Risks*; Springer Nature: Cham, Switzerland, 2014; Volume 9784431548, ISBN 9784431548041.
8. McDaniel, M.D.; Tiemann, L.K.; Grandy, A.S. Does Agricultural Crop Diversity Enhance Soil Microbial Biomass and Organic Matter Dynamics? A Meta-Analysis. *Ecol. Appl.* **2014**, *24*, 560–570. [CrossRef]
9. de Deyn, G.; Gattinger, A.; Lori, M.; Symnaczik, S.; Ma, P. Organic Farming Enhances Soil Microbial Abundance and Activity—A Meta-Analysis and Meta-Regression. *PLoS ONE* **2017**, *12*, e0180442.
10. Gwinn, K.D.; Hansen, Z.; Kelly, H.; Ownley, B.H. Diseases of Cannabis Sativa Caused by Diverse Fusarium Species. *Front. Agron.* **2022**, *3*, 796062. [CrossRef]
11. Summerell, B.A.; Botanic, R.; Sydney, G.; Wales, N.S.; Leslie, J.F. To Fusarium Identification. *Plant Dis.* **2003**, 117–128. [CrossRef]
12. Lei, S.; Wang, L.; Liu, L.; Hou, Y.; Xu, Y.; Liang, M.; Gao, J.; Li, Q.; Huang, S. Infection and Colonization of Pathogenic Fungus Fusarium Proliferatum in Rice Spikelet Rot Disease. *Rice Sci.* **2019**, *26*, 60–68. [CrossRef]
13. Sharma, K.D. Abel Piqueras Saffron (*Crocus sativus* L.) Tissue Culture: Micropropagation and Secondary Metabolite Production. *Funct. Plant Sci. Biotechnol. Saffron* **2010**, *4*, 64–73.
14. Mirghasempour, S.A.; Studholme, D.J.; Chen, W.; Zhu, W.; Mao, B. Molecular and Pathogenic Characterization of Fusarium Species Associated with Corm Rot Disease in Saffron from China. *J. Fungi* **2022**, *8*, 515. [CrossRef] [PubMed]
15. Palmero, D.; Rubio-Moraga, A.; Galvez-Patón, L.; Nogueras, J.; Abato, C.; Gómez-Gómez, L.; Ahrazem, O. Pathogenicity and Genetic Diversity of Fusarium Oxysporum Isolates from Corms of *Crocus sativus*. *Ind. Crops Prod.* **2014**, *61*, 186–192. [CrossRef]

16. Register, L.; Help, C. *Feasibility of Using Mycoherbicides for Controlling Illicit Drug Crops*; National Academies Press: Cambridge, MA, USA, 2011. [CrossRef]
17. Wang, T.; Hao, Y.; Zhu, M.; Yu, S.; Ran, W.; Xue, C.; Ling, N.; Shen, Q. Characterizing Differences in Microbial Community Composition and Function between Fusarium Wilt Diseased and Healthy Soils under Watermelon Cultivation. *Plant Soil* **2019**, *438*, 421–433. [CrossRef]
18. Leslie, J.F.; Summerell, B.A. *The Fusarium Laboratory Manual*, 1st ed.; Leslie, J.F., Summerell, B.A., Eds.; Blackwell Publishing Ltd.: Oxford, London, 2006; ISBN 9780813819198.
19. Mizrahi-Man, O.; Davenport, E.R.; Gilad, Y. Taxonomic Classification of Bacterial 16S RRNA Genes Using Short Sequencing Reads: Evaluation of Effective Study Designs. *PLoS ONE* **2013**, *8*, e53608. [CrossRef]
20. Bolyen, E.; Rideout, J.R.; Dillon, M.R.; Bokulich, N.A.; Abnet, C.C.; Al-Ghalith, G.A.; Alexander, H.; Alm, E.J.; Arumugam, M.; Asnicar, F.; et al. Reproducible, Interactive, Scalable and Extensible Microbiome Data Science Using QIIME 2. *Nat. Biotechnol.* **2019**, *37*, 852–857. [CrossRef]
21. Douglas, G.M.; Maffei, V.J.; Zaneveld, J.R.; Yurgel, S.N.; Brown, J.R.; Taylor, C.M.; Huttenhower, C.; Langille, M.G.I. PICRUSt2 for Prediction of Metagenome Functions. *Nat. Biotechnol.* **2020**, *38*, 685–688. [CrossRef]
22. Barberán, A.; Bates, S.T.; Casamayor, E.O.; Fierer, N. Using Network Analysis to Explore Co-Occurrence Patterns in Soil Microbial Communities. *ISME J.* **2012**, *6*, 343–351. [CrossRef]
23. Newman, M.E.J. The Structure and Function of Complex Networks. *SIAM Rev.* **2003**, *45*, 167–256. [CrossRef]
24. Csardi, G.; Nepusz, T. The Igraph Software Package for Complex Network Research. *InterJournal Complex Syst.* **2006**, *1695*, 1–9.
25. Shannon, P.; Markiel, A.; Ozier, O.; Baliga, N.S.; Wang, J.T.; Ramage, D.; Amin, N.; Schwikowski, B.; Ideker, T. Cytoscape: A Software Environment for Integrated Models of Biomolecular Interaction Networks. *Genome Res.* **2003**, *13*, 2498–2504. [CrossRef]
26. Assenov, Y.; Ramírez, F.; Schelhorn, S.-E.; Lengauer, T.; Albrecht, M. Computing Topological Parameters of Biological Networks. *Bioinformatics* **2008**, *24*, 282–284. [CrossRef] [PubMed]
27. Casida, L.E.J.R.; Klein, D.A.; Santoro, T. *Soil Enzymology, Soil Biology 22*; Springer: Berlin/Heidelberg, Germany, 1964; Volume 98.
28. Xie, J.; Hu, W.; Pei, H.; Dun, M.; Qi, F. Detection of Amount and Activity of Living Algae in Fresh Water by Dehydrogenase Activity (DHA). *Environ. Monit. Assess.* **2008**, *146*, 473–478. [CrossRef] [PubMed]
29. Ambardar, S.; Singh, H.R.; Gowda, M.; Vakhlu, J. Comparative Metagenomics Reveal Phylum Level Temporal and Spatial Changes in Mycobiome of Belowground Parts of *Crocus sativus*. *PLoS ONE* **2016**, *11*, e0163300. [CrossRef] [PubMed]
30. Ambardar, S.; Sangwan, N.; Manjula, A.; Rajendhran, J.; Gunasekaran, P.; Lal, R.; Vakhlu, J. Identification of Bacteria Associated with Underground Parts of *Crocus sativus* by 16S RRNA Gene Targeted Metagenomic Approach. *World J. Microbiol. Biotechnol.* **2014**, *30*, 2701–2709. [CrossRef] [PubMed]
31. Mahaffee, W.F.; Kloepper, J.W. Temporal Changes in the Bacterial Communities of Soil, Rhizosphere, and Endorhiza Associated with Field-Grown Cucumber (*Cucumis sativus* L.). *Microb Ecol* **1997**, *34*, 210–223. [CrossRef]
32. İnceoğlu, Ö.; Al-Soud, W.A.; Salles, J.F.; Semenov, A.V.; van Elsas, J.D. Comparative Analysis of Bacterial Communities in a Potato Field as Determined by Pyrosequencing. *PLoS ONE* **2011**, *6*, e23321. [CrossRef]
33. Rani, A.; Porwal, S.; Sharma, R.; Kapley, A.; Purohit, H.J.; Kalia, V.C. Assessment of Microbial Diversity in Effluent Treatment Plants by Culture Dependent and Culture Independent Approaches. *Bioresour. Technol.* **2008**, *99*, 7098–7107. [CrossRef]
34. Wang, Y.; Chen, Y.; Zhou, Q.; Huang, S.; Ning, K.; Xu, J.; Kalin, R.M.; Rolfe, S.; Huang, W.E. A Culture-Independent Approach to Unravel Uncultured Bacteria and Functional Genes in a Complex Microbial Community. *PLoS ONE* **2012**, *7*, e47530. [CrossRef]
35. Loviso, C.L.; Lozada, M.; Guibert, L.M.; Musumeci, M.A.; Sarango Cardenas, S.; Kuin, R.V.; Marcos, M.S.; Dionisi, H.M. Metagenomics Reveals the High Polycyclic Aromatic Hydrocarbon-Degradation Potential of Abundant Uncultured Bacteria from Chronically Polluted Subantarctic and Temperate Coastal Marine Environments. *J. Appl. Microbiol.* **2015**, *119*, 411–424. [CrossRef]
36. Agri, U.; Chaudhary, P.; Sharma, A.; Kukreti, B. Physiological Response of Maize Plants and Its Rhizospheric Microbiome under the Influence of Potential Bioinoculants and Nanochitosan. *Plant Soil* **2022**, *474*, 451–468. [CrossRef]
37. Deng, J.; Yin, Y.; Zhu, W.; Zhou, Y. Variations in Soil Bacterial Community Diversity and Structures Among Different Revegetation Types in the Baishilazi Nature Reserve. *Front. Microbiol.* **2018**, *9*, 2874. [CrossRef] [PubMed]
38. Yang, Y.; Viscarra Rossel, R.A.; Li, S.; Bissett, A.; Lee, J.; Shi, Z.; Behrens, T.; Court, L. Soil Bacterial Abundance and Diversity Better Explained and Predicted with Spectro-Transfer Functions. *Soil Biol. Biochem.* **2019**, *129*, 29–38. [CrossRef]
39. Zou, Z.; Yuan, K.; Ming, L.; Li, Z.; Yang, Y.; Yang, R.; Cheng, W.; Liu, H.; Jiang, J.; Luan, T.; et al. Changes in Alpine Soil Bacterial Communities With Altitude and Slopes at Mount Shergyla, Tibetan Plateau: Diversity, Structure, and Influencing Factors. *Front. Microbiol.* **2022**, *13*, 839499. [CrossRef] [PubMed]
40. Kim, H.-S.; Lee, S.-H.; Jo, H.Y.; Finneran, K.T.; Kwon, M.J. Diversity and Composition of Soil Acidobacteria and Proteobacteria Communities as a Bacterial Indicator of Past Land-Use Change from Forest to Farmland. *Sci. Total Environ.* **2021**, *797*, 148944. [CrossRef]
41. Zhou, D.; Jing, T.; Chen, Y.; Wang, F.; Qi, D.; Feng, R.; Xie, J.; Li, H. Deciphering Microbial Diversity Associated with Fusarium Wilt-Diseased and Disease-Free Banana Rhizosphere Soil. *BMC Microbiol.* **2019**, *19*, 161. [CrossRef]
42. Shen, Z.; Ruan, Y.; Chao, X.; Zhang, J.; Li, R.; Shen, Q. Rhizosphere Microbial Community Manipulated by 2 Years of Consecutive Biofertilizer Application Associated with Banana Fusarium Wilt Disease Suppression. *Biol. Fertil. Soils* **2015**, *51*, 553–562. [CrossRef]

43. Bhagat, N.; Sharma, S.; Ambardar, S.; Raj, S.; Trakroo, D.; Horacek, M.; Zouagui, R.; Sbabou, L.; Vakhlu, J. Microbiome Fingerprint as Biomarker for Geographical Origin and Heredity in *Crocus sativus*: A Feasibility Study. *Front. Sustain. Food Syst.* **2021**, *5*, 688393. [CrossRef]
44. Djebaili, R.; Pellegrini, M.; Bernardi, M.; Smati, M.; Kitouni, M.; del Gallo, M. Biocontrol Activity of Actinomycetes Strains against Fungal and Bacterial Pathogens of *Solanum lycopersicum* L. and *Daucus carota* L.: In Vitro and In Planta Antagonistic Activity. In Proceedings of the 1st International Electronic Conference on Plant Science, Online, 1–15 December 2020; MDPI: Basel, Switzerland, 2020; p. 27.
45. Wang, S.; Wang, J.; Zhou, Y.; Huang, Y.; Tang, X. Prospecting the Plant Growth–Promoting Activities of Endophytic Bacteria Franconibacter Sp. YSD YN2 Isolated from *Cyperus esculentus* L. Var. Sativus Leaves. *Ann. Microbiol.* **2022**, *72*, 1. [CrossRef]
46. Santoyo, G.; Orozco-Mosqueda, M.D.C.; Govindappa, M. Mechanisms of Biocontrol and Plant Growth-Promoting Activity in Soil Bacterial Species of Bacillus and Pseudomonas: A Review. *Biocontrol. Sci. Technol.* **2012**, *22*, 855–872. [CrossRef]
47. Donn, S.; Kirkegaard, J.A.; Perera, G.; Richardson, A.E.; Watt, M. Evolution of Bacterial Communities in the Wheat Crop Rhizosphere. *Environ. Microbiol.* **2015**, *17*, 610–621. [CrossRef] [PubMed]
48. Ofek, M.; Hadar, Y.; Minz, D. Ecology of Root Colonizing Massilia (Oxalobacteraceae). *PLoS ONE* **2012**, *7*, e40117. [CrossRef] [PubMed]
49. Raaijmakers, J.M.; Paulitz, T.C.; Steinberg, C.; Alabouvette, C.; Moënne-Loccoz, Y. The Rhizosphere: A Playground and Battlefield for Soilborne Pathogens and Beneficial Microorganisms. *Plant Soil* **2009**, *321*, 341–361. [CrossRef]
50. Bejarano-Bolívar, A.A.; Lamelas, A.; Aguirre von Wobeser, E.; Sánchez-Rangel, D.; Méndez-Bravo, A.; Eskalen, A.; Reverchon, F. Shifts in the Structure of Rhizosphere Bacterial Communities of Avocado after Fusarium Dieback. *Rhizosphere* **2021**, *18*, 100333. [CrossRef]
51. Venkateshwaran, M.; Jayaraman, D.; Chabaud, M.; Genre, A.; Balloon, A.J.; Maeda, J.; Forshey, K.; den Os, D.; Kwiecien, N.W.; Coon, J.J.; et al. A Role for the Mevalonate Pathway in Early Plant Symbiotic Signaling. *Proc. Natl. Acad. Sci. USA* **2015**, *112*, 9781–9786. [CrossRef]
52. Harris, J.A. Measurements of the Soil Microbial Community for Estimating the Success of Restoration. *Eur. J. Soil Sci.* **2003**, *54*, 801–808. [CrossRef]
53. Wu, Z.; Hao, Z.; Sun, Y.; Guo, L.; Huang, L.; Zeng, Y.; Wang, Y.; Yang, L.; Chen, B. Comparison on the Structure and Function of the Rhizosphere Microbial Community between Healthy and Root-Rot Panax Notoginseng. *Appl. Soil Ecol.* **2016**, *107*, 99–107. [CrossRef]
54. Siles, J.A.; García-Sánchez, M.; Gómez-Brandón, M. Studying Microbial Communities through Co-Occurrence Network Analyses during Processes of Waste Treatment and in Organically Amended Soils: A Review. *Microorganisms* **2021**, *9*, 1165. [CrossRef]
55. Campos, J.A.; Peco, J.D.; García-Noguero, E. Antigerminative Comparison between Naturally Occurring Naphthoquinones and Commercial Pesticides. Soil Dehydrogenase Activity Used as Bioindicator to Test Soil Toxicity. *Sci. Total Environ.* **2019**, *694*, 133672. [CrossRef]
56. Paz-Ferreiro, J.; Fu, S. Biological Indices for Soil Quality Evaluation: Perspectives and Limitations. *Land Degrad. Dev.* **2016**, *27*, 14–25. [CrossRef]
57. Aspray, T.; Gluszek, A.; Carvalho, D. Effect of Nitrogen Amendment on Respiration and Respiratory Quotient (RQ) in Three Hydrocarbon Contaminated Soils of Different Type. *Chemosphere* **2008**, *72*, 947–951. [CrossRef] [PubMed]
58. Dotaniya, M.L.; Aparna, K.; Dotaniya, C.K.; Singh, M.; Regar, K.L. Role of Soil Enzymes in Sustainable Crop Production. In *Enzymes in Food Biotechnology*; Elsevier: Amsterdam, The Netherlands, 2019; pp. 569–589.
59. Wolinska, A.; Stepniewsk, Z. Dehydrogenase Activity in the Soil Environment. In *Dehydrogenases*; InTech: London, UK, 2012.
60. Wiatrowska, K.; Komisarek, J.; Olejnik, J. Variations in Organic Carbon Content and Dehydrogenases Activity in Post-Agriculture Forest Soils: A Case Study in South-Western Pomerania. *Forests* **2021**, *12*, 459. [CrossRef]
61. Dukare, A.; Paul, S. Effect of Chitinolytic Biocontrol Bacterial Inoculation on Soil Microbiological Activities and Fusarium Population in Rhizophere of Pigeon Pea (*Cajanus cajan*). *Ann. Plant Prot. Sci.* **2018**, *26*, 98. [CrossRef]
62. Posas, M.B.; Toyota, K.; Islam, T.M. Inhibition of Bacterial Wilt of Tomato Caused by Ralstonia Solanacearum by Sugars and Amino Acids. *Microbes Environ.* **2007**, *22*, 290–296. [CrossRef]

Article

Water Regulation Ecosystem Services of Multifunctional Landscape Dominated by Monoculture Plantations

Yudha Kristanto [1,2,*], Suria Tarigan [3], Tania June [4], Enni Dwi Wahjunie [3] and Bambang Sulistyantara [5]

1. Watershed Management Science, Department of Soil Science and Land Resources Management, Faculty of Agriculture, IPB University, Bogor 16680, Indonesia
2. Natural Resources and Environmental Management Science, Graduate School, IPB University, Bogor 16129, Indonesia
3. Department of Soil Science and Land Resources Management, Faculty of Agriculture, IPB University, Bogor 16680, Indonesia; sdtarigan@apps.ipb.ac.id (S.T.); enniedw@apps.ipb.ac.id (E.D.W.)
4. Department of Geophysics and Meteorology, Faculty of Mathematics and Natural Sciences, IPB University, Bogor 16680, Indonesia; taniajune@apps.ipb.ac.id
5. Department of Landscape Architecture, Faculty of Agriculture, IPB University, Bogor 16680, Indonesia; bambang_sulistyantara@apps.ipb.ac.id
* Correspondence: kristan_yudha@apps.ipb.ac.id

Abstract: Meeting the growing demand for agricultural production while preserving water regulation ecosystem services (WRES) is a challenge. One way to preserve WRES is by adopting multifunctional landscape approach. Hence, the main objective was to evaluate the role of forest patches (FP) in preserving WRES in tropical landscapes dominated by oil palm plantations. The SWAT model was used to evaluate the essential WRES, such as water yield (WYLD), soil water (SW), surface runoff (SURQ), groundwater recharge (GWR), and evapotranspiration (AET). Due to a compaction, soils in monoculture plantation have higher bulk density and lower porosity and water retention, which decrease WRES. Conserving FP among oil palms evidently improves WRES, such as decreasing SURQ and rain season WYLD and increasing GWR, SW, AET, and dry season WLYD. FP has sponge-like properties by storing water to increase water availability, and pump-like properties by evaporating water to stabilize the microclimate. Mature oil palm also has pump-like properties to maintain productivity. However, it does not have sponge-like properties that make water use more significant than the stored water. Consequently, a multifunctional landscape could enhance WRES of forest patches and synergize it with provisioning ecosystem services of oil palm plantations.

Keywords: evapotranspiration; groundwater recharge; soil compaction; surface runoff; soil water retention

1. Introduction

Local, countrywide, and global consumption of processed palm oil products, such as food, bioenergy, and oleochemicals, is growing along with population growth. This phenomenon implies changing the tropical landscapes into oil palm plantations (*Elaeis guineensis* Jacq.) that are intensively cultivated in monocultures [1]. The area of oil palm plantations has rapidly grown and is expected to increase in the coming years [2,3]. On the one hand, the development of oil palm plantations rapidly contributes to the local, regional, and national economies [4]. In addition, oil palm plantations also positively impact increasing access to the basic needs of local communities, such as school facilities, health facilities, road networks, and electricity networks. However, if the development of oil palm plantations is not appropriately managed, it can cause the degradation of ecosystem services, and reduce the ecological function of a landscape and harms the socio-economic community [5].

Previous research has examined the transformation of natural ecosystem into agroecosystem, especially monoculture plantations on ecosystem services changes [2]. One of the noticeable changes in ecosystem services in agroecosystem is the change in water regulation ecosystem services (WRES) caused by soil compaction due to intensive land management. These phenomena encompass a decrease in infiltration [6], an increase in surface runoff in the rainy season [3], and a decrease in groundwater availability in the dry season [5]. Therefore, meeting the growing demand for oil palm production while maintaining WRES is a challenge faced by a landscape of oil palm plantations related to sustainable and climate smart agriculture. Ecosystem services support human well-being through provision, regulation, and culture formed by natural and manufactured ecosystem structures and processes. WRES are essential ecosystem services in sustainable development, referring to the quantity, quality, and time of water stored in and out of an ecosystem [7]. WRES benefits are freshwater supply, flood and drought protection, electrical power generation, irrigation, and aquatic ecosystem maintenance.

One way to balance landscape WRES is by applying a multifunctional landscape approach through preserving the remnants of forest patches among oil palm plantations. These forest patches can be in secondary dryland forests, riparian vegetation, or agroforestry systems designated as high conservation areas (HCV). A *multifunctional landscape* is a landscape that can serve multiple ecosystem services simultaneously, not only for provisioning services, but also for regulation, cultural, and support services [8]. The multifunctional landscape is a more realistic soil and water conservation approach to optimize ecosystem services, where forest patches serve water regulation and the other regulation services, while oil palm plantations serve crop production. Furthermore, the multifunctional landscape maps the potential trade-offs and win-win synergies between each ecosystem service, especially provisioning services, which often conflict with regulation, cultural, and support services [9].

Since forest patches are essential to maintain the balance of landscape-scale WRES, it is necessary to model and evaluate WRES on oil-palm-dominated landscapes assisted by dynamic models, such as the Soil and Water Assessment Tools (SWAT). The strength of SWAT in WRES simulation is that SWAT has complex parameters and model structures, which can simulate the influence of physical processes in the soil–plant–atmosphere continuum that affects WRES on the watershed scale. However, field observation cannot measure several SWAT parameters, so the uncertainty that arises from parameter justification needs to be evaluated and considered in selecting the output model to be interpreted. In addition, previous research on oil palm ecosystem services generally does not differentiate between mature and young oil palms, especially in the same tropical landscape [3,6]. Oil palm growths affect the WRES changes through soil compaction due to intensive cultivation activities and canopy development that affect hydrological parameters at the hydrologic response unit (HRU) scale.

Because of soil compaction, WRES evaluation in this study needs to consider the soil hydrological characteristics and soil water retention curve (SWRC) at the smallest analysis unit, which significantly affect the WRES due to different land use and crop dynamics. The relationship between SWRC and WRES occurs at the HRU scale, where the information from SWRC is inputted into the .sol database to update the HRU value. SWRC was closely related to calculating the retention parameter of curve number and the range of soil moisture dynamics that affect surface runoff, evapotranspiration, and soil water storage. When the HRU as the smallest unit of analysis is updated, the sub-basin parameters are automatically updated through the routing mechanism. Therefore, this study answers the main research question: how do forest patches preserve WRES in landscapes dominated by oil palm plantations? Consequently, two objectives can be drawn in this study: (i) analyze the soil water retention characteristics of each land use due to land management, and (ii) evaluate the role of forest patches among oil palm plantations in preserving landscape-scale WRES.

2. Materials and Methods

2.1. Study Area

The study area is a tropical lowland located in Jambi Province, Sumatra, Indonesia. This area has an equatorial rainfall type, characterized by two peaks of the rainy season (December and April) and one dry season (July) in one year. This area also has the same soil type, Hapludults, with a sandy clay loam texture (50% sand, 30% clay, and 20% silt) with small topography variations. The land use is dominated by oil palm plantations that are cultivated in monoculture, both mature plants (MOP) and young plants (YOP) (Figure 1). Among oil palm plantations, there are remnants of forest patches (FP) in the form of secondary dryland forest, riparian vegetation, and agroforestry plot separated from their primary ecosystem. The agroforestry plot (AGF ex-MOP) was planted in 2013 from former mature oil palm plantations.

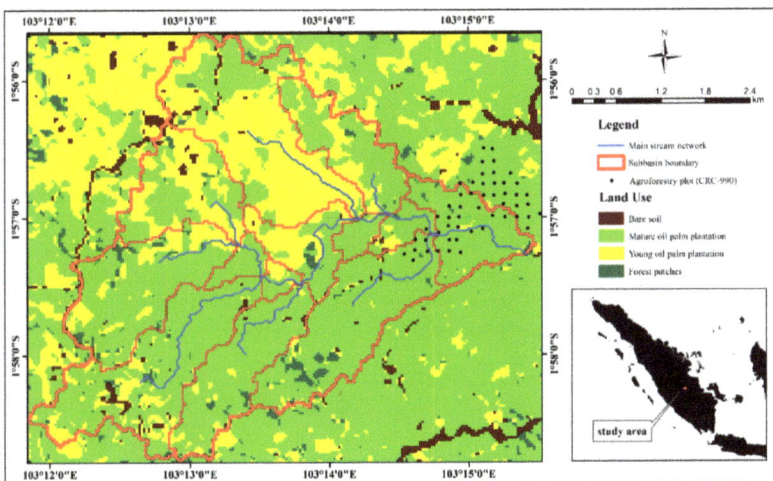

Figure 1. Study area.

A micro-watershed with 13 sub-basins is delineated as a system boundary for WRES modeling. The micro-watershed has an area of 19.1 square kilometers, and lies between 1°55′38.7″–1°58′48.3″ S and 103°11′50.1″–103°15′34.8″ E, with an elevation range of 27–106 m above sea level. An automatic water level recorder (AWLR) is installed in the watershed outlet to support model simulation. The water level data are converted into streamflow through river morphometry measurements and derived into a rating curve equation using an exponential regression model. The rating curve equation obtained from the measurement is $Q = 0.091 \exp(2.309 H)$, where Q is the river discharge and H is the water level. In addition, an automatic weather station (AWS) is installed to record hourly and daily meteorological parameters. Meteorological measurements were carried out from January 2015 to January 2021 as model input, while river discharge measurements were carried out from November 2020 to January 2021 for model calibration and validation.

2.2. Soil Water Retention Characteristics

Undisturbed soil samples at a 0–15 cm depth were collected using soil ring samples with a 7.5 cm diameter and 5 cm height. The total sample for each land use was 8, consisting of 4 different locations and 2 replications for each location. There is a 20 × 20-m plot at each location divided into four quadrants with a size of 10 × 10 m. The soil was sampled in quadrants II and IV, which were diagonal to each other so that, at each location, there were two replications. Because the landscape tends to be homogeneous in terms of elevation, topography, and soil types, we assume that differences in ecosystems are only caused by land use, so that differences in land use can represent different ecosystems at the landscape level. The undisturbed soil samples were used to determine soil hydrological properties, such as bulk density and porosity, and soil water retention characteristics at a particular suction matrix, such as pF 0 (saturated water content), pF 1, pF 2, pF 2.54 (field capacity), and pF 4.2 (permanent wilting point). Analysis of variance (ANOVA) at the 95% level was used to analyze differences in soil hydrological properties between land uses in the study area. The post hoc test with Duncan's multiple range test (DMRT) is conducted if the ANOVA showed a significant difference (p-Value < 0.05).

The observed soil hydrological properties are then used to model the soil water retention curve (SWRC). SWRC is a curve that describes the characteristics of soil water retention by defining the relationship between volumetric water content ($θ$) and matrix suction ($ψ$). SWRC is generally defined using a mathematical equation or pedotransfer function (PTF). The most used mathematical equation for SWRC modeling is the van Genuchten equation, as follows in Equation (1) [10]. The modeled SWRC was then calibrated using the observed moisture content at the pF 1, pF 2, pF 2.54, and pF 4.2.

$$θ(h) = θr + \frac{θs - θr}{(1 + |αH|^n)^m} \quad (1)$$

where $θ(h)$ is the soil moisture content (%v/v), $θr$ is the residual water content (%v/v), $θs$ is the saturated water content (value equal to the total porosity) (%v/v), $α$ is the parameter related to the air entry value into the saturated soil (pF) ($ψα$ = suction where the saturated soil goes through desaturation or air begins to enter the soil pores), H is the matrix suction in a logarithmic scale (pF), and n and m are parameters related to the slope of the curve at the inflection point ($ψ > ψα$). The slope of the curve (S) (1/pF) as a function of n and m can be calculated based on Equation (2) [11]:

$$S = -n(θs - θr)\left[1 + \frac{1}{m}\right]^{-(1+m)} \quad (2)$$

Furthermore, soil data obtained from sampling and laboratory analysis besides SWRC, such as soil permeability, texture, and soil organic matter, were used as inputs for the .sol database inside the SWAT model. The observed soil data are beneficial for WRES simulation, such as determining the initial abstraction, curve number calculation, soil moisture modeling, and evapotranspiration modeling.

2.3. Evaluation of Water Regulation Ecosystem Services

2.3.1. Simulation of Water Regulation Ecosystem Services Using SWAT Model

The Soil and Water Assessment Tools (SWAT) is a physically based, computationally efficient, spatially semi-distributed, and temporally continuous watershed-scale ecosystem services model [12]. This model was developed to simulate various essential ecosystem services related to soil and water in the watershed system, such as water regulation, nutrient retention, and erosion prevention [13,14]. Compared to other WRES models, SWAT integrates the hydrological model with soil and land management attributes, such as irrigation, drainage, fertilization, tillage, and pesticides, and integrates the hydrological model with the crop growth model to simulate dynamic WRES related to crop growth phases [15]. SWAT also simulates various soil and water conservation options and ecological disturbance scenarios, such as land-use change and climate change (LUCCC), that affect ecosystem services [16]. It makes the SWAT results widely used as decision-making tools related to natural resources and environmental management, such as flood and drought mitigation, hydroelectric power generation, and LUCCC impact assessment [15].

SWAT simulates WRES, such as soil moisture (SW), surface runoff (SURQ), lateral flow (LAT), baseflow (BFO), and actual evapotranspiration (AET) from observed precipitation (PRECIP) data based on conservation of mass [14], as described by the Equation (3):

$$SW_t = SW_0 + \sum_{i=1}^{t}(PRECIP - SURQ - AET - LAT - BFO) \qquad (3)$$

The WRES simulation follows three stages: (i) preprocess, (ii) run model, and (iii) calibration, validation, and sensitivity analysis, and (iv) model uncertainty test.

(i) Preprocess

The preprocess stage includes watersheds, sub-basin, and river networks delineation from elevation data and defining HRU from land cover, soil type, and slope class. This stage resulted in a 19.1 km² study area enclosed within the micro-watershed boundary. SWAT divides a basin into sub-watersheds and divides a sub-watershed into hydrological response units (HRU) as the smallest unit of analysis. The defined HRUs are 105 HRUs, consisting of four land covers, one soil type, three slope classes, and thirteen subbasins combinations. HRU explains the spatial heterogeneity of WRES within the watershed and improves the accuracy of WRES modeling for any combination of land use, land management, vegetation type, soil, topography, and climate [17–19].

The spatial data needed to run the SWAT model is a digital elevation model (DEM) with a resolution of 8 m from the Indonesian Geospatial Information Agency, land use derived from Landsat-8 OLI with a resolution of 30 m, and soil map unit with a scale of 1:50,000 from Indonesian Center for Agricultural Land Resources Research and Development. The temporal data needed are daily meteorological data, including rainfall, air temperature, solar radiation, wind speed, and humidity. The preprocessing stage also includes attribute data input, which includes land cover (.mgt), soil physical properties (.sol), rainfall (.pcp), and potential evapotranspiration (.pet).

(ii) Run Model

- Rainfall–Runoff Modeling

SWAT simulate WRES on each HRU and accumulate the WRES from HRU-scale to landscape-scale by a flow routing mechanism [20]. SWAT provides several methods for modeling WRES and flow routing, where users can choose which combination of methods best suits the characteristics of the study area and the availability of input data. One commonly used method for rainfall –runoff modeling related to WRES is the Soil Conservation Services-Curve Number (SCS-CN) [21]. SCS-CN is a powerful method of generating surface runoff, calculated based on Equation (4):

$$SURQ = \frac{(PRECIP - I_a)^2}{(PRECIP - I_a + S)}, \text{ where } S = 254\left(\frac{100}{CN} - 1\right) \qquad (4)$$

where SURQ is surface runoff (mm), PRECIP is precipitation (mm), S is retention parameter (mm), and Ia is an initial abstraction (mm). Initial abstraction is generally assumed 0.2 of the retention parameter, and SURQ only occurs when P > Ia (SURQ = 0 if P $\leq$ Ia). The three main processes considered in initial abstraction are rainfall interception, surface depression storage, and infiltration before the surface runoff. The retention parameter (S) can be approximated as the curve number (CN) function.

Due to land use, hydrologic soil group, and land management differences, the CN value varies spatially. The daily CN value will also vary temporally by considering the antecedent moisture condition (AMC): CN1—dry (wilting point), CN2—average moisture, and CN3—wet (saturated). The CN2 value for each land use, HSG, and land management was taken from the reference table [22], while CN1 and CN3 were calculated from CN2 [14]. SWAT provides two CN methods, original CN and modified or plants evapotranspiration curve number. In a previous study, [23] concluded that these two methods simulated streamflow with equally good performance but differed in simulating soil moisture and evapotranspiration in the lowland tropical landscape. The original CN was chosen in this study because it has a better performance in simulating various elements of WRES than CN-ET, including river discharge, soil water storage, and evapotranspiration.

- Evapotranspiration Modeling

Evapotranspiration as WRES also plays an important role in managing water resources, such as irrigation, soil–vegetation–atmosphere interactions, and spatial–temporal ecosystem productivity. The calculation of evapotranspiration by SWAT is based on the water continuity equation (Equation (3)) and potential evapotranspiration (PET). The PET model used to derive the actual evapotranspiration (AET) is the Penman–Monteith equation [14,24]. The PET calculation is automatically carried out by the SWAT model, which requires a database of maximum and minimum air temperature (.tmp), solar radiation (.slr), wind speed (.wnd), air humidity (.rhu), and crop parameters, such as leaf area index (LAI). The above meteorological datasets (.tmp, .slr, .wnd, and .rhu) were obtained from direct measurements in the field through weather monitoring with AWS. In addition, LAI data were also obtained from sampling using a hemisphere camera. After PET calculations, SWAT estimated AET as the sum of the canopy evaporation, soil evaporation, and plant transpiration [25]. Plant transpiration was calculated as a function of LAI, canopy evaporation was calculated as a function of rainfall interception, and soil evaporation was calculated as soil moisture [14].

(iii) Calibration, Validation, and Sensitivity Analysis

Simulations were carried out from 2015 to 2021, where 2015 was used to warm up the model. Calibration and validation are based on observed streamflow because they are easy to measure and cost-effective compared to soil moisture and evapotranspiration measurements [26]. The first half discharge data are used for the calibration, and the rest are used for validation. The calibrated parameters are parameters that cannot be measured directly in the field, such as groundwater and routing parameters. In contrast, parameters that can be measured directly, such as soil parameters and leaf area index are not calibrated. Compared with manual calibration, which takes a long time and fails to identify parameter sensitivity, this study uses automatic calibration based on the sequential uncertainty fitting-2 (SUFI-2) algorithm using SWAT-CUP software [27]. Sensitivity analysis was conducted to determine the response of changes in SWAT parameters to the significance of output changes and explore all possible combinations of model parameters to investigate output responses related to interactions between parameters [28]. The combination of model parameters and possible outputs are paired and sampled using the Latin hypercube sampling (LHS) to map their interactions and measure the output uncertainty caused by each parameter combination [27].

The Nash–Sutcliffe efficiency (NSE) is a selected model reliability indicator for the objective function during the parameter calibration. NSE value is used to measure how accurately the model's simulation results can describe the observation data. The NSE values range from -∞, which indicates that the model is highly inaccurate, to 1, which indicates that the model is highly accurate.

$$\text{NSE} = 1 - \left[\frac{\sum_{i=1}^{n} \left(Y_i^{obs} - Y_i^{sim} \right)^2}{\sum_{i=1}^{n} \left(Y_i^{obs} - \overline{Y^{obs}} \right)^2} \right] \qquad (5)$$

where Y^{obs} is observation data and Y^{sim} is simulation data.

The model's reliability can also be evidenced by the coefficient of determination (R-squared) value. R-squared values range from 0, indicating that the model is highly inaccurate, to 1, which indicates that the model is very accurate. There are no absolute criteria for assessing the reliability of the hydrological model described in the literature. However, some criteria are commonly used, such as the NSE criteria by Moriasi [29] and R-squared criteria by Ayele [30].

(iv) Model Uncertainty Test

Evaluation of the model reliability is not enough to ensure that the SWAT outputs are genuinely accurate and interpreted directly. Furthermore, it is also necessary to evaluate the model uncertainty that arises due to the complexity of the SWAT structures and parameters justification. Therefore, the SUFI-2 algorithm in SWAT-CUP introduces statistical indicators to investigate the structural uncertainties associated with model simulations [27]. The uncertainty that arises during the parameter calibration is measured by the p-Factor, which is the percentage of observed data that is within 95% predictive uncertainty between the 2.5 and 97.5 percentiles (95PPU), and the r-Factor, which indicates the thickness of the mean of 95PPU divided by the standard deviation of the observed data [31]. Besides looking for high NSE and R-squared, it is also necessary to obtain the largest possible p-Factor and the smallest possible r-Factor. The uncertainty of the model is acceptable if the p-Factor > 0.7 and the r-Factor < 1.5 [27].

2.3.2. Model Limitation

Parameter optimization during the calibration process can produce identical streamflow output with observational data regardless of how the best-fit parameters affect other WRES imprecision. However, because this research is related to the WRES assessment, the interpretation of the model is based not only on streamflow outputs, but also on other WRES, such as soil water storage and actual evapotranspiration. This study obtained precipitation as WRES input and other meteorological data for ETP calculation from the automatic weather station. Due to the limitations of time-series observations of soil moisture, we used soil hydrological properties and soil water retention curve (SWRC) observations from soil sampling and laboratory analysis. We linked the information from SWRC with the SWAT model by updating the .sol database for each HRU as soil moisture modeling inputs.

SWAT simulates soil moisture for each HRU as soil water storage (mm) in the range of available water content (AWC) between permanent wilting point (WP) and field capacity (FC). To obtain %v/v AWC, SWAT divides the soil water storage (mm) by soil depth (SOL_Z) and adds this result with WP. Based on the information of FC, AWC, and WP from SWRC, the results of the soil moisture from the SWAT model are still within the AWC range following AWC observations on each land use. Finally, we consider the actual evapotranspiration as the "residual" component of the modeling based on the water balance equation (AET = PRECIP $-$ Q $-$ ΔSW). Therefore, the reliability of meteorological observation, SWRC observation, streamflow modeling, and SW modeling would affect the reliability of AET. If we could appropriately simulate the streamflow and soil moisture, then the AET value can also be relied upon in the future WRES evaluation.

2.3.3. Water Regulation Ecosystem Services Indicators

One of the further challenges of evaluating WRES is determining the essential indicators based on the SWAT outputs. In general, precipitation is distributed into three flow elements: surface runoff (SURQ), groundwater recharge (GWR), and actual evapotranspiration (GWR), and the remainder is stored as soil water storage (SW).

$$PRECIP_n = SURQ_n + GWR_n + AET_n + SW_n \qquad (6)$$

where PRECIP is precipitation (mm), SURQ is surface runoff (mm), GWR is groundwater recharge (mm), AET is actual evapotranspiration (mm), SW is soil water storage (mm), and n is land use type. Water yield is also an essential WRES indicator related to the sustainable management of water resources in the study area and the key to river regime sustainability. Water yield is the amount of water from each ecosystem that enters the water body. Water yield has a complex component consisting of surface runoff with a short concentration time and lateral flow and baseflow with a longer concentration time.

$$WYLD_n = SURQ_n + LAT_n + BFO_n \qquad (7)$$

where WYLD is water yield (mm), SURQ is surface runoff (mm), LAT is lateral flow (mm), and BFO is baseflow (mm). This study also assesses WRES on an annual and seasonal scale. Temporal assessment is significant to rationally allocate water resources, especially in areas with seasonal excess water and drought. SWAT output is separated based on the monthly rainfall pattern in one year, the wet month when the rainfall is >200 mm, and the dry month when the rainfall is <100 mm.

3. Results

3.1. Soil Water Retention Due to Soil Compaction

Bulk density and porosity as indicators of soil compaction were observed in the topsoil because agricultural activities that encourage soil compaction occurred in the topsoil compared to the subsoil. The results showed that the total pore size had the following trends: FP (55.6%) > YOP (52.0%) > AGF Ex-MOP (49.3%) > MOP (49.0%), while the bulk density trends as follows: MOP (1.35 g/cm^3) > AGF Ex-MOP (1.34 g/m^3) > YOP (1.12 g/cm^3) > FP (0.91 g/cm^3). The relationship between bulk density and soil porosity is reciprocal, meaning that higher bulk density will reduce porosity and vice versa. FP has the lowest bulk density and highest porosity, while MOP has the highest bulk density and lowest porosity. Based on ANOVA and DMRT, FP bulk density was significantly the smallest ($p \leq 0.05$), and soil porosity was significantly higher ($p \leq 0.05$) than YOP, MOP, and AGF Ex-MOP (Table 1). YOP bulk density was significantly different ($p \leq 0.05$) from MOP and Ex-AGF, but YOP porosity was not significantly different ($p > 0.05$) from FP and MOP. Meanwhile, although slight differences exist, AGF Ex-MOP and MOP bulk density and porosity were not significantly different ($p > 0.05$)

Table 1. The mean values of soil porosity and bulk density.

Land Use	Total Pore Space *	Porosity (%v/v)				Bulk Density * (g/cm^3)	Soil Organic Matter (%)
		Dainage Pores		Water Holding Pores			
		pF 1 *	pF 2 *	pF 2.54 *	pF 4.2 ^{n}		
FP	55.6 ± 1.2 [a]	48.7 ± 3.4 [a]	40.0 ± 3.1 [a]	34.7 ± 3.6 [a]	17.8 ± 4.9	0.91 ± 0.06 [a]	5.45 ± 0.82 [a]
YOP	52.0 ± 3.1 [a,b]	43.8 ± 6.6 [b]	36.5 ± 4.6 [a,b]	31.6 ± 4.8 [a]	16.8 ± 4.2	1.12 ± 0.12 [b]	4.15 ± 2.39 [a,b]
MOP	49.0 ± 5.9 [b]	39.7 ± 3.9 [b]	32.8 ± 3.2 [b]	25.5 ± 2.2 [b]	17.0 ± 2.8	1.35 ± 0.16 [c]	2.49 ± 0.15 [b]
AGF Ex-MOP	49.3 ± 3.6 [b]	41.7 ± 3.6 [b]	36.1 ± 3.0 [b]	26.9 ± 2.7 [b]	17.9 ± 2.2	1.34 ± 0.09 [c]	5.19 ± 0.04 [a]

* Significant at 95% level, ^{n}: not significant at 95% level; [a,b,c] the mean value followed by the same letter does not differ according to the DMRT.

Due to soil compaction, bulk density and soil porosity changes lead to soil water retention changes. The soil water retention curve (SWRC) is presented in Figure 2, showing that the trend of soil water retention in each land use has the same pattern as the soil porosity: FP > YOP > AGF Ex-MOP > MOP. These results indicate that FP has the highest soil water retention for the same potential energy, while MOP has the lowest. The best-fit van Genuchten parameters obtained from adjustments are presented in Table 2. The SWRC of each land use has high suitability to the observed data, evidenced by a low RMSE and high R-squared. Parameter n and m are related to the slope of the retention curve after the inflection point (S) [10]. The S value is often used to describe the level of soil degradation. The results obtained indicate that MOP has the highest level of soil degradation, while FP is the lowest.

Table 2. SWRC best-fit parameters and S value.

Land Use	θs (v/v)	θr (v/v)	α (pF)	n	m	S (pF^{-1})	R-Squared	RMSE (v/v)
FP	0.556	0.000	0.110	1.142	0.124	0.0533	0.983	0.0151
YOP	0.520	0.000	0.200	1.129	0.114	0.0463	0.976	0.0153
MOP	0.490	0.000	0.437	1.120	0.107	0.0413	0.983	0.0112
AGF	0.493	0.000	0.230	1.123	0.110	0.0425	0.964	0.0172

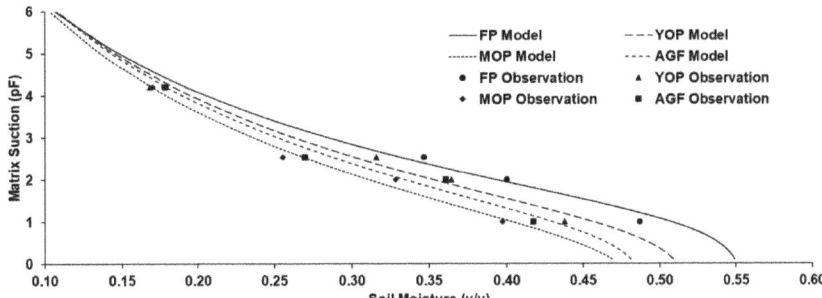

Figure 2. Soil water retention curve for each land use.

Precipitation or irrigation that infiltrates the soil column moves freely under gravitational force, and the soil matrix binds the rest by adhesive force. Gravitational water occupies drainage pores or the range of soil pores between saturated water content and field capacity (pF 0–pF 2.54). FP has higher saturated water content, field capacity, and drainage pores (Table 3), which indicates that FP has more gravitational water than YOP and MOP. Land use with high gravitational water has implications for higher percolation, lateral flow, and groundwater recharge, evidenced by SWAT simulation. Soil water bound by soil matrix is divided into available water content (AWC) or water that plant roots can still absorb and permanent wilting point (PWP) or water that plant roots can no longer absorb. AWC is the water content occupying the available water pore space (pF 2.54–pF 4.2). In contrast, PWP occupies the unavailable water pore space (pF $\geq$ 4.2). FP with high available water pore space implies that FP holds more soil water as a source for the plant uptake and evapotranspiration process.

Table 3. Soil water retention characteristic on the same potential for each land use.

Porosity (v/v)	FP	YOP	MOP	AGF Ex-MOP
Solid layer	0.45	0.49	0.53	0.52
Total pore space	**0.55**	**0.51**	**0.47**	**0.48**
pF 1	0.51	0.46	0.40	0.43
pF 2	0.39	0.35	0.31	0.33
Field capacity	0.33	0.30	0.27	0.29
Permanent wilting point	0.19	0.18	0.17	0.18
Residual pores	0.00	0.00	0.00	0.00
Drainage pores	**0.22**	**0.21**	**0.20**	**0.20**
Fast drainage pores	0.16	0.16	0.16	0.15
Low drainage pores	0.06	0.05	0.04	0.05
Water holding pores	**0.33**	**0.30**	**0.27**	**0.29**
Available water pores	0.14	0.12	0.10	0.11
Unavailable water pores	0.19	0.18	0.17	0.18

3.2. Evaluation of Water Regulation Ecosystem Services

The performance of the calibrated SWAT was evaluated quantitatively based on statistical values compared to the criteria recommended by Abbaspour [27], Moriasi [29], and Ayele [30]. The model's performance is very good for the calibration period, with NSE 0.78 and R-squared 0.88, and suitable for the validation period, with NSE 0.67 and R-squared 0.83. In addition, the p-Factor and r-Factor obtained are 0.93 and 1.09 for the calibration period and 0.75 and 0.54 for the validation period, which indicates model uncertainty is acceptable. Based on the literature review, 20 key parameters capable of capturing the main WRES were selected for calibration and 7 of them were the most sensitive parameters based on the Latin hypercube sensitivity analysis (Table 4).

Table 4. SWAT calibrated parameters with their range and best-fit values.

Parameters [a]	Range		Value			Sensitivity [b]	
	Lower Bound	Upper Bound	Calibrated Value	Initial Value	Unit	T-Stat	p-Value
v_ESCO *	0	1	0.739		-	3.721	0.000
v_EPCO	0	1	0.875		-	−0.739	0.460
v_CANMX	0	10	8.67		mm	−0.479	0.632
r_CN2 *	−0.25	0.25	1.190	[22]	-	16.053	0.000
r_OV_N	−0.2	0.2	0.842	[32]	-	−0.138	0.890
r_SOL_Z	−0.9	0.9	1.38	Obs.	mm	−0.014	0.988
r_SOL_K	−0.2	0.2	0.948	Obs.	mm/h	0.129	0.897
r_SOL_AWC	−0.2	0.2	0.932	Obs.	%	−0.650	0.516
r_SOL_CBN	−0.2	0.2	1.159	Obs.	%	−0.123	0.902
r_SOL_BD *	−0.2	0.2	1.139	Obs.	g/cm^3	−2.500	0.012
v_LAT_TTIME	0	180	63.54		day	−0.178	0.854
v_ALPHA_BF	0	1	0.347		1/h	0.230	0.818
v_GWQMN *	0	5000	3615		mm	−2.108	0.035
v_REVAPMN	0	500	296.5		Mm	0.435	0.664
v_GW_DELAY	0	300	294.9		day	−1.932	0.054
v_RCHRG_DP	0	1	0.197		-	−1.888	0.060
v_CH_N1	0	0.3	0.274		-	−1.493	0.136
v_CH_N2 *	0	0.3	0.122		-	−2.192	0.029
v_ALPHA_BNK *	0	1	0.187		day	22.880	0.000
v_CH_K2 *	0	500	31.5		mm/h	−19.43	0.000

[a] v: replace the initial value with the best fit value, r: multiply the initial value with the best fit value; [b] parameter is sensitive when p-value < 0.05 or |T-stat| > Tα. df, sensitive parameters is marked with (*).

All calibrated parameters are CN2 (curve number in average moisture conditions) and OV_N (manning "n" coefficient for overland flow) that related with surface runoff; LAT_TTIME (lateral flow travel time) that related with lateral flow; CH_N2 (manning "n" coefficient for the main channel), CH_K2 (hydraulic conductivity of the main channel), CH_N1 (manning "n" coefficient for tributary channel), and ALPHA_BNK (riverbank recession constant) that related with streamflow routing; CANMX (maximum canopy storage), ESCO (soil evaporation compensation coefficient), and EPCO (plant uptake compensation factor) that related evapotranspiration; ALPHA_BF (baseflow recession constant) and GWQMN (baseflow threshold) that related baseflow; and REVAPMN (water level threshold for "revap"), GW_DELAY (groundwater delay), and RCHRG_DP (deep aquifer recharge proportion) that related groundwater recharge. Furthermore, all observed parameters are SOL_BD (bulk density), SOL_Z (soil depth), SOL_AWC (available water content), and SOL_CBN (soil carbon content) that related with soil water storage; and SOL_K (soil permeability) that related with lateral flow. The initial value of curve number (CN2) and manning "n" coefficient for overland flow was based on frequently used and reliable literature.

The three parameters related to streamflow routing are sensitive parameters, such as ALPHA_BNK, CH_K2, and CH_N2. ALPHA_BNK is the most sensitive parameter, indicated by the highest |T-stat|. The sensitivity of ALPHA_BNK, CH_K2, and CH_N2 indicates that the flow routing mechanism influenced by these parameters greatly determines the streamflow dynamics. The sensitivity of CH_K2 shows that streamflow is strongly influenced by two-way interactions between rivers and shallow aquifers, where this interaction only occurs in intermittent rivers. Rivers receive water from shallow aquifers when the water table level exceeds the riverbed (rainy season) and lose water when the water table level is less than the riverbed (dry season). The velocity of streamflow filling and loss is closely related to the hydraulic conductivity of the soil layer (CH_K2) between the riverbed and the shallow aquifer. CH_N2 is also a sensitive parameter, which means that the flow velocity greatly determines the streamflow dynamics. Higher CH_N2 is associated with lower flow rates, while lower CH_N2 is associated with higher flow rates.

The other sensitive parameters are CN2, ESCO, SOL_BD, and GWQMN. As a component that dominates streamflow during the rainy period, the magnitude of surface runoff calculated from the CN2 value also dramatically determines the streamflow dynamics. Previous studies have shown that CN2 is always a sensitive parameter when the CN method is chosen for rainfall–runoff modeling [33]. CN2 has a value range of 0 to 100, but the often-used values are in the range of 25 to 98. The greater the CN2 value, the higher the surface runoff generated from rainfall. SOL_BD is a parameter that determines the soil water retention, which then implies soil moisture dynamics. Soil moisture dynamics are necessary to determine surface runoff, lateral flow, groundwater recharge, and actual evapotranspiration. The last, GWQMN, is a parameter that determines the amount of baseflow, where baseflow only appears if the water table exceeds the GWQMN value.

The SWAT model used for evaluating WRES has been updated by adding soil water retention characteristics in the HRU scale. The soil water retention characteristics are inputted into soil attributes and affect the WRES value for each HRU after HRU definition. Soil water retention characteristics are related to calculating curve number retention parameter and defining soil moisture ranges that affect surface runoff, evapotranspiration, and soil water storage. Precipitation that reaches soil surface is generally allocated as surface runoff (SURQ), groundwater recharge (GWR), actual evapotranspiration (AET), and soil water storage (SW) (Figure 3). In forest patches (FP), 43% of rainfall is distributed as SURQ, 26% as AET, 25% for GWR, and 6% for SW. On the other hand, on mature (MOP) and young (YOP) oil palm plantations, rainfall is allocated as SURQ by 56% and 73%, respectively, AET by 25% and 20%, and GWR 15% and 6%. The temporal WRES of each land use is presented in Figure 4. The area of agroforestry plots on former oil palm plantations (AGF ex-MOP) are very narrow (<0.01% of the watershed area) and do not significantly affect the landscape-scale water regulation. Therefore, these agroforestry plots are not included in the SWAT simulation.

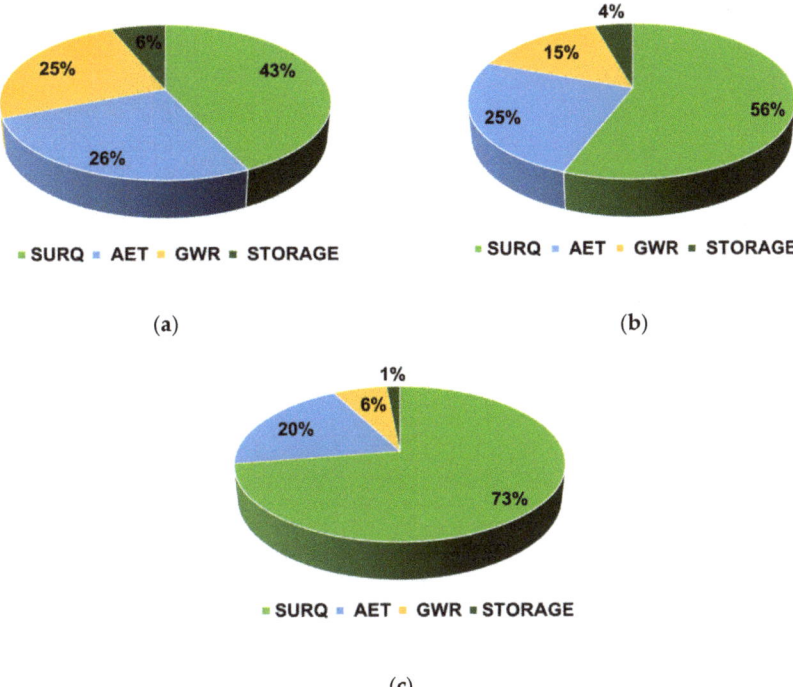

Figure 3. Distribution of water regulation ecosystem service elements in forest patches (**a**), mature oil palm (**b**), and young oil palm (**c**). **Notes: SURQ:** surface runoff, **AET:** actual evapotranspiration, **GWR:** groundwater recharge, **STORAGE:** soil water storage.

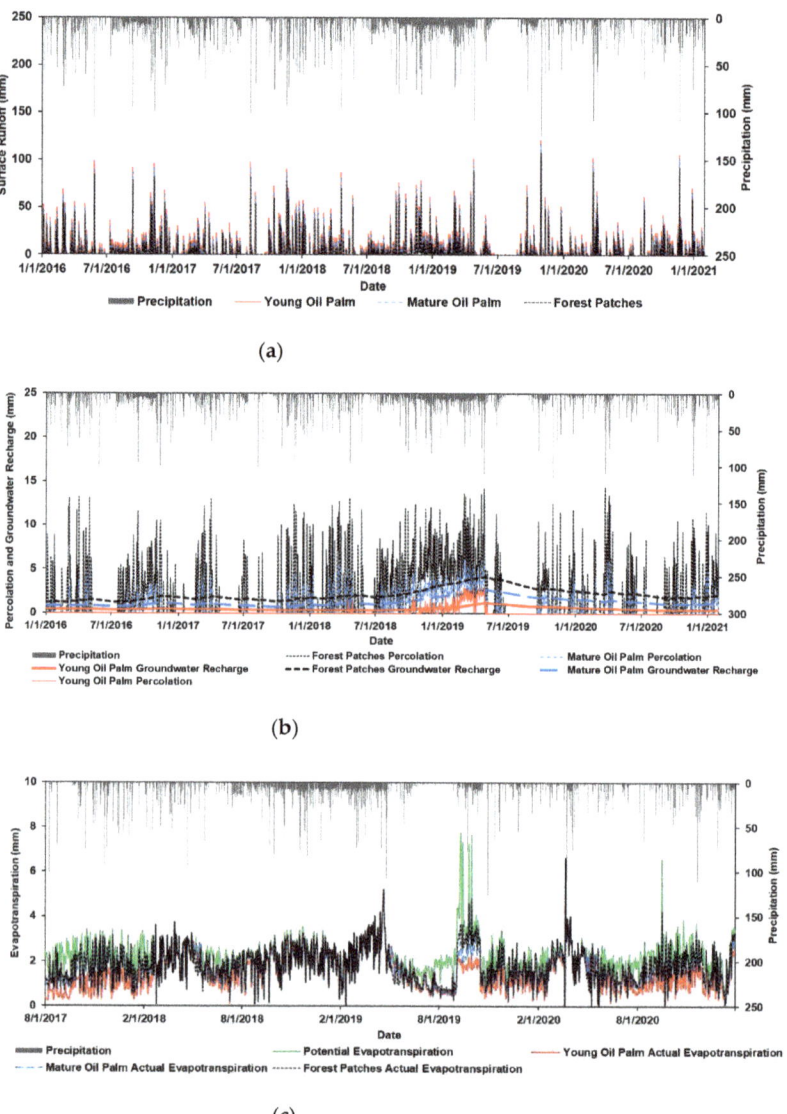

Figure 4. Hydrograph of water regulation ecosystem service elements: surface runoff (**a**), groundwater recharge (**b**), actual evapotranspiration (**c**).

Forest patches have better annual and monthly water regulation ecosystem services than young and mature oil palm plantation, evidenced by lower surface runoff, higher groundwater recharge, higher actual evapotranspiration, higher soil water storage, lower water yield in wet months, and higher water yield in dry months (Figure 5). The annual water yield of forest patches is 2222 mm, while the average water yield in the wet and dry months is 220 mm and 95 mm. On the other hand, the annual water yield of young and mature oil palms were 2465 mm and 2298 mm, wet month water yields were 261 mm and 237 mm, and dry month water yields were 55 mm and 72 mm. The annual and seasonal water yield difference between forest patches and young oil palms is very high and decreases as the oil palm grows [2]. An increasing proportion of surface runoff and decreasing lateral flow and base flow to water yield in mature and young oil palm led to more significant water yield seasonal variability despite the higher annual water yield (Figure 6). It means that oil palm water yield became concentrated in the rainy season and decreased in the dry season, while forest patch water yield is less concentrated in the rainy season and more available in the dry season.

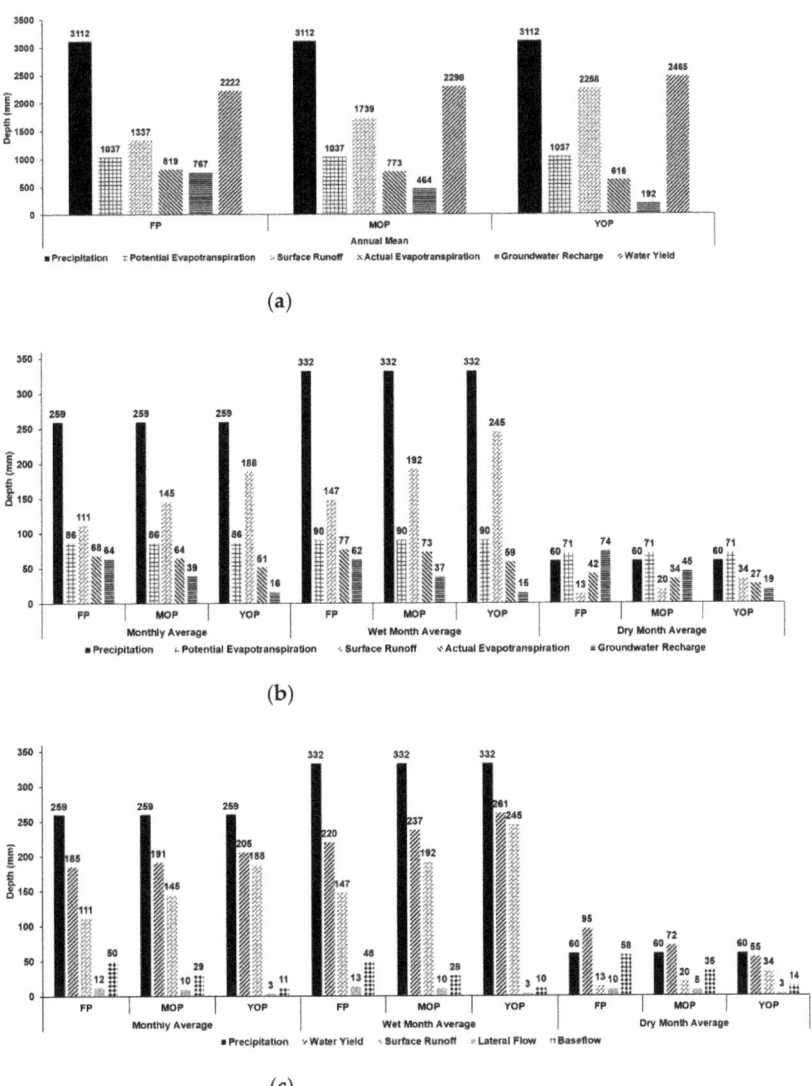

Figure 5. Annual (**a**) and seasonal mean of water regulation ecosystem services (**b**) and water yield (**c**) elements.

Actual evapotranspiration (AET) is also an essential component of water regulation ecosystem services because it plays a role in crop production (water use) and microclimate regulation. The annual actual evapotranspiration of forest patches was 819 mm, with a monthly average of 77 mm in the wet months and 42 mm in the dry months. Meanwhile, the annual actual evapotranspiration of young and mature oil palms was 773 mm and

616 mm, with an average of 73 mm and 59 mm in the wet months and 34 mm and 27 mm in the dry months. Potential evapotranspiration (PET) and soil moisture at available water content (AWC) are the main limiting factors for AET. PET limits AET in saturated soil, while AWC limits AET in unsaturated soil. FP with higher AWC implied higher AET than YOP and MOP. AET values were also controlled by canopy cover, quantified by the leaf area index (LAI) by direct measurement using hemispherical photos. FP with LAI of 3.1 ± 0.63 had a higher AET than MOP with LAI of 1.27 ± 0.19. FP has an AET/PET coefficient of 0.79, while YOP and MOP are 0.59 and 0.75. This value means that, from 100% of the potential energy allocated for the evapotranspiration, FP uses 79% of that energy to evaporate water, while YOP uses 59%, and MOP uses 75% of potential energy.

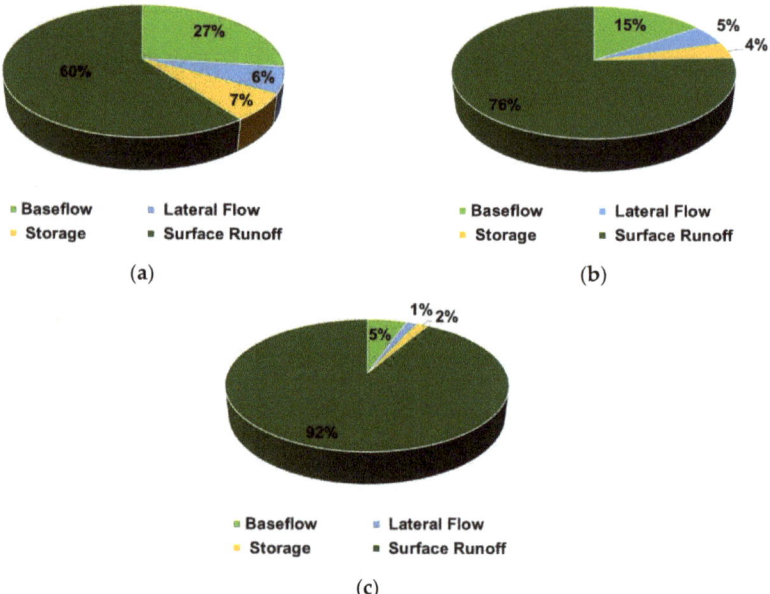

Figure 6. Distribution of water yield elements in forest patches (a), mature oil palm (b), and young oil palm (c).

4. Discussion

Water regulation through the soil layer is the primary process determining how much water flows above the soil surface as surface runoff, returns to the atmosphere through evapotranspiration, percolates into the aquifer, and is stored in the soil pores. Water regulation is highly dependent on soil quality, which is indicated by soil physical, chemical, and biological characteristics that vary in response to soil type, land use, and topography [34]. Water regulation through the soil layer also determines the retention and movement of dissolved nutrients, such as nitrogen, phosphorus, and other nutrients. Therefore, good soil quality needs to be maintained to support sustainable agricultural production. However, a decline in soil quality has led to soil degradation mainly caused by anthropogenic pressures exerted on the soil beyond its carrying capacity [35].

Soil degradation in the form of soil compaction is a consequence that must be accepted due to natural ecosystem changes to agroecosystems, especially intensive monoculture-cultivated agriculture, such as oil palm plantation [34,36,37]. Soil compaction in the oil palm plantation is characterized by increased soil bulk density and decreased porosity. On the other hand, ecosystem restoration, such as preserving forest patches or constructing agroforestry islands inside oil palm plantations, will improve soil structure indicated by increased porosity and reduced bulk density. Soil compaction in oil palm plantations is mainly caused by soil structure detachment by heavy equipment pressure during soil tillage and harvesting, intensive inorganic fertilization, and decreased soil organic matter and vegetation cover [38].

Soil compactions have negative effects on the soil hydrological process, such as reduction in infiltration capacity, permeability, and soil water retention that are related with reduction in soil porosity [39]. This phenomenon, in turn, has implications for the WRES degradation. For example, [5,6] reported that the study area experienced increasing problems related to water resources due to soil compaction in the oil palm plantations area, especially the prolonged water shortage during the dry season. Based on soil water retention curve interpretation in MOP and YOP, the soil is degraded and less structured, has less organic matter, and has fewer pores. Meanwhile, FP has more structured soil with more organic matter and pores. According to [10], the SWRC slope (S value) correlates with soil bulk density, porosity, and soil organic matter content. The S value will decrease along with the increase in bulk density, decrease in porosity, and decrease in soil organic matter.

The higher the level of soil compaction causes the macropores to be further reduced, which causes the field capacity, saturated water content, and gravitational water to decrease. The decrease in gravitational water can be seen from the reduction in lateral flow, percolation, and groundwater recharge. On the other hand, soil compaction effect on water retention characteristics is almost nonexistent at very high matrix suction (pF > 4.2) [40,41]. Water retention at a high matrix suction tends to be influenced by textural pores associated with clay content. The higher the clay content, the higher the water retained by the textural pores [42,43]. Soil compaction only changes the structural pores and does not change the textural pores. For the same soil type, although there are differences in AWC and gravitational water associated with different land management, the water content in the high suction matrix tends to be the same [41]. The modeling result proves that each SWRC in the study area tends to coincide when the suction matrix gets bigger, considering that the soil in the study area has relatively the same clay content based on soil texture data from ground survey and laboratory analysis.

The SWAT model assisted in WRES upscaling from plot scale to landscape and time-series scale. Nevertheless, before further interpretation, it is necessary to evaluate the reliability and uncertainty of the SWAT model through the NSE, R-squared, p-Factor, and r-Factor values. The objective function used during the calibration process is NSE, meaning that the value of each parameter will be optimized from its initial value until it reaches the desired NSE value. When the system calculates the NSE value, other statistical values, such as R-squared, will be adjusted automatically. In addition, evaluation of model uncertainty is vital because SWAT provides various combinations of parameters and different modules that produce one of the same outputs but produce another very different output. A good model has a p-Factor value close to 1 and an r-Factor close to 0. However, to achieve a p-Factor close to 1, it is necessary to sacrifice the r-Factor value away from 0 and vice versa. Therefore, the best simulation is defined as a balanced p-Factor and r-Factor when it reaches the highest NSE or R-squared value during the calibration and validation periods. To obtain a balanced p-Factor and r-Factor, we must arrange each model parameter's lower and upper bound.

The calibrated results are still in the reliable category, and the model's uncertainty is still acceptable, although the model's performance for the validation period is not as good as the calibration period. The high values of R2 and NSE in the calibration and validation periods indicate that the combination of calibrated parameters can capture the impact of daily meteorological input variations on daily WRES. Moriasi [29] stated that the model reliability criteria could be used for model evaluation on a monthly and daily scale. However, with the same criteria, model evaluation for the daily scale is generally more robust than the monthly one because daily output captures variations and inaccuracies arising from parameter uncertainty and daily input data (SWAT input must be daily, while the output can be daily or monthly). More accurate modeling of WRES for micro-watersheds can serve as complementary information for environmental restoration planning at a local scale. A calibrated and validated SWAT model with acceptable reliability and uncertainty was then used to evaluate WRES of forest patches among oil palm plantations.

Due to differences in soil water characteristics, WRES varies significantly between land uses and management, so changes in land use and management will impact the change in WRES at the landscape scale by routing mechanism. Meanwhile, oil palm is the dominant vegetation in the study area, so the oil palm WRES dominates the landscape-scale water balance. In addition, given the dynamic nature of oil palm plantations that are cleared and replanted every 20–30 years, it is necessary to understand oil palm WRES when they are young and mature (yielding plants). The transition from YOP to MOP occurs 8–9 years after planting, when the canopy cover reaches its maximum value. The simulation consistently shows that preserving forest patches among oil palm plantations has implications for decreasing SURQ, increasing SW, and increasing GWR. These advantages have consequences for the FP water yield, which is higher in the dry month and lower in the wet month so that water is still available in the dry season and does not overflow in the rainy season. The lower WYLD FP compared to oil palm plantations is supported by [44], which states that the response to WYLD in agroecosystems, especially oil palm plantations, tends to be higher than forest.

SURQ occurs when rainfall exceeds infiltration capacity, where lower infiltration capacity in oil palm due to soil compaction causes more rainfall to turn into SURQ. Two factors caused the higher SURQ, and more insufficient water storage in oil palm plantations: (1) decreased canopy and ground cover in YOP and (2) intensive soil compaction due to MOP harvesting. The decrease in canopy and ground cover causes the SURQ rate to be faster, time concentration to be shorter, and decreases in interception and evapotranspiration. The rough surface of the FP due to more complex canopy stratification, cover crops, and forest litter, besides lowering SURQ, also slows the SURQ rate on its way to water bodies. High infiltration capacity also increases LAT, BFO, and GWR. LAT and BFO are relatively stable WYLD parts because they have a slower rate to reach water bodies. LAT and BFO indicate the river regime's sustainability as it maintains water availability on a day without rain. An increase in SURQ and a decrease in LAT, BFO, and GWR lead to an increased risk of flooding, drought, and water scarcity. Changes in local water resources had become a significant concern for residents in the study area, including the faster shallow aquifer depletion during the dry season and high fluctuations in streamflow between the rainy and dry season [5].

High canopy cover in FP also causes soil water storage to be more compensated for the AET besides reducing SURQ. High AET increases the initial abstraction and decreases rainfall proportion for SURQ. Three factors limit AET: PET as an energy source, AWC as a water source, and stomatal conductance (correlated with LAI) as the water pump from the soil to the atmosphere. A high AET causes soil moisture to decrease faster so that the soil water changes more quickly from saturated to unsaturated conditions. Unsaturated soils have higher infiltration rates and lower SURQ rates than saturated soils, so, in this case, AET plays a role in reducing runoff. AET is also an essential element of surface energy balance related to microclimate regulation. A higher AET for the same net radiation will reduce the proportion of energy for atmosphere heating (sensible heat). The increase in air temperature in the study area is evidence of the relationship between lower AET in oil palm plantations (especially YOP) and atmospheric warming. Beside depletion in groundwater, the residents in the study area also feel the air has become much warmer since oil palm plantations have dominated the landscape [5]. In addition, the comparison of lower PET and AET in oil palm plantations (especially in the dry season) indicates that these land uses have a higher water deficit than FP.

FP has properties like a sponge and a pump in the hydrological cycle. Meanwhile, MOP only has pump properties, and YOP does not have these two properties. FP act like sponges by increasing soil water retention and releasing it slowly through LAT and BFO because it has more soil pores. Water is stored and maintained on days without rain and even remains available until the dry season through these properties. On the other hand,

FP also act like pumps by evaporating a large amount of soil water into the atmosphere. The proof that MOP only has pump properties is that MOP evaporates a large amount of water, although not as much as FP, but cannot store large amounts of soil water. YOP does not have sponge and pump properties, indicated by low AET and low soil water storage. The unavoidable soil degradation due to soil compaction is the leading cause of oil palm plantations losing their sponge properties. The loss of sponge properties due to soil compaction has three consequences: a decrease in infiltration leading to an increase in SURQ, a reduction in the GWR, and a decrease in AWC leading to a reduction in AET.

The anecdotal information that oil palm plantations are "water-greedy crops" [5,45] because they have higher AET and are associated with water scarcity is inaccurate. The MOP AET tends to be less than equal to FP [2], while YOP AET is much lower. More scientific evidence for this water scarcity case is that oil palm plantations encounter soil compaction so that more rainfall flows become SURQ than stored in soil pores. Oil palm water use is greater than the stored water because of improper management, where high AET is not accompanied by high AWC and GWR, as is the case with FP. Higher AET also reduces shallow aquifers through the capillary water movement to plant roots when soil moisture is insufficient to compensate for AET, which causes the groundwater to be dwindled and become unavailable during the dry season.

WRES sustainability is mainly determined by land use and management, which changes the soil and surface and affects rainfall distribution into SURQ, GWR, and AET [46]. According to locals' information that there is a groundwater decline during the dry season as the primary water source, the desired alternative for water management is to reduce SURQ and increase GWR. Landscape SURQ can be reduced and GWR can be increased through a multifunctional landscape, i.e., retaining the remaining FP among oil palm plantations. By maintaining FP or agroforestry as a high conservation area with an optimal area, soil and water conservation becomes a more suitable alternative for improving landscape WRES while maintaining oil palm productivity. FP with good soil structure and more complex canopy stratification enhance the WRES, so their existence in a landscape dominated by oil palm plantations is vital to maintaining the watershed's ecological integrity. In addition, the presence of FP inside oil palm plantations can synergize the provisioning and regulating services, where oil palm provides provision ecosystem services and FP provide regulation ecosystem services. The recommended multifunctional landscape is to maintain FP or create agroforestry islands separately around oil palm plantations. If oil palm is mixed with forest vegetation, there will be competition for light and water, which hinders the productivity of the entire vegetation.

Multifunctional landscapes also enhance other ecosystem services besides WRES, such as biodiversity conservation and erosion prevention [47]. Multifunctional landscapes are also an effort to adapt to the negative impacts of climate change. Changes in rainfall patterns due to climate change, where rainfall is predicted to increase in wet months and decrease in dry months causes WRES variations to become more extreme [48]. Further research is needed regarding the spatial configuration of multifunctional landscapes with the most optimum ecosystem functions and economic benefits based on ecosystem services trade-offs. Another aspect that needs to be considered is that the chosen multifunctional landscapes policies must be site-specific and examine the local biophysical characteristics of the landscape, such as soil type, topography, and elevation [49,50]. Even more broadly, they need to include social and economic factors. For example, soil type in the study area belongs to the hydrologic soil group (HSG) C based on the observation of soil permeability, where the runoff coefficient of each land use is higher than HSG A or HSG B. In addition, the topography in the study area belongs to the relatively flat areas, where the runoff coefficient is lower than the steeper slope. Environmental planners require these site-specific quantitative relationships to balance landscape-scale ecological and socio-economic functions.

5. Conclusions

Multifunctionality landscapes approach through maintaining forest patches between oil palm plantations, can improve landscape WRES, shown by a decrease in surface runoff, an increase in groundwater recharge, an increase in soil water storage, and an increase in actual evapotranspiration. As a result, water is not concentrated in the rainy season and remains available in the dry season. The forest patches can improve landscape WRES because they have good soil hydrological characteristics, so the existence of forest patches is essential to maintain the ecological integrity of the watershed. Soil hydrological characteristics in forest patches are indicated by the lowest bulk density and the highest soil porosity compared to oil palm plantations. Good soil hydrological characteristics have implications for increasing soil water retention, imply more soil water is stored and available to plants (available water content), and flows through soil pore spaces to fill aquifers and river networks (gravitational water). Meanwhile, soil compaction increases bulk density, decreases porosity, and decreases soil water retention in oil palm plantation.

The calibrated SWAT is reliable and acceptable model, shown by the high NSE and R-squared value and balanced p-Factor and r-Factor values. The SWAT model consistently proves that forest patches have sponge and pump properties in the hydrological cycle. The sponge properties are related to the optimal distribution of soil porosity so that water is stored in the rainy season and flows slowly during the dry season. Meanwhile, the pump properties are related to plant tissue, which plays a role in absorbing and returning soil water to the atmosphere to maintain microclimate stability. Mature oil palm can only evaporate large amounts of water like forest patches but cannot retain some soil water because of degraded soil pores due to soil compaction. The pump properties, which are not accompanied by the sponge properties, cause the water use to be greater than the stored water. Therefore, the multifunctional landscape approach by conserving forest patches between oil palm plantations is one approach that can improve the sustainability of oil palm plantations. The multifunctional landscape can synergize provisioning services from oil palm plantations and regulating services from forest patches.

Author Contributions: Y.K. and S.T. designed the research. Y.K. wrote the manuscript. S.T., T.J., E.D.W. and B.S. reviewed the manuscript. All authors have read and agreed to the published version of the manuscript.

Funding: This research was funded by PMDSU (Program Pendidikan Magister Menuju Doktor untuk Sarjana Unggul) scholarship from Ministry of Education, Culture, Research, and Technology, Republic of Indonesia.

Data Availability Statement: The datasets presented in the study are included in the article material; further inquiries can be directed to the corresponding author/s.

Acknowledgments: This research was supported by CRC-990 EFForTS (Collaborative Research Center 990 Ecological and Socio-economic Functions of Tropical Lowland Rainforest Transformation System) in form of access to the study site and provision of daily meteorological data.

Conflicts of Interest: The authors declare no conflict of interest.

References

1. Ewers, R.M.; Scharlemann, J.P.W.; Balmford, A.; Green, R.E. Do increases in agricultural yield spare land for nature? *Glob. Chang Biol.* **2009**, *15*, 1716–1726. [CrossRef]
2. Dislich, C.; Keyel, A.C.; Salecker, J.; Kisel, Y.; Meyer, K.M.; Auliya, M.; Barnes, A.D.; Corre, M.D.; Darras, K.; Faust, H.; et al. A review of the ecosystem functions in oil palm plantations, using forests as a reference system. *Biol. Rev.* **2017**, *92*, 1539–1569. [CrossRef] [PubMed]
3. Tarigan, S.; Wiegand, K.; Sunarti; Slamet, B. Minimum forest cover required for sustainable water flow regulation of a watershed: A case study in Jambi Province, Indonesia. *Hydrol. Earth Syst. Sci.* **2018**, *22*, 581–594. [CrossRef]
4. Sharma, S.K.; Baral, H.; Laumonier, Y.; Okarda, B.; Komarudin, H.; Purnomo, H.; Pacheco, P. Ecosystem services under future oil palm expansion scenarios in West Kalimantan, Indonesia. *Ecosyst. Serv.* **2019**, *39*, 100978. [CrossRef]
5. Merten, J.; Röll, A.; Guillaume, T.; Meijide, A.; Tarigan, S.D.; Agusta, H.; Dislich, C.; Dittrich, C.; Faust, H.; Gunawan, D.; et al. Water scarcity and oil palm expansion: Social views and environmental processes. *Ecol. Soc.* **2016**, *21*, 5. [CrossRef]

6. Tarigan, S.; Stiegler, C.; Wiegand, K.; Knohl, A.; Murtilaksono, K. Relative contribution of evapotranspiration and soil compaction to the fluctuation of catchment discharge: A case study from a plantation landscape. *Hydrol. Sci. J.* **2020**, *65*, 1239–1248. [CrossRef]
7. Millenium Ecosystem Assessment (MEA). *Ecosystem and Human Well-Being: Synthesis*; Island Press: Washington, DC, USA, 2005.
8. Bolliger, J.; Battig, M.; Gallati, J.; Klay, A.; Stauffacher, M.; Kienast, F. Landscape multifunctionality: A powerful concept to identify effects of environmental change. *Reg. Environ. Chang.* **2011**, *11*, 203–206. [CrossRef]
9. Rallings, A.M.; Smukler, S.M.; Gergel, S.E.; Mullinix, K. Towards multifunctional land use in an agricultural landscape: A trade-off and synergy analysis in the Lower Fraser Valley, Canada. *Landsc. Urban Plan.* **2019**, *184*, 88–100. [CrossRef]
10. Dexter, A.R. Soil physical quality part I: Theory, effects of soil texture, density, and organic matter, and effects on root growth. *Geoderma* **2004**, *120*, 201–214. [CrossRef]
11. van Genuchten, M.T. A closed-form equation for prediction the hydraulic conductivity of unsaturated soils. *Soil Sci. Soc. Am. J.* **1980**, *4*, 892–898. [CrossRef]
12. Wei, Z.; Zhang, B.; Liu, Y.; Xu, D. The application of a modified version of the SWAT model at the daily temporal scale and the hydrological response unit spatial scale: A case study is covering an irrigation district in the Hei River basin. *Water* **2018**, *10*, 1064. [CrossRef]
13. Arnold, J.G.; Srinivasan, R.; Muttiah, R.S.; Williams, J.R. Large area hydrologic modeling and assessment part I: Model development. *JAWRA J. Am. Water Resour. Assoc.* **1998**, *34*, 73–89. [CrossRef]
14. Neitsch, S.L.; Arnold, J.G.; Kiniry, J.R.; Williams, J.R. *Soil and Water Assessment Tool Theoretical Documentation Version 2009*; Texas A&M University: Commerce, TX, USA, 2011.
15. Dash, S.S.; Sahoo, B.; Raghuwanshi, N.S. A novel embedded pothole module for soil and water assessment tool (SWAT) improving streamflow estimation in paddy dominated catchments. *J. Hydrol.* **2020**, *588*, 125103. [CrossRef]
16. Zhang, H.; Wang, B.; Liu, D.L.; Zhang, M.; Leslie, L.M.; Yu, Q. Using an improved SWAT model to simulate hydrological responses to land use change: A case study of a catchment in tropical Australia. *J. Hydrol.* **2020**, *585*, 124822. [CrossRef]
17. Sajikumar, N.; Remya, R.S. Impact of land cover and land-use change on runoff characteristics. *J. Environ. Manag.* **2015**, *161*, 460–468. [CrossRef]
18. Wang, Y.; Jiang, R.; Xie, J.; Zhao, Y.; Yan, D.; Yang, S. Soil and water assessment tool (SWAT) model: A systematic review. *J. Coast. Res.* **2019**, *93*, 22–30. [CrossRef]
19. Zhang, D.; Lin, Q.; Chen, X.; Chai, T. Improved curve number estimation in SWAT by reflecting the effect of rainfall intensity on runoff generation. *Water* **2019**, *11*, 163. [CrossRef]
20. Gao, X.; Chen, X.; Biggs, T.W.; Yao, H. Separating wet and dry years to improve the calibration of SWAT in Barret Watershed, Southern California. *Water* **2018**, *10*, 274. [CrossRef]
21. Hawkins, R.H.; Theurer, F.D.; Rezaeianzadeh, M. Understanding the basis of the curve number method for watershed models and TMDLs. *J. Hydrol. Eng.* **2019**, *24*, 06019003. [CrossRef]
22. Arsyad, S. *Soil and Water Conservation*; IPB Press: Bogor, ID, USA, 2009.
23. Kristanto, Y.; Tarigan, S.D.; June, T.; Wahjunie, E.D. Evaluation of different runoff curve number (CN) approaches on water regulation ecosystem services assessment in intermittent micro catchment dominated by oil palm plantation. *Agromet* **2021**, *35*, 73–88. [CrossRef]
24. Allen, R.G.; Pereira, L.S.; Raes, D.; Smith, M. Crop evapotranspiration-guidelines for computing crop water requirements. In *FAO Irrigation and Drainage Paper 56*; Food and Agriculture Organization: Rome, IT, USA, 1998.
25. Dash, S.S.; Sahoo, B.; Raghuwanshi, N.S. How reliable are the evapotranspiration estimates by soil and water assessment tool (SWAT) and variable infiltration capacity (VIC) models for catchment-scale drought assessment and irrigation planning? *J. Hydrol.* **2021**, *592*, 125838. [CrossRef]
26. Dile, Y.T.; Karlberg, L.; Srinivasan, R.; Rockstrom, J. Investigation of the curve number method for surface runoff estimation in tropical regions. *JAWRA J. Am. Water Resour. Assoc.* **2016**, *52*, 1155–1169. [CrossRef]
27. Abbaspour, K.C. *SWAT-CUP: SWAT Calibration and Uncertainty Programs—A User Manual*; Swiss Federal Institute of Aquatic Science and Technology: Dübendorf, CH, USA, 2015.
28. Muleta, M.; Nicklow, J. Sensitivity and uncertainty analysis coupled with automatic calibration for a distributed watershed model. *J. Hydrol.* **2005**, *306*, 127–145. [CrossRef]
29. Moriasi, D.N.; Arnold, J.G.; van Liew, M.W.; Bingner, R.L.; Harmel, R.D.; Veith, T.L. Model evaluation guidelines for systematic quantification of accuracy in watershed simulation. *Trans. ASABE* **2007**, *50*, 885–900. [CrossRef]
30. Ayele, G.T.; Teshale, E.Z.; Yu, B.; Rutherfurd, I.D.; Jeong, J. Streamflow and sediment yield prediction for watershed prioritization in the Upper Blue Nile River basin Ethiopia. *Water* **2017**, *9*, 782. [CrossRef]
31. Singh, V.; Bankar, N.; Salunkhe, S.S.; Bera, A.K.; Sharma, J.R. Hydrological stream flow modeling on Tungabhadra catchment: Parameterization and uncertainty analysis using SWAT CUP. *Curr. Sci.* **2013**, *104*, 1187–1199.
32. Jung, I.K.; Park, J.Y.; Park, G.A.; Lee, M.S.; Kim, S.J. A grid-based rainfall-runoff model for flood simulation including paddy fields. *Paddy Water Environ.* **2011**, *9*, 275–290. [CrossRef]
33. van Griensven, A.; Meixner, T.; Grunwald, S.; Bishop, T.; Diluzio, M.; Srinivasan, R. A global sensitivity analysis tool for the parameters of multi-variable catchment models. *J. Hydrol.* **2005**, *324*, 10–23. [CrossRef]

34. Nanganoa, L.T.; Okolle, J.N.; Missi, V.; Tueche, J.R.; Levai, L.D.; Njukeng, J.N. Impact of different land-use system on soil physicochemical properties and macrofauna abundance in the humid tropics of Cameroon. *Appl. Environ. Soil Sci.* **2019**, *2019*, 5701278. [CrossRef]
35. Arthur, M.D.; Asamoah, E.F. Soil compaction under three different land use systems within the semi-deciduous agroecological zone of Ghana. *Int. J. Plant Soil Sci.* **2016**, *13*, 1–12. [CrossRef]
36. Liu, X.; Herbert, S.J.; Hashemi, A.M.; Zhang, X.; Ding, G. Effects of agricultural management on soil organic matter and carbon transformation-a review. *Plant Soil Environ.* **2006**, *52*, 531–543. [CrossRef]
37. Khormali, F.; Ajami, M.; Ayoubi, S.; Srinivasarao, C.; Wani, S.P. Role of deforestation and hillslope position on soil quality attributes of loess-derived soils in Golestan province, Iran. *Agric. Ecosyst. Environ.* **2009**, *134*, 178–189. [CrossRef]
38. Hamza, M.A.; Anderson, W.K. Soil compaction in cropping systems: A review of the nature, causes, and possible solutions. *Soil Tillage Res.* **2005**, *82*, 121–145. [CrossRef]
39. Liu, C.; Tong, F.; Yan, L.; Zhou, H.; Hao, S. Effects of porosity on soil-water retention curve: Theoretical and experimental aspects. *Geofluids* **2020**, *2020*, 6671479. [CrossRef]
40. Dorner, J.; Sandoval, P.; Dec, D. The role of soil structure on the pore functionality of an ultisol. *J. Soil Sci. Plant Nutr.* **2010**, *10*, 495–508. [CrossRef]
41. Fashi, F.H.; Gorji, M.; Shorafa, M. Estimation of soil hydraulic parameters for different land-uses. *Modeling Earth Syst. Environ.* **2016**, *2*, 170. [CrossRef]
42. Gupta, S.C.; Sharma, P.P.; de Franchi, S.A. Compaction effects on soil structure. *Adv. Agron.* **1989**, *42*, 311–338.
43. Assouline, S. Modeling the relationship between soil bulk density and the water retention curve. *Vadose Zone J.* **2006**, *5*, 554–563. [CrossRef]
44. Goeking, S.A.; Tarboton, D.G. Forest and water yield: A synthesis of disturbance effects on streamflow and snowpack in western coniferous forests. *J. For.* **2020**, *118*, 172–192. [CrossRef]
45. Manoli, G.; Meijide, A.; Huth, N.; Knohl, A.; Kosugi, Y.; Burlando, P.; Ghazoul, J.; Fatichi, S. Ecohydrological changes after tropical forest conversion to oil palm. *Environ. Res. Lett.* **2018**, *13*, 064035. [CrossRef]
46. Ouyang, L.; Liu, S.; Ye, J.; Liu, Z.; Sheng, F.; Wang, R.; Lu, Z. Quantitative assessment of surface runoff and base flow response to multiple factors in Pengchongjian small watershed. *Forest* **2018**, *9*, 533. [CrossRef]
47. Tarigan, S.; Buchori, D.; Siregar, I.Z.; Azhar, A.; Ullyta, A.; Tjoa, A.; Edy, N. Agroforestry inside oil palm plantation for enhancing biodiversity-based ecosystem functions. In Proceedings of the International e-Conference on Sustainable Agriculture and Farming System, Bogor, Indonesia, 24–25 September 2020; Volume 694, p. 012058.
48. Tarigan, S.; Kristanto, Y. Assessment of water security in Indonesia considering future trends in land use change and climate change. In *Water Security in Asia*; Babel, M., Haarstrick, A., Ribbe, L., Shinde, V.R., Dichtl, N., Eds.; Springer: Cham, Switzerland, 2021.
49. Tarigan, S.; Wiegand, K.; Dislich, C.; Slamet, B.; Heinonen, J.; Meyer, K. Mitigation options for improving the ecosystem function of water flow regulation in a watershed with rapid expansion of oil palm plantation. *Sustain. Water Qual. Ecol.* **2016**, *8*, 4–13. [CrossRef]
50. Tarigan, S.; Zamani, N.P.; Buchori, D.; Kinseng, R.; Suharnoto, Y.; Siregar, I.Z. Peatlands are more beneficial if conserved and restored than drained for monoculture crops. *Front. Environ. Sci.* **2021**, *9*, 749279. [CrossRef]

Article

Topsoil Seed Bank as Feeding Ground for Farmland Birds: A Comparative Assessment in Agricultural Habitats

Aikaterini Voudouri, Evgenia Chaideftou and Athanassios Sfougaris *

Crop Production and Rural Environment, Laboratory of Ecosystems and Biodiversity Management, Department of Agriculture, University of Thessaly, Fytokou Str., N. Ionia, GR-38446 Volos, Greece; k.voudouri@hotmail.com (A.V.); eugeniachd@gmail.com (E.C.)
* Correspondence: asfoug@agr.uth.gr

Abstract: The topsoil seed bank was studied in four types of agricultural bird habitats: fields with cereals, maize, clover and tilled fields of a Mediterranean plain to determine the potentially richest habitat based on food supply for the wintering farmland birds. The diversity and abundance of topsoil seeds differed between seasons but did not differ significantly between habitats. The cereal habitat was the richest in food supply for the overwintering of farmland birds. The topsoil seed bank was dominated by *Chenopodium album*, *Polygonum aviculare* and *Amaranthus retroflexus*. The findings of this study provide insight for low-intensity management of higher-elevation mount agricultural areas of southern Mediterranean by preserving seed-rich habitats for farmland avifauna.

Keywords: topsoil seed bank; farmland bird diet; agricultural ecosystem; biodiversity; habitat

Citation: Voudouri, A.; Chaideftou, E.; Sfougaris, A. Topsoil Seed Bank as Feeding Ground for Farmland Birds: A Comparative Assessment in Agricultural Habitats. *Land* **2021**, *10*, 967. https://doi.org/10.3390/land10090967

Academic Editors: Chiara Piccini and Rosa Francaviglia

Received: 13 August 2021
Accepted: 3 September 2021
Published: 14 September 2021

Publisher's Note: MDPI stays neutral with regard to jurisdictional claims in published maps and institutional affiliations.

Copyright: © 2021 by the authors. Licensee MDPI, Basel, Switzerland. This article is an open access article distributed under the terms and conditions of the Creative Commons Attribution (CC BY) license (https://creativecommons.org/licenses/by/4.0/).

1. Introduction

The management of agricultural land has greatly changed over recent decades. This has resulted in different physiognomy and a reduction of agricultural biodiversity and heterogeneity [1–4]. The depletion of the natural transient soil seed bank during cultivation was one of the changes (e.g., [5]). Shifts in agricultural management also led to decline of rural birds [6–13].

Mosaics of low-intensity cultivation in the Mediterranean areas may preserve high diversity of bird species, but intensification or land abandonment probably do not benefit biodiversity [4,14]. Reviews identified that agricultural intensification [15] and concomitant abandonment [16] remain the major threat to agricultural ecosystems of the 21st century across Europe and elsewhere (e.g., [17]) with many ecological and biodiversity impairments. As a consequence, investigation of floristic and seed diversity and abundance, along with the physiognomy of the rural landscape, is necessary for identifying the most interesting bird habitats. These habitat features, i.e., high quality or food resources or aboveground floristic components like stubbles or semi-natural with natural habitats suitable for breeding (low intensity farmland with steppe-like vegetation), which can be proved to be beneficial for birdlife per case facilitate biodiversity maintenance [6,18].

Approximately 30% of the bird species being "Species of European Conservation Concern" exploit agricultural ecosystems [19,20]. In rural landscape across Europe, food availability (plant and seed food items) is reduced especially during winter [3,21], and nesting habitats in spring are deficient for many bird species that have declined over recent decades [22]. For instance, seed-eating birds face the risk of limited accessibility to preferred seeds when vegetation is dense in uncultivated areas close to farmlands [23] or they feed in stubble of reduced quality due to modernized techniques in cereals harvested as arable silage [24].

Especially in winter the most important food resource in the rural landscape for birds is the soil seed bank (e.g., [3,25]). However, research has been focused on the impacts of agricultural practices on the seed bank composition, abundance and vertical distribution of weeds in the rural landscape regarding their persistence (e.g., [26]), fertilization (e.g., [27]) crop rotation or varied tillage systems (e.g., [28,29]). As a result, the seed bank fraction on topsoil of agricultural habitats that serves as food resource to bird seed-eaters is rarely, if at all, studied.

Food resource provision [30] is crucial for wintering seed-eaters especially them of a high conservation interest. However, it is less studied in the Mediterranean regions compared to northern Europe. For conservation of bird populations which are exclusively or partially seed foragers, such as *Passer montanus* (Eurasian tree sparrow), *P. domesticus* (house sparrow), *Fringilla coelebs* (common chaffinch), *Carduelis chloris* (European greenfinch), *C. carduelis* (European goldfinch), *Miliaria calandra* (corn bunting), *Turdus merula* (common blackbird) and *Emberiza* spp. (bunting birds), a precise "instruction" of the most proper habitat, i.e., seed-rich in winter [31], is not yet defined [32].

A seed-eating bird may have preference on the seeds of particular plant species for their diet [3,22] thus seed consumption can cause seed limitation of these particular species. As a result, determination of the value a habitat carries to supply winter food to birds [33] is critical in decision-making for maintaining diversity in agricultural ecosystems [21,34,35]. Therefore, there is scope for further consideration of how we manage areas of former traditional low-intensity agriculture in Europe, supported by European subsidies [24], and principally seed-rich habitats such as cereal stubbles in certain seasons of the year (e.g., [36,37]).

This study aimed at determining the potentially richest habitat (for food supply) for the wintering of rural avifauna in the Dolichi plain of Elassona region. To this aim, the study investigated the effect of the type of crop (habitat for birds) on the topsoil seed bank in four arable fields undergoing post-dispersal consumption of seeds by farmland birds. This topsoil seed bank entails a fraction that serves as food source especially to the seed-eaters.

2. Materials and Methods

2.1. Study Area

The study area covers an area of 40 km^2, where the settlement of Dolichi and the municipality of Livadi are situated. Livadi is located at an altitude of 1100 m. Dolichi lies 5 km from the foothills of Mount Olympus and 21 km from Elassona, at an altitude of 590 m asl. The inhabitants of Dolichi, numbering 473 (in 2001) are principally involved in land cultivation and animal husbandry.

The settlement of Dolichi is located in the center of a cultivated plain and is surrounded by hilly and mountainous natural ecosystems. The main crops are cereals, with wheat (*Triticum* sp.) dominating over the barley (*Hordeum* sp.). Previously, the second largest crop was tobacco, but due to the regime of European subsidies tobacco cultivation was abandoned, therefore the second rank is currently held by maize crops (*Zea mays*). Legumes and vegetables are cultivated on a smaller scale. Clover (*Trifolium* sp.) in particular is cultivated in a mixed agricultural-livestock farming system. In 2005, vetch crops exceptionally dominated due to compulsory crops rotation in line with the Codes of Good Agricultural Practice. Ecosystems with a high plant cover such as grasslands, hedgerows, uncultivated vegetation strips and riparian zones are largely present in the area. The cultivation system is mechanized, but clearly less intensive than the one of the main plain of Thessaly. In the area many plantations of locust (*Robinia pseudoacacia*) have been established through subsidies of the agri-environmental measure "Afforestation of agricultural land".

The landscape characteristics in this area were appropriate for the research. Approximately two thirds (2/3) of the study area are covered by anthropogenic ecosystems and only one third by natural ecosystems. Among natural ecosystems, grasslands (including ecotones) hold the largest area. Cereals comprise 90% of anthropogenic ecosystems while legumes only 5%. The remaining 5% is covered by other types of agricultural ecosystems.

In this area of study, fields are undergoing post-dispersal consumption of seeds by farmland birds, having thus a potential as feeding sources to them, with non-cultivated and cultivated plant species ([38]; Table A1 of Appendix A). The study took place in four selected habitat types which actually are arable cropped fields: cereals, maize, clover and tillage, with average surface area of 2.35 ± 0.3 ha ([38]; Table A2 of Appendix A). According to [38], within this area the above-ground non-cultivated species richness and (%) plant cover differed from the respective ones of plant species serving as food sources to farmland birds among the four studied habitats (Table A2). This is not true though for the field physiognomy (see Table A2).

The avian diversity in this area was also proper for the research aims. Overall, 33 bird species were recorded in the study area (unpublished data, see Table A3 in Appendix A). A high majority, that is, 26 bird species comprising 79% of the recorded bird species in the study area, are classified as seed-eaters and all birds listed in Table A3 are present in all studied habitat types (namely the crop types: cereals, maize, clover and tillage).

2.2. Research Design

The sampling area is shown in Figure 1. Four types of fields were sampled and analyzed between September 2006 and March 2007 in the current study: cereals, maize, clover and tillage (bare soil during winter). Since these fields include 'micro-sites' providing refuge and food resources to rural bird populations, they are referred to here as "habitats". There were crops and stubbles in the maize and cereal fields during winter.

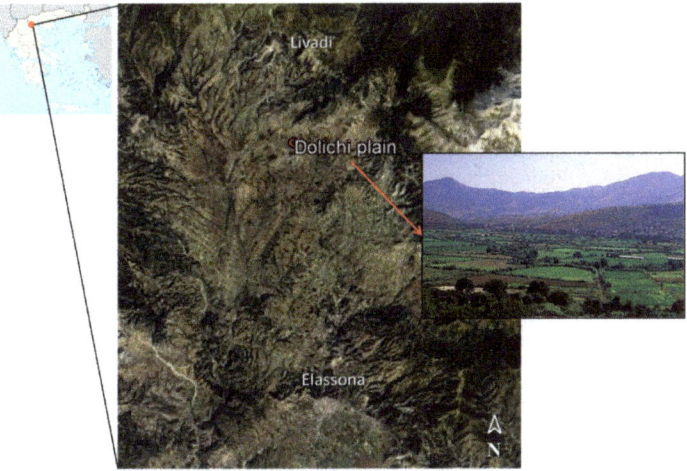

Figure 1. Map of Dolichi plain in the area of Elassona (study area). Image and photo sources: The blank map of Greece (on the top left) is by Lencer, CC BY-SA 3.0, https://commons.wikimedia.org/w/index.php?curid=4432468. The satellite map image is extracted and edited from Google Earth (https://earth.google.com/) on 22 November 2017. The landscape photo of the Dolichi plain is taken from [38].

Plots of approximately 20 m^2 each were randomly chosen so that the major species are represented in the cultivated area. A total of 36 plots, in total 846.7 m^2, were recorded and sampled. The number of studied plots (n) per crop type was: n = 10 in winter cereal fields (197.5 m^2), n = 10 in maize fields (274.5 m^2), n = 6 in clover fields (171.5 m^2) and n = 10 in tillage (203.2 m^2) (Table A4).

2.3. Topsoil Seed Bank Sampling, Seed Extraction and Identification

During fall-end and winter-start of 2006, within 21 plots (cereal n = 6, maize n = 6, clover n = 3 and tillage n = 6) randomly selected out of the total of 36, soil cores were sampled (soil corer of 15 cm diameter, and 1 cm depth). In randomly selected plots (at every second measurement of plant cover), 10 soil samples were collected (R = 10) across the diagonal of the plot and were placed into encoded polyethylene bags that were transferred to the laboratory. The second soil core sampling took place in spring of 2007 following exactly the same protocol, though the number of plots differed due to the seasonal change of landscape in Dolichi plain. A total of 12 plots were totally studied in spring: cereal (n = 4), maize (n = 2), clover (n = 3) and tillage (n = 3). Seeds up to one centimeter of soil depth have been sampled. Therefore, the seeds and seed bank are referred in this article as 'topsoil'.

Soil core samples were retained at 4 °C for 24 h to avoid seed germination. Then seeds were isolated from soil phase using sieves and were identified at species level in petri dishes under stereoscope and magnifying lens using a series of keys (Appendix B). In addition, plant specimens were collected in spring and autumn of 2006 so that all plant species the seeds of which are potentially present in the topsoil seed bank of autumn 2007, are included in a reference plant collection that facilitated the seed identification (sources are listed in Appendix B). Classification of plant species on the basis of their significance to the farmland bird diet is presented in Tables A1 and A5 of Appendix A.

2.4. Data Analysis

The species richness of the seed bank was estimated as the number of species per m^2. The Shannon index (entropy) was also estimated for the topsoil seed bank. The seed abundance of the topsoil seed bank was estimated as the average number of seeds per square meter (m^2).

The Shannon index (entropy) and the seed abundance of the topsoil seed bank were tested by estimating the differences between habitats (cereal, maize, clover and tillage) and seasons (winter and spring) using generalized linear mixed (GLM) effect models. For Shannon index, the GLM for Gaussian family with random intercept of plot were used. For seed abundance, the GLM for Poisson family (with log link function) with random intercept of plot were employed. For model selection, model with and without interaction between season and habitat were compared with simple generalized linear model (for Gaussian family in the case of Shannon Index; for Poisson family (with log link function) in the case of seed abundance). In each case, all three models were compared using p-value of ANOVA and Akaike Information Criterion (AIC). Signs of heteroscedasticity (residuals vs. fitted plot) and normality of residuals (q-q plot) were also tested for identifying the best performing model. Overdispersion of the Poisson model was also checked and, if needed, the analysis was reconducted using generalized linear (simple or mixed effect) model with negative-binomial family with log link function.

Predictions were generated with and without inclusion of random effects. The 95% confidence intervals were estimated with bootstrapped simulation (n = 1000) using the bootMer function. Post-hoc (Tukey all-pairs) comparisons were conducted.

The data were processed in R 4.1.0 [39] using the packages: broom.mixed [40], dplyr [41], ggplot2 [42], lme4 [43], lmerTest [44], MASS [45], multcomp [46], and tidyr [47].

3. Results

3.1. Composition of the Topsoil Seed Bank as Food Source to Farmland Birds

Overall, the soil seed bank sampling and seed identification resulted in a total of 66 plant species, 64 of which are non-cultivated and the other two are the cultivated species *Triticum aestivum* and *Zea mays*. Out of these 66 plant species, 49 species were classified as having a level of significance as food items to the farmland birds (Table A5).

Soil seed bank sampling during winter in 36 fields resulted in a total of 62 plant species, 15 of which were present in all habitats while the soil seed bank sampling during spring in 12 fields resulted in 39 identified species. A total of 26 and 21 species of seeds serving as food source to farmland avifauna were identified in winter and spring topsoil seed bank, respectively (Table A5).

The highest number of species with significance as food sources to birds was recorded in cereals (34 species) while the lowest in tillage (23 species) (see Table A5). The habitats can be ranked on the basis of the number of significant species in serving as food sources to birds as follows: tillage (23) < clover (25) < maize (31) < cereal (34).

The winter seed bank was dominated in all studied habitats by *Chenopodium album*, *Polygonum aviculare* and *Amaranthus retroflexus*. The last two species also dominated the spring seed bank. Apart from these two species, the following species predominated in the spring seed bank: *Lithospermum arvense* in cereals, *Amaranthus albus* in maize, *Echinochloa crus-galli* in clover and *Digitaria sanguinalis* in tillage (Table A5).

Commonly, 35 species were present in both spring and winter while only seven species (*Brassica juncea*, *Sinapis arvensis*, *Silene dioica*, *Chenopodium album*, *Polygonum aviculare*, *Portulaca oleracea*, *Amaranthus retroflexus*) were present in all habitats in both seasons (Table A5).

These common and dominant plant species, with the exception of *Amaranthus* sp. and *P. oleracea*, are classified to highest significance as food sources to farmland birds (Table A5).

3.2. Shannon Entropy and Seed Abundance of the Topsoil Seed Bank

3.2.1. Model Selection

Mixed effect model without interaction did not differ significantly from model with interaction for both seed abundance and Shannon index (entropy). However, it differed in both cases significantly from model without random intercept. The model without interaction also performed best in terms of AIC in the case of seed abundance, and similar to model with interactions in the case of Shannon entropy. Visual inspection showed no clear signs of heteroscedasticity nor deviation from normality of residuals.

However, in the case of seed abundance, the model showed significant overdispersion (dispersion ratio = 12.97, Pearson's Chi-Squared = 2931.16, $p < 0.001$). Thus, instead of using Poisson family model, negative binomial family was used for model estimation. The new model also performed better in case of AIC (Tables 1 and 2). Visual inspection showed that normality of residuals was slightly worse than in Poisson family model.

Table 1. Generalized Linear Model comparison for Shannon entropy and seed abundance in the topsoil seed bank.

Variable contrast	npar	logLik	deviance	Chisq	Df	Pr (>Chisq)
Shannon Entropy						
season + habitat	6	−127.29	254.57			
season + habitat + (1 | plots)	7	−121.41	242.83	11.7466	1	0.0006095 **
season * habitat + (1 | plots)	10	−118.27	236.53	6.2956	3	0.0980824
Seed Abundance						
season + habitat	6	−1099.8	2199.5			
season + habitat + (1 | plots)	7	−1054.5	2109	90.5848	1	$<2 \times 10^{-16}$ **
season * habitat + (1 | plots)	10	−1053.5	2107.1	1.8701	3	0.5998

+ indicates it is additive model (y = spring + winter + tillage + clover + maize + cereals). * indicates there is interaction between habitat and season (y = spring + winter + tillage + clover + maize + cereals + spring * tillage + spring * clover + spring * maize + spring * cereals + winter * tillage). ** indicates statistical significance, $p < 0.05$.

Table 2. Generalized Linear Model comparison for Shannon entropy and seed abundance in the topsoil seed bank on the basis of AIC.

	Shannon Entropy		
Variable contrast		df	AIC
season + habitat		6	2211.542
season + habitat + (1 | plots)		7	2122.957
season * habitat + (1 | plots)		10	2127.087
	Seed Abundance		
Variable contrast		df	AIC
season + habitat		6	2211.542
season + habitat + (1 | plots)		7	2122.957
season * habitat + (1 | plots)		10	2127.087

+ indicates it is additive model. * indicates there is interaction between habitat and season.

The calculated parameters of the model for Shannon entropy and seed abundance are summarized in Table 3.

Table 3. Model summary for Shannon entropy and seed abundance.

				Shannon Entropy				
Effect	Term	Estimate	Std. Error	Statistic	df	p Value	Conf. Low	Conf. High
fixed	(Intercept)	4.713	0.442	16.516	34.476	6.01×10^{-18}	3.921	5.665
fixed	Season winter	0.874	0.080	−1.462	38.114	0.152	0.731	1.047
fixed	Habitat Clover	1.019	0.121	0.157	27.754	0.876	0.807	1.287
fixed	Habitat Maize	0.877	0.095	−1.211	26.078	0.236	0.709	1.085
fixed	Habitat Tillage	0.861	0.096	−1.341	32.468	0.189	0.692	1.071
				Seed Abundance				
Effect	Term	Estimate	Std. Error	Statistic		p Value	Conf. Low	Conf. High
fixed	(Intercept)	72.366	18.403	16.837		1.30×10^{-63}	43.962	119.123
fixed	Season winter	0.434	0.106	−3.415		0.001	0.268	0.700
fixed	Habitat Clover	1.444	0.485	1.093		0.275	0.747	2.789
fixed	Habitat Maize	1.061	0.327	0.190		0.849	0.579	1.942
fixed	Habitat Tillage	0.600	0.182	−1.679		0.093	0.331	1.089

3.2.2. Shannon Entropy of the Topsoil Seed Bank

The Shannon entropy of the topsoil seed bank did not differ significantly between seasons or between habitats (Table 4).

Table 4. Tukey all-pairs comparisons for Shannon entropy of the topsoil seed bank.

		Post-Hoc Comparisons				
Term	Contrast	Null. Value	Estimate	Std. Error	Statistic	Adj. p Value
Season	Winter—Spring	0	−0.134	0.092	−1.461	0.144
Habitat	Clover—Cereals	0	0.019	0.119	0.157	0.999
	Maize—Cereals	0	−0.131	0.108	−1.211	0.619
	Tillage—Cereals	0	−0.149	0.111	−1.342	0.535
	Maize—Clover	0	−0.150	0.126	−1.188	0.633
	Tillage—Clover	0	−0.168	0.128	−1.311	0.554
	Tillage—Maize	0	−0.018	0.117	−0.154	0.999

In spring, the estimated Shannon entropy was higher in cereal than in the other three habitats; while in maize it was the lowest (Table 5). However, these differences were not statistically different.

Table 5. Model predictions compared to Shannon entropy (mean and median) of the topsoil seed bank.

Habitat	Season	Mean Shannon Entropy	Median Shannon Entropy	Prediction Adjusted	Predicted Unadjusted
Cereals	spring	1.69	1.63	1.63	1.55
Cereals	winter	1.34	1.39	1.37	1.56
Clover	spring	1.53	1.58	1.56	1.42
Clover	winter	1.47	1.63	1.46	1.40
Maize	spring	1.08	1.20	1.29	1.42
Maize	winter	1.35	1.37	1.32	1.43
Tillage	spring	1.43	1.44	1.41	1.28
Tillage	winter	1.21	1.31	1.22	1.27

Table 5 presents the model predictions, both unadjusted (not including random effect) and adjusted (including random effect), compared to the original-data mean and median.

3.2.3. Seed Abundance of the Topsoil Seed Bank

Significant differences in the seed abundance of the topsoil seed bank were observed only between seasons (winter-spring; see Post-Hoc comparisons in Table 6). In tillage the seed abundance was lower than in the other three habitats, and this difference was more pronounced with clover (Table 7). It is noted, however, that these differences in the seed abundance between habitats were insignificant and this is also confirmed by the Post-Hoc comparisons (Table 6).

Table 6. Tukey all-pairs comparisons for seed abundance of the topsoil seed bank.

		Post-Hoc Comparisons				
Term	Contrast	Null. Value	Estimate	Std. Error	Statistic	Adj. p Value
Season	Winter—Spring	0	−0.835	0.245	−3.415	0.0006 *
Habitat	Clover—Cereals	0	0.367	0.336	1.093	0.693
	Maize—Cereals	0	0.059	0.309	0.190	0.997
	Tillage—Cereals	0	−0.510	0.304	−1.679	0.334
	Maize—Clover	0	−0.308	0.356	−0.867	0.821
	Tillage—Clover	0	−0.877	0.350	−2.505	0.059
	Tillage—Maize	0	−0.569	0.320	−1.775	0.284

* indicates statistical significance, $p < 0.05$.

Table 7. Model predictions compared to seed abundance (mean and median) of the topsoil seed bank.

Habitat	Season	Mean Seed Abundance	Median Seed Abundance	Prediction Adjusted	Predicted Unadjusted
Cereals	spring	89.7	87	87.7	72.4
Cereals	winter	41.5	28	39.6	31.4
Clover	spring	96.5	90	96.9	104.5
Clover	winter	52.7	45	51.9	45.3
Maize	spring	125.3	104.5	111.1	76.7
Maize	winter	35.1	26	34.5	33.3
Tillage	spring	37.2	37	38.6	43.4
Tillage	winter	22.1	12	19.7	18.8

Table 7 shows the model predictions, both unadjusted (not including random effect) and adjusted (including random effect), compared to the original-data mean and median.

4. Discussion

4.1. Effect of Habitat (Crop) Type on the Topsoil Seed Bank

The differences in Shannon entropy of the topsoil seed bank were insignificant for the tested samples of this study. The effect of habitat (crop) type in such landscapes require, to our view, further coordinated research that would also include samples of different size, and consideration of a soil-property matrix [29].

The winter and spring Shannon entropy were lower in tillage though not significantly. This difference, although insignificant, could be explained by the widespread practice of autumn tillage which buries the surface seeds [48] thus reducing their availability in fields [49]. Tillage techniques prevent vegetation growth, seed germination and seedling growth [50] and temporarily enrich the topsoil with seeds [51]. Topsoil seeds are easily depleted from the soil surface also because they are consumed by high numbers of birds using stubbles as feeding grounds [52]. Moreover, in all studied fields where tillage was employed the seed abundance of dominant species was very high, as tillage decrease seed diversity [53]. [29] stressed the significance of such practices to preserve biodiversity in crop fields, and the complexity of it, as continuous tillage was found to have increased the soil seedbank diversity and density under specific soil conditions.

The winter topsoil seed bank is dominated only by *Chenopodium album*, *Polygonum arviculare* and *Amaranthus retroflexus* that have long-lived seeds according to [54]. By contrast, the spring topsoil seed bank reveals a different picture since apart from *Chenopodium album*, and *Amaranthus retroflexus*, four other species are prominently present: *Lithospermum arvense*, *Amaranthus albus*, *Echinochloa crus-galli* and *Digitaria sanguinalis* in cereal, maize, clover and tillage respectively. The above species, apart from *Lithospermum arvense*, are also reported to have seeds of high longevity [54]. However, note that the seeds of *Echinochloa crus-galli* found in clover and *Digitaria sanguinalis* found in tillage in our case, are classified as very-short lived for untilled systems by [55].

Since in this study only the topsoil seed bank has been investigated, no conclusion on seed persistence per species under heavy disturbance can be given. Regarding seed availability as food sources to farmland birds it should be considered that in no- or low-tillage fields where the soil is not heavily disturbed, seed predation is enhanced [56,57].

In this study, the genus Amaranthus dominated [58] maize crops. *Chenopodium album* dominated the topsoil seed bank of all habitats either in spring or in winter. This is consistent with the findings of [59], who detected high seed abundances of this annual broadleaved species in the upper 5 cm of soil irrespective of barley tillage treatments in Alaska, as well as of other authors [58,60]. *Polygonum arviculare* was dominant in winter in the topsoil seed bank, implying that autumn tillage did not bury the bulk of its surface seeds. [59] detected higher seed density of *Polygonum arviculare*, only for medium-intensity tillage treatments (disc once) in the upper 5 cm of soil.

Ref. [61] found in fields of Poland that the base of the winter diet of reed bunting *Emberiza schoeniclus* are seeds of *Chenopodium album*, *Amaranthus retroflexus*, *Setaria viridis*, *Stellaria media* and *Fumaria officinalis*. It should be underlined that the aforementioned differences in dominant seeds are consistent with the high spatial variability of seeds predated, such as *Chenopodium album*, given that some birds of the study area may count on alternate food resources, have preference to specific species and respond differently to different plant cover [62] and landscape composition in winter [35].

Species present in the aboveground flora and linked to disturbance in agricultural soils are *Amaranthus* spp., *Chenopodium album* and *Echinochloa crus-galli*, while *Digitaria sanguinalis*, of which seeds were dominant in clover, are linked to undisturbed soils [54] and these species have over 3-year seed longevity [54].

4.2. Agricultural Habitats with a Topsoil Seed Bank Serving as Food Source to Farmland Birds

A total of 26 and 21 species of seeds serving as food source to farmland avifauna were identified in winter and spring, respectively. Differences between spring and winter seed abundance are mostly attributed to seed consumption of species with high significance to farmland bird diet in this study. The farmland birds use stubbles more frequently in winter [63,64]. The main food source of seed-eating birds during winter is the soil seed bank [3,22,25,65].

Cereal seeds would rather show a higher potential to positively impact rural bird diversity in the studied landscape, while the structural characteristics in clover habitats might also favor birds' presence, but these are objectives of future study in a more systems' thinking approach, beyond single-farm scales [21]. [66] underlined the importance of features like hedgerows in diversifying habitats associated with many farmland bird benefits. [67] proposed that the best option for birds in winter are the seed-rich habitats while in the summer structurally and floristically rich habitats. The results of this study would rather support the findings of [18] that highlighted the importance of the presence of suitable breeding habitats in mixed landscape for farmland birds. Furthermore, in our case, whether differences between spring and winter seed abundance are attributed to seed consumption of species with high significance to farmland bird diet needs further investigation. From this viewpoint, more thorough investigation of the relationship between the richness and abundance of bird fauna and the respective parameters of seeds is necessary in the future.

Seed-eating birds are important topsoil-seed consumers inferring quantitative and qualitative changes in soil seed bank, especially in winter when plants serving as food sources to avifauna are highly reduced, compared to spring in the same fields [65]. Conversely, seed predators also have a determining role in plant communities at landscape level by impacting the abundance of specific plants of their preference, thus inferring floristic variations even at areas that are distanced in the landscape [68]. Consequently, it could not be disregarded that reduced seed availability can be a limiting factor to wintering birds, a fact that highlights the importance of interspecific competition of avian communities [69]. As such, neither the preference of seed foragers for seeds of varied seed sizes of specific annual plant species at landscape patches is to be overlooked in current and future agro-ecological management [70] nor the importance of the minimum distance of available food-resource patches in the rural landscape [31].

In this respect, as [71] supported, the landscape heterogeneity may benefit generalist birds but may mean habitat loss and fragmentation for specialists, and therefore management should not include unique standalone measures. Fragmentation and land-use changes in rural landscape also influence the soil seed bank in terms of size and composition [72]. These, and the current study findings, highlight the importance of habitat provision for farmland birds during winter and breeding seasons [21].

Author Contributions: Conceptualization, A.V. and A.S.; methodology A.V. and A.S.; software, A.V. and E.C.; validation, A.V., E.C. and A.S.; formal analysis, E.C.; data curation, E.C.; writing—original draft preparation, A.V., E.C. and A.S.; writing—review and editing, E.C. and A.S.; visualization, E.C.; supervision, A.S. All authors have read and agreed to the published version of the manuscript.

Funding: This research received no external funding.

Data Availability Statement: Data available on request from the authors.

Acknowledgments: The authors would like to particularly acknowledge the contribution of Martin Hermy, at KU Leuven, for his critical review comments on a previous version of this manuscript. Special appreciation to Michał Czyż, at Coding Manatee Ninja, for advice and verification of the data analyses conducted. We also wish to thank the reviewers for their improvements on a previous version of this manuscript.

Conflicts of Interest: The authors declare no conflict of interest.

Appendix A

Table A1. Presence (+), total species richness and phenology of 61 herb-layer non-cultivated species of 25 families (11 grasslike species and 50 broadleaf (4 legumes and 46 forbs) herb-layer species) in Cereals, Maize, Clover, and Tillage in Dolichi plain during the growing season of 2006 (from [38]).

Family	Plant Species	Phenology *		Habitat			
		Life-Cycle	BG (Life-Form)	Cereals	Maize	Tillage	Clover
Poaceae	*Alopecurus myosuroides*	A	G (The)	+			
	Avena spp.	A	G (The)	+			
	Bromus tectorum	A	G (The)				+
	Bromus spp.	A	G (The)	+			
	Cynodon dactylon	P	G (The/Geo/Hem)	+	+	+	
	Hordeum murinum	A	G		+		+
	Lolium multiflorum	A	G (The)	+			
	Lolium rigidum	A	G	+	+		+
	Phalaris brachystachys	A	G	+			
	Sorghum halepense	A	G (Cha/Geo/The)		+		
Cyperaceae	*Cyperus glomeratus*	A or P	G (The)	+			
Amaranthaceae	*Amaranthus blitoides*	A	F (The)		+	+	
	Amaranthus retroflexus	A	F (The)		+	+	
Asteraceae	*Anthemis altissima*	A	F (Pha)				+
	Anthemis arvensis	A	F (Pha)	+			+
	Sonchus arvensis	P	F (Geo)	+	+	+	+
	Sonchus asper	A or B	F (Hem/The)	+		+	
	Sonchus oleraceus	WA	F (The)	+	+	+	+
	Taraxacum officinale	WA	F (Cha/Hem)	+			+
	Tragopogon longifolius	P	F	+			
	Tragopogon pratensis	B	F				+
	Xanthium spinosum	A	F (The)		+		
Apiaceae	**Bifora radians**	A	F	+	+		
	Caucalis platycarpos	A	F (The)	+			
	Scandix pecten-veneris	A	F (The)	+			
Boraginaceae	*Echium italicum*	B	F	+			
	Lithospermum arvense	A	F (The)		+		
Brassicaceae	*Capsella bursa-pastoris*	A	F (Hem/The)		+		
	Cardaria draba	A	F (The)	+			
	Neslia paniculata	A	F (Hem)				+
	Sisymbrium altissimum	A or B	F (The)				+
	Sisymbrium irio	A	F (The)		+		
Campanulaceae	*Legusia spegulum veneris*	A	F	+			
Caryophyllaceae	*Agrostemma githago*	A	F	+			
	Dianthus armeria var. *uniflorus*	A or B	F	+			
	Silene inflate	A	F (Cry/Hem)	+			
Chenopodiaceae	**Chenopodium album** var. *viride*	A	F (The)	+		+	
Convolvulaceae	*Convolvulus arvensis*	P	F (The/Geo/Cli)		+	+	
Euphorbiaceae	*Chrozophora tinctonia*	A	F (The)		+		
Fabaceae	*Lathyrus aphaca*	A	L (The/Cli)				+
	Trifolium striatum	A	L (Cha)	+			
	Medicago spp.	A	L (Hem/The)				+
	Vicia spp.	A or P	L (The/Cli)	+			+
Fumariaceae	*Fumaria capreolata*	A	F (Cli)				+
	Fumaria officinalis	A	F	+			+
Geraniaceae	*Geranium purpureum*	A	F (The)				
Zygophyllaceae	*Tribulus terrestris*	A	F (The)		+		

Table A1. Cont.

Family	Plant Species	Phenology *		Habitat			
		Life-Cycle	BG (Life-Form)	Cereals	Maize	Tillage	Clover
Lamiaceae	*Lamium amplexicaule*	A	F (The)		+		+
Malvaceae	*Malva sylvestris*	B	F (Hem)	+			
Papaveraceae	*Papaver hybridum*	A	F (The)				
	Papaver rhoeas	A	F (The)	+	+		
Polygonaceae	*Bilderdykia convolvulus*	A	F (The)	+	+		
	Polygonum aviculare	A	F (Cry/The)	+	+		+
Portulacaceae	*Portulaca oleracea*	A	F (The)		+		
Ranunculaceae	*Adonis aestivalis*	A	F	+			+
	Consolida regalis	A	F	+			
	Delphinium orientale	A	F	+			
	Ranunculus spp.	A	F (The/Hem)	+			
Rubiaceae	*Galium* spp.	A	F (The)	+	+		
Scrophulariaceae	*Veronica persica*	A	F (The)	+			
Solanaceae	*Solanum nigrum*	P	F (Hem/The)		+		
Total number (species richness) of the overall 61 species per habitat:				19	23	8	19

* Phenological classes of herb-layer species: according to life cycle: A = Annual, B = Biennial, P = Perennial; according to biological group (BG): G = Grass, L = Legume, F = Forb, B = Bulbous (geophyte). Classification of plant life form in line with [73] in parenthesis: The = Therophyte; Hem = Hemicryptophyte; Pha = Phanerophyte; Cha = Chamaephyte; Cry = Cryptophyte; Cli = Climber. G: Cyperus is considered grass-like. Related background references: [54,74–77]. In **bold** are shown species of (least to highest) significance as food items to rural birds, while significant and highly significant species are besides underlined; classification followed [22,78,79], as well as field observations.

Table A2. Mean (±Standard Deviation) above-ground variable estimations in the studied habitats [38]. Plant cover was recorded following [64]. The Field Physiognomy Index estimation followed [80].

Variable [1]	Habitat			
	Cereals	Maize	Tillage	Clover
Field surface area (in habitat)	19.75 ± 3.46	27.45 ± 5.78	20.32 ± 3.95	28.58 ± 14.16
Field Physiognomy Index	0.42 ± 0.22	0.22 ± 0.11	0.04 ± 0.02	0.59 ± 0.41
% bare soil	56.8 ± 1.96	94.3 ± 0.76	98 ± 0.41	76.3 ± 2.04
Number of non-cultivated species	46.1 ± 1.99	5.61 ± 1.07	17.2 ± 3.5	12.6 ± 1.73
% plant cover of non-cultivated species	33.8 ± 1.84	2.38 ± 0.55	1.82 ± 0.4	1.74 ± 0.47
Number of species serving as food items to birds	21 ± 1.26	0.22 ± 0.07	0.95 ± 0.15	0.58 ± 0.35
% plant cover of species serving as food items to birds	25.9 ± 1.76	0.19 ± 0.11	1.01 ± 0.31	0.01 ± 0
% plant cover of cultivated species	9.22 ± 0.88	3.32 ± 0.33	0.21 ± 0.1	22 ± 1.97

[1] Species richness of non-cultivated species and of species serving as food items to birds differ with habitat (physiognomy), with the highest values in cereals (1-way ANOVA; LSD; $p < 0.001$; unpublished data from [38]).

Recordings were conducted using quadrat (1 m^2) at sampling points of a plot diagonal (see also design in Table A4). The recorded variables were: 1. The total number of non-cultivated plant species. 2. The percentage of the sampling plot area (1 m^2) that is covered by non-cultivated plant species (%) percent of non-cultivated plant cover). 3. The number of plant species serving as food items (namely species producing seeds where birds feed on) to rural bird (food resources); this classification was based on [22,78,79] and observations in the field. 4. The (%) percentage of food items from the total surface non-cultivated plant cover serving as food items to rural birds (%) percent of non-cultivated plant cover serving as foot item to birds. 5. The (%) percentage covered by each crop species (cereals, maize, clover and residue from the previous crop field for tillage), respectively, for cereals, maize, clover and tillage fields (% plant cover of each crop species).

Table A3. Bird species recorded in the study area (in bold exclusively or partially seed-eaters). *: The bird species experienced decline in Europe [6,22]; F = farmland specialist, W = primarily woodland species that commonly use farmland [6].

Bird Species	Habitat	Bern Convention	79/409 EC Directive	SPEC	Bonn Convention
Accipiter nisus		II			
Alauda arvensis *	F	III	II/2	3	
Anthus pratensis		II			
Athene noctua		II		3	
Buteo buteo		II			
Carduelis cannabina *	F	II		4	
Carduelis carduelis *	F	II			
Carduelis chloris	F	II		4	
Circus cyaneus		II	I		II
Coccothraustes coccothraustes		II			
Corvus corone					
Corvus monedula	F			4	
Dendrocopos syriacus		II	I	4	
Emberiza cirlus		II			
Erithacus rubecula	W	II		4	II
Falco columbarius		II	I		
Falco tinnunculus *	F	II		3	II
Fringilla coelebs	W	III		4	
Galerida cristata		III		3	
Garrulus glandarius					
Melanocorypha calandra		II	I	3	
Miliaria calandra *	F	III		4	
Parus major	W	II			
Passer domesticus *					
Passer montanus *	F	III			
Phoenicurus ochruros		II			
Pica pica					
Pluvialis apricaria		III	I-II/2		II
Prunella modularis *		II		4	
Streptopelia decaocto		III	II/2		
Sturnus vulgaris *	F				
Turdus merula *	W	III	II/2	4	II
Turdus philomelos *	W	III	II/2	4	II

1. For the species listed in Annex II, states that have signed the treaty are required to take the necessary measures for the protection and conservation of these species and their habitats; for the species listed in Annex III, states that have signed the treaty are required to regulate the exploitation of wild fauna and prevent illegal means of capture and killing. 2. I: Species that will be subject of special conservation measures taking into account their habitat to ensure their survival and reproduction in the area of their dispersal; II/1: Species that can be hunted in the geographical sea and land where this Directive applies; II/2: Species that can be hunted only in the Member States, having regard to local laws. 3. 1: Species of global interest for their conservation; 2: Concentrated in Europe and with an unfavorable conservation status; 3: Not concentrated in Europe, but with an unfavorable conservation status; 4: Concentrated in Europe and with a favorable conservation status; w: Category related to winter populations; 4. I: Species with risk of total or at large extent extinction; II: Species that can benefit from the international cooperation for their conservation and management.

Table A4. Summary of the studied fields, sampling plots and methods [38].

Season		Sampling/Parameter	Habitat (i.e., Crop)	Number of Fields	Area (Hectares)	Replicates (R)	Materials		Methods
fall/winter 2006	i.	(a) physiognomy index	cereals	10	19.75	20	Quadrat 1 × 1 m²	i.	[80]
			clover	6	17.15	20		ii.	[64]
			tillage	10	20.31	20			
	ii.	plant cover	maize	10	27.45	20			
		(b) soil cores sampling/soil seed bank abundance & diversity	cereals	6	19.75	10	i. Cylindrical ring, 1 cm high		[64]
			clover	3	17.15	10			
			tillage	6	20.31	10			
			maize	6	27.45	10			
spring 2007			cereals	4	-	10	ii. Sweep		
			clover	3	-	10	iii. Squirrel		
			tillage	3	-	10			
			maize	2	-	10			

Table A5. The total number of seeds per m² estimated for each species of the topsoil seedbank in each studied habitat for winter (left columns) and spring (right columns), respectively.

Family	Plant Species	Cereals		Maize		Tillage		Clover	
		Winter	Spring	Winter	Spring	Winter	Spring	Winter	Spring
Amaranthaceae	Amaranthus albus *	-	112.99	112.36	10,112.99	56.18	112.99	-	1977.40
Amaranthaceae	Amaranthus blitoides *	56.18	-	-	-	0.00	112.99	1629.21	-
Amaranthaceae	Amaranthus retroflexus *,‡	10,224.72	6271.19	28,146.07	10,000.00	3539.33	4124.29	12,078.65	21,807.91
Asteraceae	Lactuca serriola	-	-	-	-	-	-	56.18	-
Asteraceae	Sonchus asper	-	-	11123.60	-	-	-	-	-
Apiaceae	Aethusa cynapium *	112.36	2316.38	-	-	-	-	0.00	4802.26
Apiaceae	Bifora radians *	1460.67	112.99	-	-	-	-	-	-
Apiaceae	Torilis nodosa	-	-	56.18	-	-	-	-	-
Boraginaceae	Lithospermum arvense *	2415.73	8644.07	112.36	56.50	112.36	-	1910.11	338.98
Brassicaceae	Brassica juncea *,‡	786.52	1186.44	1460.67	56.50	1235.96	169.49	449.44	1581.92
Brassicaceae	Brassica nigra	-	-	-	-	-	-	-	1016.95
Brassicaceae	Brassica rapa	-	-	617.98	-	-	-	280.90	-
Brassicaceae	Brassica sp.	-	-	-	-	56.18	-	-	-
Brassicaceae	Camelina microcarpa *	-	169.49	112.36	-	-	-	-	56.50
Brassicaceae	Capsella bursa-pastoris *	112.36	-	56.18	-	-	-	280.90	56.50
Brassicaceae	Sinapis arvensis *,‡	3483.15	790.96	2191.01	1468.93	3370.79	508.47	617.98	1186.44
Caryophyllaceae	Silene dioica *,‡	1348.31	2881.36	4438.20	903.95	1179.78	225.99	4438.20	1920.90
Caryophyllaceae	Stellaria media	-	-	-	-	-	1468.93	-	-
Chenopodiaceae	Chenopodium album *,‡	23,651.69	8135.59	24,887.64	14,689.27	19,157.30	3276.84	22,134.83	6610.17
Chenopodiaceae	Chenopodium vulvaria *	2134.83	-	393.26	2259.89	2303.37	1977.40	4213.48	-
Convolvulaceae	Ipomoea hederacea	-	-	-	-	-	-	56.18	-
Euphorbiaceae	Chrozophora tinctoria	168.54	-	-	-	56.18	-	-	-
Euphorbiaceae	Euphorbia spp. *	1741.57	790.96	337.08	-	561.80	-	1460.67	-
Fabaceae	Juncus sp. *	3707.87	1751.41	-	-	561.80	-	-	-
Fabaceae	Medicago mimina	56.18	-	-	-	-	-	-	-
Fabaceae	Medicago sativa	-	-	-	-	-	-	168.54	-
Fabaceae	Medicago polymorpha	168.54	-	-	-	112.36	-	-	-
Geraniaceae	Geranium lucidum	674.16	-	56.18	-	-	-	224.72	-
Geraniaceae	Geranium pusillum	56.18	-	-	-	-	-	-	-
Lamiaceae	Lamium amplexicaule *	112.36	1129.94	224.72	-	112.36	-	617.98	960.45
Malvaceae	Abutilon theophrasti *	-	169.49	56.18	395.48	-	56.50	-	-
Malvaceae	Malva sylvestris	-	-	-	-	-	-	112.36	-
Papaveraceae	Papaver rhoeas *	3876.40	4124.29	2640.45	-	1348.31	-	393.26	1412.43
Plantaginaceae	Plantago lanceolata	-	-	-	-	-	-	112.36	-
Poaceae	Alopecurus myosuroides	168.54	-	-	-	-	-	-	-
Poaceae	Alopecurus pratensis	-	-	56.18	-	-	-	-	-
Poaceae	Apera spica-venti	-	-	337.08	-	-	-	-	-
Poaceae	Avena nuda	-	5.00	-	-	-	-	-	-
Poaceae	Avena sterillis	1011.24	-	-	-	224.72	-	-	-
Poaceae	Cynodon dactylon	56.18	-	-	-	-	-	-	-
Poaceae	Digitaria sanguinalis *	-	56.50	5112.36	-	337.08	3276.84	-	2824.86
Poaceae	Echinochloa crus-galli *	-	-	674.16	112.99	-	-	-	16,045.20
Poaceae	Zea mays	-	-	1797.75	-	-	-	-	-
Poaceae	Panicum repens *	561.80	-	-	-	-	-	-	56.50
Poaceae	Setaria pumila *	56.18	225.99	4213.48	-	280.90	1242.94	337.08	225.99
Poaceae	Setaria spp. *	-	56.50	1966.29	-	449.44	56.50	-	-
Poaceae	Sorghum halepence *	-	-	4269.66	56.50	-	-	-	-
Poaceae	Triticum aestivum *	56.18	225.99	-	-	56.18	-	-	-

Table A5. Cont.

Family	Plant Species	Habitat							
		Cereals		Maize		Tillage		Clover	
		Winter	Spring	Winter	Spring	Winter	Spring	Winter	Spring
Polygonaceae	*Bilderdykia convolvulus* *	6460.67	2542.37	786.52	395.48	730.34	-	-	169.49
Polygonaceae	*Eriogonum racemon*	-	-	-	-	-	-	168.54	-
Polygonaceae	*Polygonum aviculare* *,‡	38,202.25	35,593.22	14,213.48	903.95	6516.85	1129.94	16,629.21	2937.85
Polygonaceae	*Polygonum lapathifolium*	-	-	-	-	337.08	-	-	-
Polygonaceae	*Polygonum persicaria*	-	56.50	-	-	-	-	-	56.50
Polygonaceae	*Rumex sanguineus* *	-	10,677.97	56.18	-	-	225.99	-	-
Polygonaceae	*Rumex* sp. *	56.18	169.49	112.36	-	617.98	56.50	-	56.50
Portulacaceae	*Portulaca oleracea* *,‡	9662.92	19,491.53	2921.35	734.46	898.88	564.97	10,224.72	4519.77
Primulaceae	*Anagallis arvensis*	-	-	1348.31	-	-	-	-	-
Rosaceae	*Rubus* spp.	56.18	-	168.54	-	112.36	-	-	-
Ranunculaceae	*Consolida regalis* *	1910.11	1186.44	-	-	-	-	-	-
Rubiaceae	*Galium aparine* *	505.62	112.99	56.18	-	168.54	-	-	-
Scrophulariaceae	*Veronica arvensis.*	56.18	-	-	-	-	-	-	-
Scrophulariaceae	*Veronica hederifolia.*	56.18	-	-	-	-	-	-	-
Scrophulariaceae	*Veronica persica* *	14,719.10	6214.69	393.26	-	1460.67	225.99	3595.51	225.99
Solanaceae	*Datura stramonium* *	-	112.99	-	-	-	-	337.08	-
Solanaceae	*Solanum nigrum* *	224.72	960.45	898.88	338.98	1123.60	112.99	337.08	-
Zygophyllaceae	*Tribulus terrestris*	-	-	-	-	-	-	56.18	-

Out of the 66 identified species, 35 were commonly detected in both seasons (winter and spring) and are marked with an asterisk (*); seven species found across all habitats and seasons are additionally shown with the double cross (‡) indicates absence. In **bold** are shown species of (least to highest) significance as food items to rural birds, while significant and highly significant species are besides underlined; classification followed [22,78,79], and field observations.

Appendix B

List of sources used for seed and plant specimen identification.

Seed identification:

- Flood, R.J. and Gates, S.C., 1986. Seed Identification Handbook, Official Seed Testing Station. National Institute Agricultural Botany. Publishing, Cambridge, UK.
- Lola P., 2003. Weeds Weed-Herbicides. Fate and behavior in the environment. Publications Modern Education.
- Seed collection of the Weed Laboratory of Department of Agriculture Crop Production and Rural Environment. University of Thessaly. (Professor P. Lolas).
- Seeds collected in the field
- Plant specimen and seed collections

Websites:

- Scottish Crop Research Institute
- University of Abertay Dundee
- ASIS Arable Seed Identification System

  http://asis.scri.ac.uk/

- The Ohio State University. Department of Horticulture and Crop Science. Seed IDWorkshop

  http://www.oardc.ohio-state.edu/seedid/

- University of Missouri Extension. Missouri Weed Seeds. Department of Agronomy Fred Fishel Kevin Bradley

  http://extension.missouri.edu/explore/agguides/pests/ipm1023.htm

- Seeds of Success Collections at the Bend Seed Extractory

  http://www.nps.gov/plants/sos/bendcollections/index.htm

- The seed identification web page. Paleoethnobotany Project

  http://www.oldthingsforgotten.com/seeds/seeds.htm

- Visual Identification of Small Oilseeds and Weed Seed Contaminants Grain Biology Bulletin No. 3

http://www.grainscanada.gc.ca/Pubs/Grainbio/bulletin3/sows_03-e.htm

Plant specimen identification:

- Kavvadas S., 1956. Illustrated Botany—Botanic Dictionary, Volumes 1–9. Pegasus Publications, Athens.
- Vardavaki M. Zouzouli D., 2003. Anatomy and Morphology of plants. Ziti, Thessaloniki.
- Lola P., 2003. Weeds Weed-Herbicides. Fate and behavior in the environment. Publications Modern Education.
- The growers weed identification Handbook. Collective work. Publisher University of California, Division of Agriculture and Natural Resources.
- Flowers of Greece and the Balkans, A field Guide. Collective work. Publisher Oxford University.
- Bonnier G., 1989. La Grande Flora En Couleurs, Volumes 1–2. Publications Delachaux et Niestle.

Websites (online databases):

- SRI Ilinois Council on food and Agricultural Research

  http://weedid.aces.uiuc.edu/

- United States Department of Agriculture

  http://plants.usda.gov/classification.html

- Weed Identification and Descriptions

  http://twig.tamu.edu/weedid.htm

- Utah State University extension. The weed web

  http://extension.usu.edu/weedweb/ident/ID.htm

- University of California, Agriculture and Natural Resources, Statewide IPM Program

  http://www.ipm.ucdavis.edu/PMG/WEEDS/low_amaranth.html

References

1. Sotherton, N.W. Land use changes and the decline of farmland wildlife: An appraisal of the set-aside approach. *Biol. Conserv.* **1998**, *83*, 259–268. [CrossRef]
2. Benton, T.G.; Vickery, J.A.; Wilson, G.D. Farmland biodiversity: Is habitat heterogeneity the key? *Trends Ecol. Evol.* **2003**, *18*, 182–188. [CrossRef]
3. Robinson, R.A.; Hart, J.D.; Holland, J.M.; Parrott, D. Habitat use by seed-eating birds: A scale-dependent approach. *Ibis* **2004**, *146*, 87–98. [CrossRef]
4. Pedersen, C.; Krøgli, S.V. The effect of land type diversity and spatial heterogeneity on farmland birds in Norway. *Ecol. Indic.* **2017**, *75*, 155–163. [CrossRef]
5. Römermann, C.; Dutoit, T.; Poschlod, P.; Buisson, E. Influence of former cultivation on the unique Mediterranean steppe of France and consequences for conservation management. *Biol. Conserv.* **2005**, *121*, 21–33. [CrossRef]
6. Chamberlain, D.E.; Fuller, R.J.; Bunce, R.G.H.; Duckworth, J.C.; Shrubb, M. Changes in the abundance of farmland birds in relation to the timing of agricultural intensification in England Wales. *J. Appl. Ecol.* **2000**, *37*, 771–788. [CrossRef]
7. Donald, P.F.; Pisano, G.; Rayment, M.D.; Pain, D.J. The Common Agricultural Policy, EU enlargement and the conservation of Europe's farmland birds. *Agric. Ecosyst. Environ.* **2002**, *89*, 167–182. [CrossRef]
8. Stephens, P.A.; Freckleton, R.P.; Watkinson, A.R.; Sutherland, W. Predicting the response of farmland bird populations to changing food supplies. *J. Appl. Ecol.* **2003**, *40*, 970–983. [CrossRef]
9. Vickery, J.A.; Bradbury, R.B.; Henderson, I.G.; Eaton, M.A.; Grice, P.V. The role of agri-environment schemes and farm management practices in reversing the decline of farmland birds in England. *Biol. Conserv.* **2004**, *119*, 19–39. [CrossRef]
10. Wretenberg, J.; Pärt, T.; Berg, Å. Changes in local species richness of farmland birds in relation to land-use changes and landscape structure. *Biol. Conserv.* **2010**, *143*, 375–381. [CrossRef]
11. Sálek, M.; Havlícek, J.; Riegert, J.; Nesporc, M.; Fuchs, R.; Kipson, M. Winter density habitat preferences of three declining granivorous farmland birds: The importance of the keeping of poultry dairy farms. *J. Nat. Conserv.* **2015**, *24*, 10–16. [CrossRef]
12. Calvi, G.; Campedelli, T.; Florenzano, T.G.; Rossi, P. Evaluating the benefits of agri-environment schemes on farmland bird communities through a common species monitoring programme. A case study in northern Italy. *Agric. Syst.* **2018**, *160*, 60–69. [CrossRef]

13. Gayer, C.; Kurucz, K.; Fischer, C.; Tscharntke, T.; Batáry, P. Agricultural intensification at local and landscape scales impairs farmland birds, but not skylarks (*Alauda arvensis*). *Agric. Ecosyst. Environ.* **2019**, *277*, 21–24. [CrossRef]
14. Suarez-Seoane, S.; Osborne, P.E.; Baudry, J. Responses of birds of different biogeographic origins and habitat requirements to agricultural land abandonment in northern Spain. *Biol. Conserv.* **2002**, *105*, 333–344. [CrossRef]
15. Lewis-Phillips, J.; Brooks, S.; Sayer, C.D.; McCrea, R.; Siriwardena, G.; Axmacher, J.C. Pond management enhances the local abundance and species richness of farmland bird communities. *Agric. Ecosyst. Environ.* **2019**, *273*, 130–140. [CrossRef]
16. Stoate, C.; Beja, B.P.; Boatman, N.D.; Herzon, I. Ecological impacts of early 21st century agricultural change in Europe—A review. *J. Environ. Manag.* **2009**, *91*, 22–46. [CrossRef]
17. Stanton, R.L.; Morrissey, C.A.; Clark, R.G. Analysis of trends and agricultural drivers of farmland bird declines in North America: A review. *Agric. Ecosyst. Environ.* **2018**, *254*, 244–254. [CrossRef]
18. Tarjuelo, R.; Benítez-López, A.; Casas, F.; Martín, C.A.; García, J.T.; Vinuela, J.; Mougeot, F. Living in seasonally dynamic Farmland: The role of natural and semi-natural habitats in the movements and habitat selection of a declining bird. *Biol. Conserv.* **2020**, *251*, 108794. [CrossRef]
19. Perkins, A.J.; Whittingham, M.J.; Bradbury, R.B.; Wilson, J.D.; Morris, A.J.; Barnett, P.R. Habitat characteristics affecting use of lowland agricultural grassland by birds in winter. *Biol. Conserv.* **2000**, *95*, 279–294. [CrossRef]
20. Perkins, A.J.; Maggs, H.E.; Wilson, J.D. Winter bird use of seed-rich habitats in agri-environment schemes. *Agric. Ecosyst. Environ.* **2008**, *126*, 189–194. [CrossRef]
21. Redhead, J.W.; Hinsley, S.A.; Beckmann, B.C.; Broughton, R.K.; Pywell, R.F. Effects of agri-environmental habitat provision on winter and breeding season abundance of famrland birds. *Agric. Ecosyst. Environ.* **2018**, *251*, 114–123. [CrossRef]
22. Marshall, E.J.P.; Brown, V.K.; Boatman, N.D.; Lutman, P.J.W.; Squire, G.R.; Ward, L.K. The role of weeds in supporting biological diversity within crop fields. *Weed Res.* **2003**, *43*, 77–89. [CrossRef]
23. Butler, S.J.; Bradbury, R.B.; Whittingham, M.J. Stubble height affects the use of stubble fields by farmland birds. *J. Appl. Ecol.* **2005**, *42*, 469–476. [CrossRef]
24. Hancock, M.H.; Duffield, S.; Boyle, J.; Wilson, J.D. The effect of harvest method on cereal stubble use by seed-eating birds in a High Nature Value farming system. *Agric. Ecosyst. Environ.* **2016**, *219*, 119–124. [CrossRef]
25. McHugh, N.M.; Prior, M.; Grice, P.V.; Leather, S.R.; Holland, J.M. Agri-environmental measures and the breeding ecology of a declining farmland bird. *Biol. Conserv.* **2017**, *212*, 230–239. [CrossRef]
26. Scherner, A.; Melander, B.; Kudsk, P. Vertical distribution and composition of weed seeds within the plough layer after eleven years of contrasting crop rotation and tillage schemes. *Soil Tillage Res.* **2016**, *161*, 135–142. [CrossRef]
27. Lal, B.; Gautam, P.; Raja, R.; Tripathi, R.; Shahid, M.; Mohanty, S.; Panda, B.B.; Bhattacharyya, P.; Nayak, A.K. Weed seed bank diversity and community shift in a four-decade-old fertilization experiment in rice–rice system. *Ecol. Eng.* **2016**, *86*, 135–145. [CrossRef]
28. Hosseini, P.; Karimi, H.; Babaei, S.; Mashhadi, H.R.; Oveisi, M. Weed seed bank as affected by crop rotation and disturbance. *Crop Prot.* **2014**, *64*, 1–6. [CrossRef]
29. Santín-Montanyá, M.I.; Martín-Lammerding, D.; Zambrana, E.; Tenorio, J.L. Management of weed emergence and weed seed bank in response to different tillage, cropping systems and selected soil properties. *Soil Tillage Res.* **2016**, *161*, 38–46. [CrossRef]
30. Chamberlain, D.E.; Vickery, J.A.; Glue, D.E.; Robinson, R.A.; Conway, G.J.; Woodburn, R.J.W.; Cannon, A.R. Annual and seasonal trends in the use of garden feeders by birds in winter. *Ibis* **2005**, *147*, 563–575. [CrossRef]
31. Siriwardena, G.M.; Calbrade, N.A.; Vickery, J.A.; Sutherland, W.J. The effect of the spatial distribution of winter seed food resources on their use by farmland birds. *J. Appl. Ecol.* **2006**, *43*, 628–639. [CrossRef]
32. Siriwardena, G.M.; Stevens, D.K. Effects of habitat on the use of supplementary food by farmland birds in winter. *Ibis* **2004**, *146*, 144–154. [CrossRef]
33. Henderson, I.G.; Vickery, J.A.; Carter, N. The use of winter bird crops by farmland birds in lowland England. *Biol. Conserv.* **2004**, *118*, 21–32. [CrossRef]
34. Marshall, E.J.P.; West, T.M.; Kleijn, D. Impacts of an agri-environment field margin prescription on the flora and fauna of arable farmland in different landscapes. *Agric. Ecosyst. Environ.* **2006**, *113*, 36–44. [CrossRef]
35. Geiger, F.; de Snoo, G.R.; Berendse, F.; Guerrero, I.; Morales, M.B.; Onate, J.J.; Eggers, S.; Part, T.; Bommarco, R.; Bengtsson, J.; et al. Landscape composition influences farm management effects on farmland birds in winter: A pan-European approach. *Agric. Ecosyst. Environ.* **2010**, *139*, 571–577. [CrossRef]
36. Moreira, F.; Beja, P.; Morgado, R.; Reino, L.; Gordinho, L.; Delgado, A.; Borralho, R. Effects of field management and landscape context on grassland wintering birds in Southern Portugal. *Agric. Ecosyst. Environ.* **2005**, *109*, 59–74. [CrossRef]
37. Hyvönen, T.; Huusela-Veistola, E. Impact of seed mixture and mowing on food abundance for farmland birds in set-asides. *Agric. Ecosyst. Environ.* **2011**, *143*, 20–27. [CrossRef]
38. Voudouri, A. The Value of Agro-Ecosystems for Biodiversity Outside the Growing Season: Comparative Assessment of Crops in the Elassona Region. Master's Thesis, University of Thessaly, Volos, Greece, 2008; 113p.
39. R Core Team. R: A Language and Environment for Statistical Computing. R Foundation for Statistical Computing: Vienna, Austria. Available online: https://www.R-project.org/ (accessed on 15 June 2021).
40. Bolker, B.; Robinson, D. Broom. Mixed: Tidying Methods for Mixed Models. R Package Version 0.2.6. 2020. Available online: https://CRAN.R-project.org/package=broom.mixed (accessed on 15 June 2021).

41. Wickham, H.; François, R.; Henry, L.; Müller, K. Dplyr: A Grammar of Data Manipulation, R Package Version 1.0.6. 2021. Available online: https://CRAN.R-project.org/package=dplyr (accessed on 15 June 2021).
42. Wickham, H. *Ggplot2: Elegant Graphics for Data Analysis*; Springer: New York, NY, USA, 2016.
43. Bates, D.; Maechler, M.; Bolker, B.; Walker, W. Fitting Linear Mixed-Effects Models Using lme4. *J. Stat. Softw.* **2015**, *67*, 1–48. [CrossRef]
44. Kuznetsova, A.; Brockhoff, P.B.; Christensen, R.H.B. lmerTest Package: Tests in Linear Mixed Effects Models. *J. Stat. Softw.* **2017**, *82*, 1–26. [CrossRef]
45. Venables, W.N.; Ripley, B.D. *Modern Applied Statistics with, S*, 4th ed.; Springer: New York, NY, USA, 2002; ISBN 0-387-95457-0.
46. Hothorn, T.; Bretz, F.; Westfall, P. Simultaneous Inference in General Parametric Models. *Biom. J.* **2008**, *50*, 346–363. [CrossRef]
47. Wickham, H. tidyr: Tidy Messy Data, R Package Version 1.1.3. 2021. Available online: https://CRAN.R-project.org/package=tidyr (accessed on 15 June 2021).
48. Ball, D.A. Weed seed bank response to tillage, herbicides, and crop rotation sequence. *Weed Sci.* **1992**, *40*, 654–659. [CrossRef]
49. Cunningham, H.M.; Bradbury, R.B.; Chaney, K.; Wilcox, A. Effect of non-inversion tillage on field usage by UK farmland birds in winter. *Bird Study* **2005**, *52*, 173–179. [CrossRef]
50. Shrestha, A. *Weed Seed Banks and Their Role in Future Weed Management*; UC Statewide IPM Program: Parlier, CA, USA, 2001.
51. Whittingham, M.J.; Devereux, C.L.; Evans, A.D.; Bradbury, R.B. Altering perceived predation risk food availability: Management prescriptions to benefit farmland birds on stubble fields. *J. Appl. Ecol.* **2006**, *43*, 640–650. [CrossRef]
52. Donald, P.F.; Sanderson, F.J.; Burfield, I.J.; van Bommel, F.P.J. Further evidence of continent-wide impacts of agricultural intensification on European farmland birds, 1990–2000. *Agric. Ecosyst. Environ.* **2006**, *116*, 189–196. [CrossRef]
53. Wilson, J.D.; Whittingham, M.J.; Bradbury, R.B. The management of crop structure: A general approach to reversing the impacts of agricultural intensification on birds. *Ibis* **2005**, *147*, 453–463. [CrossRef]
54. Zanin, G.; Otto, S.; Riello, L.; Borin, M. Ecological interpretation of weed flora dynamics under different tillage systems. *Agric. Ecosyst. Environ.* **1997**, *66*, 177–188. [CrossRef]
55. Ghersa, C.M.; Martínez-Ghersa, M.A. Ecological correlates of weed seed size and persistence in the soil under different tiling systems: Implications for weed management. *Field Crops Res.* **2000**, *67*, 141–148. [CrossRef]
56. Baskin, C.C.; Baskin, J.M. *Seeds. Ecology, Biogeography, and Evolution of Dormancy and Germination*; Academic Press: San Diego, CA, USA, 1998; 666p.
57. Field, R.H.; Benke, S.; Bádonyi, K.; Bradbury, R.B. Influence of conservation tillage on winter bird use of arable fields in Hungary. *Agric. Ecosyst. Environ.* **2007**, *120*, 399–404. [CrossRef]
58. Reuss, S.A.; Buhler, D.D.; Gunsolus, J.L. Effects of soil depth and aggregate size on weed seed distribution and viability in a silt loam soil. *Appl. Soil Ecol.* **2001**, *16*, 209–217. [CrossRef]
59. Conn, J.S. Weed seed bank affected by tillage intensity for barley in Alaska. *Soil Tillage Res.* **2006**, *90*, 156–161. [CrossRef]
60. Carter, M.R.; Ivany, J.A. Weed seed bank composition under three long-term tillage regimes on a fine sandy loam in Atlantic Canada. *Soil Tillage Res.* **2006**, *90*, 29–38. [CrossRef]
61. Orłowski, G.; Czarnecka, J. Winter diet of reed bunting *Emberiza schoeniclus* in fallow and stubble fields. *Agric. Ecosyst. Environ.* **2007**, *118*, 244–248. [CrossRef]
62. Marino, P.C.; Westerman, P.R.; Pinkert, C.; van der Werf, W. Influence of seed density and aggregation on post-dispersal weed seed predation in cereal fields. *Agric. Ecosyst. Environ.* **2005**, *106*, 17–25. [CrossRef]
63. Robinson, R.A.; Sutherland, W.J. The winter distribution of seed-eating birds: Habitat structure, seed density and seasonal depletion. *Ecography* **1999**, *22*, 447–454. [CrossRef]
64. Moorcroft, D.; Whittingham, M.J.; Bradbury, R.B.; Wilson, J.D. The selection of stubble fields by wintering granivorous birds reflects vegetation cover food abundance. *J. Appl. Ecol.* **2002**, *39*, 535–547. [CrossRef]
65. Marone, L.; Rossi, B.E.; De Casenave, J.L. Granivore impact on soil-seed reserves in the central Monte desert. *Argentina Funct. Ecol.* **1998**, *12*, 640–645. [CrossRef]
66. Broughton, R.K.; Chetcuti, J.; Burgess, M.D.; Gerard, F.F.; Pywell, R.F. A regional-scale study of associations between farmland birds and woody networks of hedgerows and trees. *Agric. Ecosyst. Environ.* **2021**, *310*, 107300. [CrossRef]
67. Vickery, J.A.; Feber, R.E.; Fuller, R.J. Arable field margins managed for biodiversity conservation: A review of food resource provision for farmland birds. *Agric. Ecosyst. Environ.* **2009**, *133*, 1–13. [CrossRef]
68. Orrock, J.L.; Levey, D.J.; Danielson, B.J.; Damschen, E.I. Seed predation not seed dispersal explains the landscape-level abundance of an early-successional plant. *J. Ecol.* **2006**, *94*, 838–845. [CrossRef]
69. Dunning, J.B.; Brown, J.H. Summer rainfall and winter sparrow densities: A test of the food limitation hypothesis. *Auk* **1982**, *99*, 123–129. [CrossRef]
70. Lortie, C.L.; Ganey, D.T.; Kotler, B.P. The effects of gerbil foraging on the natural seedbank and consequences on the annual plant community. *Oikos* **2000**, *90*, 399–407. [CrossRef]
71. Teillard, F.; Antoniucci, D.; Jiguet, F.; Tichit, M. Contrasting distributions of grassland and arable birds in heterogeneous farmlands: Implications for conservation. *Biol. Conserv.* **2014**, *176*, 243–251. [CrossRef]
72. Sanou, L.; Salvadogo, P.; Zida, D.; Thiombiano, A. Contrasting land use systems influence the soil seed bank composition and density in a rural landscape mosaic in West Africa. *Flora* **2019**, *250*, 79–90. [CrossRef]
73. Raunkiaer, C. *The Life Form of Plants and Statistical Plant Geography*; Clarendon Press: Oxford, UK, 1934.

74. Alsherif, E.A.; Ayesh, A.M.; Rawi, S.M. Floristic composition, life form and chorology of plant life at Khulais region, Western Saudi Arabia. *Pak. J. Bot.* **2013**, *45*, 29–38.
75. Carni, A.; Matevski, V.; Silc, U.; Custerevska, R. Early spring ephemeral therophytic non-nitrophilous grasslands as a habitat of various species of Romulea in the southern Balkans. *Acta Bot. Croat.* **2014**, *73*, 107–129.
76. Mehrvarz, S.S.; Nodehi, M.A. A floristic study of the Sorkhankol Wildlife Refuge, Guilan Province. *Iran. Caspian J. Environ. Sci.* **2015**, *13*, 183–196.
77. Sultan-Ud-Din, A.H.; Haidar, A.; Hamid, A. Floristic composition and life form classes of district Shangla, Khyber Pakhtunkhwa, Pakistan. *J. Biodivers. Environ. Sci.* **2016**, *8*, 187–206.
78. Wilson, J.D.; Morris, A.J.; Arroyo, B.E.; Clark, S.C.; Bradbury, R.B. A review of the abundance and diversity of invertebrate and plant foods of granivorous birds in northern Europe in relation to agricultural change. *Agric. Ecosyst. Environ.* **1999**, *75*, 13–30. [CrossRef]
79. Holland, J.M.; Hutchison, M.A.S.; Smith, B.; Aebischer, N.J. A review of invertebrates and seed-bearing plants as food for farmland birds in Europe. *Ann. Appl. Biol.* **2006**, *148*, 49–71. [CrossRef]
80. Wilson, J.D.; Evans, J.; Browne, S.J.; King, J.R. Territory distribution breeding success of skylarks Alauda arvensis on organic intensive farmland in southern England. *J. Appl. Ecol.* **1997**, *34*, 1462–1478. [CrossRef]

MDPI
St. Alban-Anlage 66
4052 Basel
Switzerland
www.mdpi.com

Land Editorial Office
E-mail: land@mdpi.com
www.mdpi.com/journal/land

Disclaimer/Publisher's Note: The statements, opinions and data contained in all publications are solely those of the individual author(s) and contributor(s) and not of MDPI and/or the editor(s). MDPI and/or the editor(s) disclaim responsibility for any injury to people or property resulting from any ideas, methods, instructions or products referred to in the content.